Contents

AF870215

Vertebrates

A LABORATORY TEXT
SECOND EDITION

Editors

Norman K. Wessells
Stanford University

Elizabeth M. Center
College of Notre Dame

Illustrator

Nina Shapley

Authors

Henry B. Kistler, Jr.
Neil J. Baker
David Anderson
Mary Beth Bryan McNabb
Mary Marshall Cooke
Nathan Schafer

With a Section on The Mudpuppy by

Carl Gans and
Ronald A. Nussbaum
The University of Michigan

Jones and Bartlett Publishers
BOSTON LONDON

Editorial, Sales, and Customer Service Offices

Jones and Bartlett Publishers
One Exeter Plaza
Boston, MA 02116
617-859-3900
1-800-832-0034

Jones and Bartlett Publishers International
7 Melrose Terrace
London W6 7RL
England

ISBN 0-86720-853-8

Printed in the United States of America
98 97 96 95 10 9 8 7 6 5 4 3 2

Preface

This is a laboratory text written largely by new kinds of authors: undergraduates writing for other undergraduates! In addition, most of the illustrations were drawn by an undergraduate student. Why is this important in the development of a text? Because these students (like undergraduates everywhere) have had to struggle through textbooks written from the point of view of the professor or other experts—not from the perspective of students delving into sharks or cats for the first time. In addition, poor organization, inadequate illustrations, and other shortcomings of many textbooks frustrate learning. Recognizing these problems from firsthand experience over their dissecting trays, these student authors were challenged to do better. This book is the result.

At Stanford University, undergraduates serve as laboratory instructors in vertebrate biology. Each such instructor, having taken the course during the previous year, has new ideas about its strengths and weaknesses and wants to improve on past performances. So the first edition of this text, compiled by six authors and one artist, represented a distillation of many people's efforts. An essential aspect of the book's development was the preparation of a completely new set of illustrations drawn from specimens, with the author of each chapter looking over the artist's shoulder. New perspectives and realistic illustrations will, we hope, facilitate the learning experience.

Significantly, this second edition contains a new section of chapters on the mudpuppy, *Necturus*, an amphibian. It was prepared by two experienced teachers and authorities on vertebrate biology, Carl Gans and Ronald A. Nussbaum of The University of Michigan. Their chapters maintain the spirit of the manual and fill a gap in the coverage of the first edition. New figures for the section on the mudpuppy were prepared by Karen Kvenvold Bailey, a professional biological artist.

The authors' goal in this text is not to cover all aspects of a wide variety of vertebrates, nor to follow a complete evolutionary sequence. Instead, the text concentrates on important structures and functions of (1) a small group of evolutionarily important organisms—tunicates, amphioxus, and lampreys—which give hints about vertebrate origins; (2) the shark, which has a simple body plan and is relatively easy to dissect; (3) the mudpuppy, an amphibian that does not undergo metamorphosis and grows large enough to permit successful dissection; and (4) the cat, which is a mammal similar in structure and physiology to human beings. The latter point does not mean that the text is oriented toward premedical students or premedical eduction. Rather, the inherent curiosity of all human beings about their bodies amply justifies the time and attention students devote to the study of a mammal.

This book is unique in that it treats the nervous system of the shark and the brain of a representative mammal—the sheep—together. In this way, the transition from one class of organism to another can be made more easily. Similarly, the skeleton and the muscles of the cat are discussed as one integrated unit. Besides helping to clarify the function of the two systems, this strategy decreases the tedium of learning the bones and the muscles, one by one, as an unending series of separate units. This approach also helps students think of bones and associated muscles as integrated functional units.

Another unique feature is a section of color plates at the end of the book. These figures repeat key numbered figures in the main body of the text, but with color added for clarity. The plates are perforated for removal in order that they may be readily at hand during dissections.

Our emphasis on the functional, integrated approach reflects our concern that too often in vertebrate biology—or in all biology, for that matter—the naming of a structure is taken as a meaningful explanation of that structure. To be sure, names of structures are useful, as are names of people or chemical compounds, but assigning names to the parts of an organism is only the beginning. For the study of vertebrate biology to be of real educational value, one needs to question why a structure exists in a particular form and what the function of that structure might be.

With these concepts in mind, the authors of this manual have tried to provide information on both evolution and function of body structures. Instructors and students will find here more than a dissection guide and atlas. Thus the introductions and discussions in each chapter, amplified with information from supplied references, may add liveliness and breadth to the laboratory experience and to a concurrent lecture course. In this new second edition, notes relating the anatomy of the animals dissected to other aspects of their biology and to other vertebrates have been updated and increased in number throughout the text. We would urge, in particular, that instructors consider incorporating classroom reports and discussions into the laboratory sessions, since such activities can add substantially to the pleasure in the course and to the educational experience for students. The information provided by our authors may serve as a source of ideas for such reports. The student should be aware, however, that our treatment of various topics is necessarily brief and that many fascinating aspects of vertebrate animals are not mentioned here. Books that may be helpful for amplification include: *Biomechanics*, by C. Gans (The University of Michigan Press); *Analysis of Vertebrate Structure*, by M. Hildebrand (John Wiley and Sons); *Vertebrate Life*, by W. N. McFarland, F. H. Pough, T. J. Cade, and J. B. Heiser (Macmillan); and *Vertebrates: Physiology* plus *Vertebrates: Adaptation*, both edited by N. K. Wessells (W. H. Freeman and Co.).

We, as instructors in charge of the vertebrate course at Stanford, could find no greater reward for our teaching efforts than the achievement of a group of undergraduates in producing the first edition of this book. In its new edition, the coverage of *Necturus* by Professors Gans and Nussbaum improves the balance of topics covered and should permit the manual to be used in an even broader array of courses on vertebrates. In preparation for the second edition, Drs. Ellen F. Porzig and Marilyn Anderson provided valuable suggestions for improvements; Dr. Jane A. Peterson of the University of California, Los Angeles, contributed to the expansion and clarification of dissection instructions for the chapters on the dogfish shark and the cat; and, finally, Dr. Carl Gans, one of the new authors, reviewed the entire book for accuracy and for additional opportunities to make meaningful, interesting comments on the animals under examination. Users of the second edition are encouraged to report to us on their experiences in the laboratory.

Elizabeth M. Center

Norman K. Wessells

General Instructions and Terminology

HENRY B. KISTLER, JR.

1

General Instructions and Terminology

GENERAL INSTRUCTIONS FOR THE LABORATORY

1. Before each laboratory session, familiarize yourself with the objectives and procedures of each dissection. This will help to prevent accidental disruption of organ systems to be studied later.

2. Learn the vocabulary of dissection. Terms describing planes of orientation and general locations are defined in this chapter. For other terms, use the Glossary at the end of this book. Learning is often easier if the roots and derivations of words are understood.

3. Careful dissection is time-consuming. Though there are many ways to dissect a given organ system, the sequence of the dissections and other instructions used in this book are the cumulative result of the experience of many students and instructors. Haste will lead only to frustrating difficulties and irreparable damage to your specimens.

4. Locate, identify, and evaluate the function of each structure described in the identification section of each chapter. Learn the relationships of structures to other members of the same system. Do not simply compare your preparation with those diagrammed in the text; use the diagrams and text as a guide, but use *your specimen* for learning the actual body parts.

5. Make every attempt to perform the dissection and identify structures without the instructor's assistance. Consult with fellow students and compare your specimen with theirs. You may occasionally find features not described in the text or identified in the diagrams. Seek the assistance of the instructor for identification of structures not listed.

6. The anatomical arrangement of certain organ systems, especially the circulatory systems, may vary from one specimen to the next. Systems also vary with the sex and age of the specimen. When you discover an oddity in your preparation, show the instructor and your fellow students. Do not hesitate to observe the specimens of other students and to offer your own for comparison. Such an atmosphere of sharing benefits everyone.

7. Preserve your specimen carefully. Cover all exposed surfaces, such as muscles and digestive organs, with wet paper towels; then secure the towels with rubber bands. Place the specimens in airtight plastic bags with a small amount of water, then seal tightly. While dissecting, never expose tissue to close direct light without repeatedly wetting the surface to prevent desiccation.

TERMINOLOGY USED IN DISSECTION (Fig. 1-1)

Anterior. **(Cranial; cephalic; rostral)** Situated toward or located in the specimen's head or front end. In the human vascular system, *superior* is used synonymously with *anterior.*

Distal. Peripheral to a point of origin or attachment; farther from the median line or reference point.

Dorsal. Situated toward or located in the specimen's upper surface or back area.

Inferior. Situated toward or located on the under surface.

Lateral. Located away from the specimen's midsagittal plane.

Medial. Located at or toward the midsagittal plane.

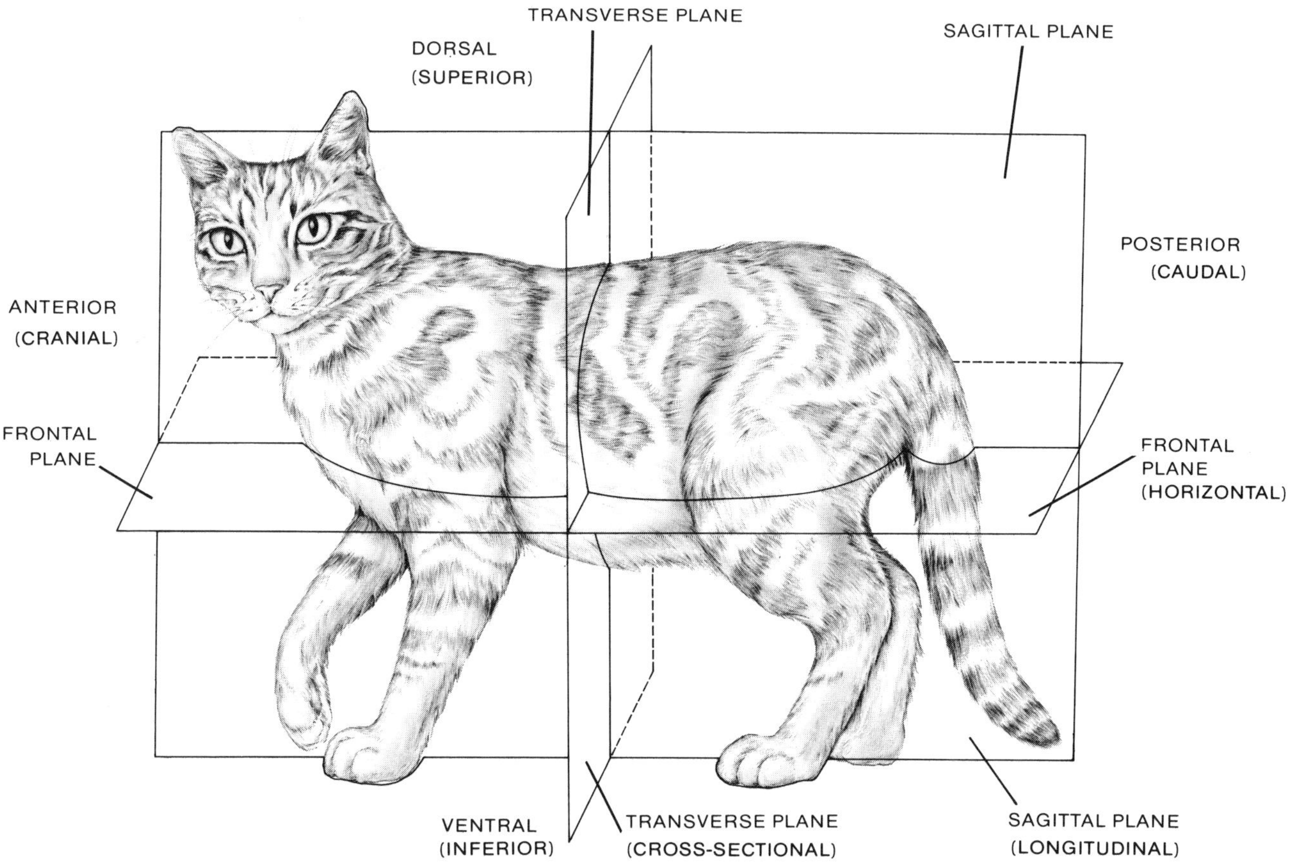

Figure 1-1. Lateral view of a cat, illustrating planes of dissection and general regional terminology.

Posterior. (**Caudal**) Situated toward or located in the specimen's hind end.

Proximal. Situated closer to the midline, axis, or other reference point.

Superior. Situated toward or located in the specimen's upper surface or head.

Ventral. Situated toward or located in the specimen's under surface or belly area.

PLANES OF ORIENTATION (Fig. 1-1)

Midsagittal. This longitudinal plane separates the body into right and left halves. Sagittal planes are those parallel to the midsagittal plane.

Transverse. (**Cross-sectional**) This plane passes dorsoventrally and from left to right, separating the body into cranial and caudal parts.

Frontal. (**Horizontal** or **coronal**) This plane passes craniocaudally and from left to right, separating the body into dorsal and ventral parts.

MATERIALS NEEDED FOR VERTEBRATE DISSECTION

½ dozen disposable gloves (optional)
1 small scalpel handle with replaceable blades
1 large scalpel handle with replaceable blades
1 blunt forceps
1 needlenosed forceps
2 dissecting needles
1 blunt probe
1 fine scissors
1 heavy scissors
1 laboratory coat (optional)

COMMENTS

The following suggestions will assist you in laboratory dissections.

1. Tools should be cleaned after use.
2. Scalpel blades must be very sharp. Replace or sharpen them often.
3. Wear gloves whenever handling specimens for a prolonged period. Rubbing powder on the hands

before putting on gloves makes it easier to remove them later.

4. Bone shears and a sharp knife will be needed but will usually be provided by the instructor.

5. Preservatives such as formalin will stain clothes. Pay special attention to long-sleeved shirts, sweaters, and coats. Lab coats are desirable.

6. Long hair should be tied back and protected.

7. Scrub your hands with strong soap and water after taking off your gloves to remove traces of preservative solution.

8. Formalin (formaldehyde) is a potential carcinogen and should be avoided.

The Hemichordates and The Lower Chordates

MARY BETH BRYAN McNABB

2

Lower Chordates: General Introduction

Our understanding of relationships between vertebrates and other animal groups is based in large part on analysis of basic body organization in embryos and adults. For instance, all chordate embryos show certain similarities in morphological development that set this group off from all the invertebrates (or, more precisely, from the nonchordates). A point of confusion to some students is that the lower chordates, and also most primitive vertebrates, do not possess vertebrae or a true vertebral column. Instead, a notochord acts as the structural supportive element down the axis of the body. In this category are urochordates, cephalochordates, and members of the phylum Chordata, to which we belong. Still more primitive animals, the hemichordates, lack a notochord and are probably best placed into a separate phylum.

Many questions about the origin of vertebrates can be approached by studying living organisms which, because of their simplicity, give hints about the possible morphology of the earliest vertebrates. We will examine three such organisms: (1) the tunicate larva, a urochordate, (2) a cephalochordate, and (3) an ammocoete, the larval form of the cyclostome, a primitive vertebrate. All three share certain characteristics that were probably evolutionary inventions of the vertebrate stock. These are: (1) a skeletal notochord at some stage in the life cycle; (2) a dorsally situated nerve tube; (3) a tail that extends posteriorly beyond the anus and which, like more anterior body regions, contains blocks of muscle tissue called myomeres; and (4) slit-like perforations in the walls of the pharynx.

All these features help to define a chordate, since they are quite different from anything found in the invertebrate phyla. They do not, however, help to explain the origin of the vertebrates. On the basis of other data, several theories have been proposed. The presence of similar protective skeletal plates in early vertebrate fishes and in early arachnids led to an "arachnid theory" of vertebrate origins. The segmentation of the annelids (such as earthworms) suggested to some that those organisms might be ancestral to vertebrates with their segmented myomeres. Perhaps the most widely quoted theory is that the early chordates arose from relatives of echinoderms—the starfishes, sea urchins, and their kind.

The "echinoderm theory" is based on embryological similarities between some echinoderms and the lower chordates. Both groups have indeterminate, radial cleavage, a similar type of body cavity formation, and a mouth that develops as a second opening instead of from the transformed blastopore. Echinoderms also have a blastula stage similar to that of urochordates, though the larvae which form are quite different.

Some biologists believe that the larvae of early chordates may be the key to the evolutionary origins of free-swimming adult chordates. Larval stages serve to disperse members of a species during the reproductive phase of the life cycle. There is no guarantee, however, that the presence of a larva means that adults of the species are necessarily sedentary or sessile. It has recently been argued that the sessile adult urochordate living today is a secondary adaptation to life at the bottom of the sea, one that arose from free-swimming creatures reminiscent of the tadpoles. According to this theory, the early chordate stock was distinguished by unique gill slits that allowed a hitherto unknown type of filter feeding to go on. Thus, among others, the marine zoologist Donald Abbott argues that a free-swimming adult stage with gill slits, a notochord, segmental myomeres, and a postanal tail gave rise to the modern urochordates as well as to the first jawless fish. In a sense, we might view the urochordate tadpole and the cyclostome tadpole as representative of different levels of complexity that may have characterized our ancient relatives. We must not be fooled, however, by the seeming simplicity of these various little organisms. Each is highly adapted to its particular niche; the pressures of natural selection must be such in the particular environments in which the organisms live that evolutionary modification is extremely conservative. Consequently, the creatures may serve as models for the original vertebrates.

MARY BETH BRYAN McNABB

3

The Hemichordates

The Hemicordata are a difficult group to interpret, since two distinct types of animals are included. *Pterobranchs* are colonial deep-sea organisms with tentacles for use during feeding; *enteropneusts*, or acorn worms, are elongate, marine creatures with the pharyngeal gill slits characteristic of chordates. Because of this and other factors, hemichordates are considered as relatives of both echinoderms and chordates.

EVOLUTION

Familial relationships between hemichordates and chordates are quite vague. No group of invertebrate animals possesses pharyngeal gill slits or homologous structures. However, other features of hemichordate bodies are quite different from vertebrate structures, such as the anus opening at the tip of the tail. The sessile *Pterobranchs* are believed by some biologists to be the source of the free-swimming ancestor of urochordates, cephalochordates, and jawless vertebrates.

STRUCTURE AND FUNCTION (Fig. 3-1)

Enteropneusts, the most plentiful form of the *Hemichordata*, are the simplest to understand and so are generally used for teaching purposes in comparisons of hemichordates and chordates. Representative enteropneusts, the acorn worms (also called intestinal worms or tongue worms), generally live on sandy or muddy sea bottoms. They employ a very specialized feeding mechanism that is somewhat different from the filter-feeding found in urochordates and cephalochordates.

Figure 3-1. External view of a stylized hemichordate (scale: approximately ×3).

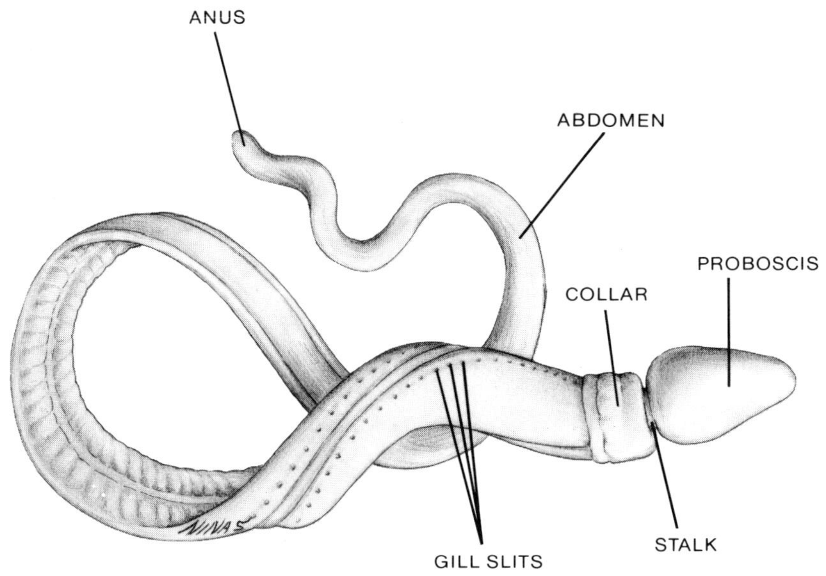

A major instrument for feeding is the long ***proboscis***, covered with cilia and mucus-secreting cells and stiffened by a diverticulum of the gut called the ***stomochord***. The mucus-trapped food particles are swept back into the mouth by beating cilia. A stream of water that is carried into the mouth along with the food-laden mucus escapes the body of the animal through 100 to 150 pairs of ***gill slits***. Food and mucus pass into the simple ***alimentary canal***, which is lined with a ciliated epithelium that carries out the digestive and absorptive functions.

To carry out this sort of feeding, the enteropneust uses a complex set of motions employing its musculature to burrow through the detritus-laden sea bottom. The muscles are not arranged segmentally as in chordates, but rather are circular fibers. A few longitudinal fibers are present; they seem to be used for escape reactions. In order to feed, the enteropneust expands its relaxed muscles by pumping sea water into the ***collar chamber***. This expanded chamber anchors the animal while the ***proboscis*** stretches forward. This use of water pressure in coelomic chambers to create turgidity is typical of echinoderms. In this respect, as well as in the general correspondence of internal body cavities, echinoderms and hemichordates are similar.

MARY BETH BRYAN McNABB

4

The Urochordates

Though Aristotle, in a moment of insight, recognized the tough-skinned, seemingly formless urochordates as animals, it was not until the larval forms were discovered in 1866 that their significance to vertebrates became known. Named for their thick, cellulose-based outer covering, the "tunic," tunicates have a mobile larval form adapted for seeking out new sites for adult habitation. The larva is equipped with a ***notochord***, a ***nerve cord***, ***muscle masses*** in a ***post-anal tail***, and the ***gill slit structures*** characteristic of chordates. (Note, however, that the muscle cells are not segregated into discrete blocks [myomeres] as is the case in other chordates, including vertebrates.)

There is a great deal of difference between the life-styles of larval and adult tunicates. The larva (Fig. 4-1) is constructed so it can find a site suitable for the sessile adult organism. Therefore the larva's anatomy and chemistry are specialized for locomotion and for sensing pertinent aspects of the environment. The chordate tail makes the larva an efficient swimmer; its sensory mechanisms are suited to dispersal and habitat selection. Shortly after hatching from spawned eggs, the larvae swim toward the light. As a result, the little animals are likely to disperse widely from their sessile parents. When the larva is ready for metamorphosis—the change to the adult state—its sensory system, composed of a ***unicellular otolith***, specialized for the detection of gravity, and an ***ocellus*** or eye spot with three ***unicellular lenses*** in the cerebral vesicle, cause the animal to move downward and away from light. Certain species have more elaborate behavioral patterns to insure placement of the adult in specific locales.

Its brief existence and limited functions permit the larva to remain small and to have simplified mechanisms for respiration, food gathering, excretion of wastes, and blood circulation.

It could well be that the absence of these capabilities may not be a "primitive" condition, but instead may reflect an evolutionary "simplification." This would be the case if the urochordates are, as discussed earlier, descendents of free-swimming adult creatures that reproduced while in possession of a vertebrate-like tail. If this speculation is correct, the modern urochordate tadpole appears to have eliminated extraneous organs and functions to adapt itself for one job: finding a suitable substratum fast!

EVOLUTION

Upon first examination of the tunicate, it is difficult to appreciate the degree of specialization and the amount of evolutionary innovation found in animals of this subphylum. The degree of specialization is exemplified by the lack of some commom chordate structures and the presence of other structures not found in other chordates. The excretory system of the urochordates provides examples. Some tunicates have a renal vesicle which accumulates and holds the by-products of nucleic acid metabolism until death of the organism. The blood of the urochordate *Ciona* (sea squirt), on the other hand, possesses certain cells called nephrocytes that may store crystals of uric acid, a nitrogenous waste product; or such cells may manufacture ammonia, a compound that may diffuse from the body at the gill slits (ammonia arises as a product of protein breakdown). In any event, tunicates do not have the glomerular structures used for excretion by hemichordates and by cephalochordates. Since these groups in this regard are more complex than tunicates, the absence of glomeruli in urochordates may reflect a loss during the evolution of the tunicate line.

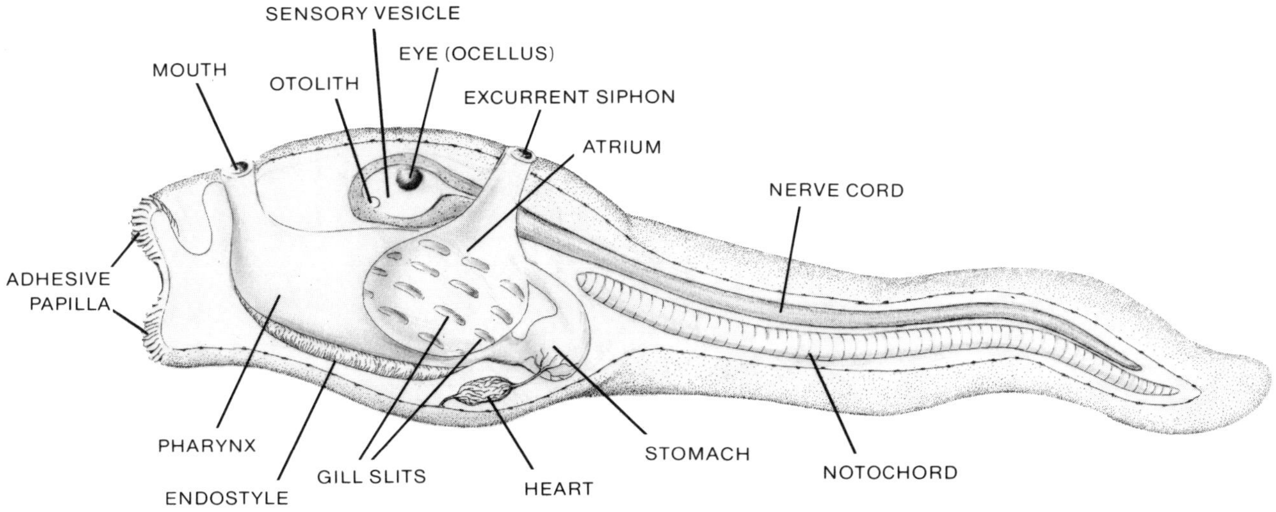

Figure 4-1. Lateral view of an idealized larval tunicate, showing internal structures.

EXAMINATION OF THE ADULT UROCHORDATE,

Ciona intestinalis (sea squirt)

An examination of the intact animal shows it to be a translucent, water-filled pouch. The external surface may be encrusted with sponges and algae. The exterior may be cleaned with fingers or tweezers until the adult tunicate with its ***excurrent*** and ***incurrent siphons*** is obvious. ***Ciona*** is usually pear-shaped and from one to three inches long. The axes of the animal are determined by designating the ***incurrent siphon*** as ***anteriormost***. The smaller ***excurrent siphon*** is sometimes distinguishable by six small tentacles located around the opening.

With the scissors, snip the posterior side of the excurrent siphon; extend the cut around the base of the animal, stopping short of the larger incurrent siphon. Evert the outer tunic, exposing the ***inner visceral mass****. In some preserved specimens, the preservatives may have gelled inside the tunic, making a few more snips necessary to free the visceral mass. A few extra cuts may also be needed around the openings of the siphons if the visceral mass is too securely attached to the outer tunic. When utilizing living specimens, the dissection should be carried out in a small dish of sea water. Living specimens should be narcotized in order to prevent contractions that distort internal organs. Add a few crystals of magnesium sulfate to the sea water a few hours prior to dissection. A narcotized* Ciona *does not respond to touch. Such animals may be fixed in alcohol or, if necessary, formalin (be careful).*

Refer to Figure 4-2 for names and locations of organs and structures.

Atrial mantle. After releasing the inner visceral mass of the tunicate, you will see that it is enclosed in a clear atrial mantle or body wall. Embedded in the mantle are whitish-yellow fibers running the length of the body, from the incurrent siphon to the base of the tunicate. These are muscles that aid in the expulsion of filtered water from the atrial cavity, which occupies the space between the pharynx and the mantle. Through the transparent wall of the atrial mantle, the ***intestines***, ***gonad***, ***genital duct***, and ***stomach*** can be seen.

Siphons. The siphons should be readily identifiable due to yellow or orange pigmentation around the actual openings.

Alimentary tract. The alimentary tract of the tunicate begins with the incurrent siphon, where food-laden water is brought in. The water passes through the ***atrial basket*** or ***pharynx***, the walls of which are perforated by ***ciliated gill slits***, and the food is trapped by mucus produced by the ***endostyle***, a groove running down one side of the pharynx. The water passes outward through the gill slits, and is subsequently squirted out of the excurrent siphon. The mucus spreads around the inside of the atrial basket and collects in a groove known as the ***dorsal lamella***. The collected "chain" of mucus follows the lamella through a short ***esophagus*** into the "stomach" of the tunicate, which is enzymatically comparable to the ***small intestine*** of other chordates. The stomach

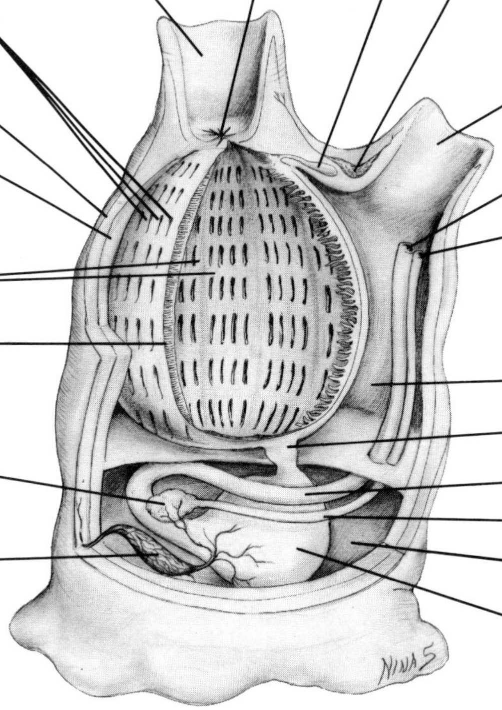

Figure 4-2. Lateral view of an idealized adult tunicate. The mantle, tunic, and part of the branchial basket have been cut away. (Modified from Storer and Usinger.)

leads to the intestine, which twists and turns in the epicardial cavity until it opens at a spot under the ***excurrent siphon***.

Gonadal ducts and gonads. The gonadal ducts originate at the gonads, which are tucked among the coils of the intestines. The more compact, larger gonad is the ovary, which has a white genital duct—the oviduct—leading from it to an opening beneath the excurrent siphon. The testis consists of small white tubules on the surface of the ovary and intestine. A thin vas deferens runs from the testis paralleling the oviduct.

Heart. The heart, located next to the stomach, is difficult to identify in a preserved specimen because the blood of the tunicate, the cardiac vessel, and the pericardium (membrane covering the heart) are all transparent. In a live specimen it is easy to find because of the waves of contraction that pass along the cardiac vessel.

Pharyngeal (gill) slits. A piece of the atrial basket should be excised and examined under a dissection microscope in order to observe the pharyngeal slits, which represent a major evolutionary advance in feeding.

FUNCTION

A unique and puzzling feature of the tunicate is its vascular system. The tunicate heart, enclosed in the only possible coelomic cavity in the organism, beats alternately in two directions! First, a pacemaker at one end of the heart initiates the waves of contraction for a period. Then it appears to become physiologically fatigued, allowing another pacemaker at the opposite end of the heart to initiate beating in the opposite direction. The blood pumped is remarkable in that it shows no evidence of a respiratory pigment.

The tunic is of interest because it contains ***tunicin***, a branched carbohydrate resembling the cellulose of plants.

The central nervous system of the tunicate consists of one nerve ganglion, located on the dorsal side of the animal between the siphons. Nerve axons from this ganglion reach most of the atrial wall musculature. There are sensory receptor cells in the walls of the pharyngeal gill basket which detect water current changes and tactile stimulation. Axons from these cells run to the nerve ganglion to initiate appropriate motor reponses. Under the nerve ganglion is the subneural gland, which produces a substance similar to the vertebrate pituitary hormone oxytocin.

So in the tunicate we have a unique heart, blood constitution, and excretory system, along with the capacity for nervous coordination and perhaps for endocrine hormonal control. The presence of these organ systems in the adult tunicate raises intriguing questions about the degree to which these and other features of our ancient relatives were adapted and modified in vertebrates themselves.

MARY BETH BRYAN MCNABB

5

The Cephalochordates

The two genera in the subphylum Cephalochordata are *Branchiostoma* and *Assymetron*.

The best-known member of the first of these genera is the amphioxus, more properly called *Branchiostoma*. This small, relatively simple chordate is found around the world. It may exist in sufficient quantities to serve as a food for humans in some areas. *Branchiostoma* suggests many evolutionary questions, but provides few answers.

Rarely more than 60 mm long, *Branchiostoma* is a burrowing filter-feeder that seems to have been diverted from a more mobile life for which its tail and musculature seem adapted. The nervous system of this animal is quite complex, coordinating various internal ciliary movements and offering a variety of behaviors to cope with the environment. However, there is no brain and only a hint of an ***adenohypophysis***, or anterior pituitary gland, a structure associated with brain tissue in chordates. The animal's excretory system is composed of special excretory cells called ***solenocytes***. These cells are of a unique type not found in other invertebrates, echinoderms, or chordates. In *Branchiostoma* the circulatory system lacks a respiratory pigment, but it is a closed system except in the tissues, where the blood bathes the cells directly.

Although the blood passes along the branchial bars, respiratory exchange there is minimal; the animal relies instead upon simple diffusion of respiratory gases across its body surfaces. Despite all these peculiarities and specializations, *Branchiostoma* is very definitely a chordate animal.

EVOLUTION

Branchiostoma was long thought to be the source of vertebrates. Today many scientists think *Branchiostoma* originated as a more mobile animal that assumed its burrowing habit in the sandy bottom of the seashore secondarily. However, hard evidence for or against such a possibility is lacking. We can tentatively conclude that, in the evolutionary scheme, the cephalochordates are probably an offshoot from a line of organisms that yielded the urochordates and the vertebrates. Important diagnostic characters include a segmented postanal tail; a dorsal, hollow nerve tube; a notochord; and pharyngeal gill slits. The vertebrates may be another offshoot of that stock. The close relationship of urochordates and cephalochordates is evident in their similar methods of feeding and larval behavior.

EXAMINATION OF *BRANCHIOSTOMA*—WHOLE MOUNT AND REPRESENTATIVE SECTIONS (Figs. 5-1, 5-2, 5-3)

Branchiostoma is a torpedo-shaped animal with a slightly barbed tail and a tentacled head. Its color results in part from the lack of red hemoglobin in the blood. The animal's actual color is that of the muscle masses and associated connective tissues as seen through the transparent skin. The fins of *Branchiostoma* are not especially well defined, but dorsal, ventral, and caudal fins can be distinguished.

A cleared whole mount and representative sections of *Branchiostoma* reveal the animal's inner structures without having to resort to dissection. With standard tissue stains, gill slits, notochord, nerve cord, wheel organ, and alimentary tract can be seen.

Notochord and nerve cord. At the anterior end, the notocord extends beyond the dorsally located ***nerve cord*** and the tentacled ***mouth region***.

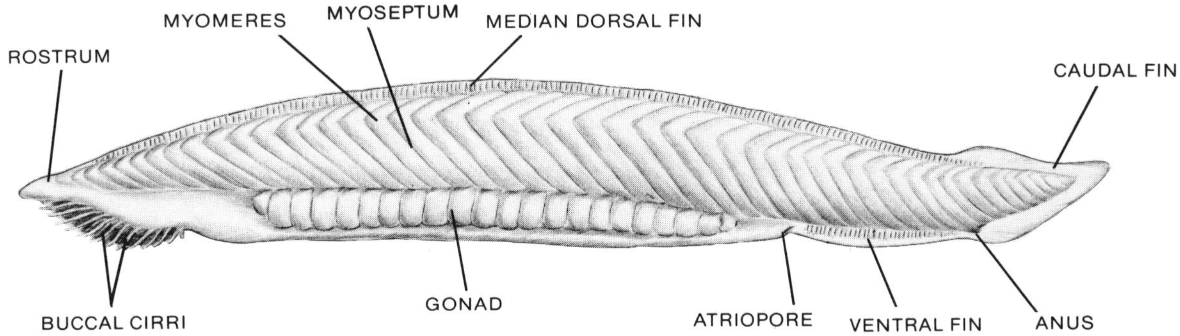

Figure 5-1. Lateral view of *Branchiostoma* (amphioxus), showing external morphology.

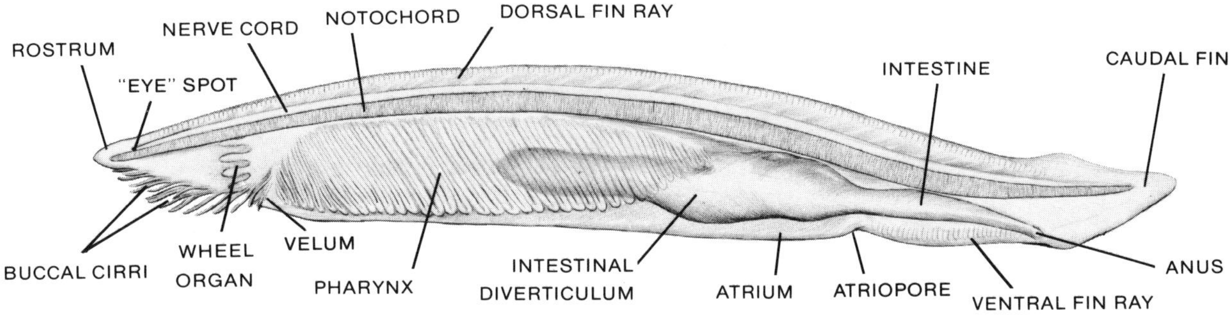

Figure 5-2. Lateral view of a cleared whole mount of *Branchiostoma*, showing internal structures.

Pigment (eye) spot. The nerve cord terminates anteriorly near a pigment spot, sometimes called an eyespot. Its association with neurosecretory cells suggests that this spot may have some endocrine function.

Myomeres and myosepta. The nerve cord and notochord are flanked laterally by the myomeres, segmented muscle masses which are separated from each other by connective tissue sheets called myosepta. Each myomere is V-shaped when viewed laterally (Fig. 5-1). As a result, it extends along a larger proportion of the side, so that contraction of the myomere produces curvature along an equivalent distance.

Gonads. These are a series of oblong structures along the ventral margin of the myomeres.

Buccal cirri. The mouth of the *Branchiostoma* is surrounded by buccal cirri, tentacles which prevent large particles from entering the alimentary tract.

Wheel organ. On the inner walls of the oral hood—which encloses the cavity behind the buccal cirri—is the wheel organ. This ciliated structure funnels the food particles and water stream through the curtain-like ***velum***. Since the buccal cirri are folded over to act as a sieve during feeding, the current of water in the oral hood is prone to stagnation. The cilia of the wheel organ keep this water moving to bring suspended food particles into contact with mucus produced and collected in ***Hatschek's pit***, a depression in the roof of the oral hood. The food and mucus, swept into a chain-like band by the wheel organ, move through the velum into the pharyngeal gill region.

Branchial basket. The ***pharyngeal slits*** in the pharynx or branchial basket do not open directly to the outside. Instead the water swept in with the food passes first into a chamber, or atrium, between the walls of the pharynx and the outer body wall. This water moves posteriorly and escapes the body via an ***atriopore***, which can be seen as an indentation anterior to the musculature of the tail.

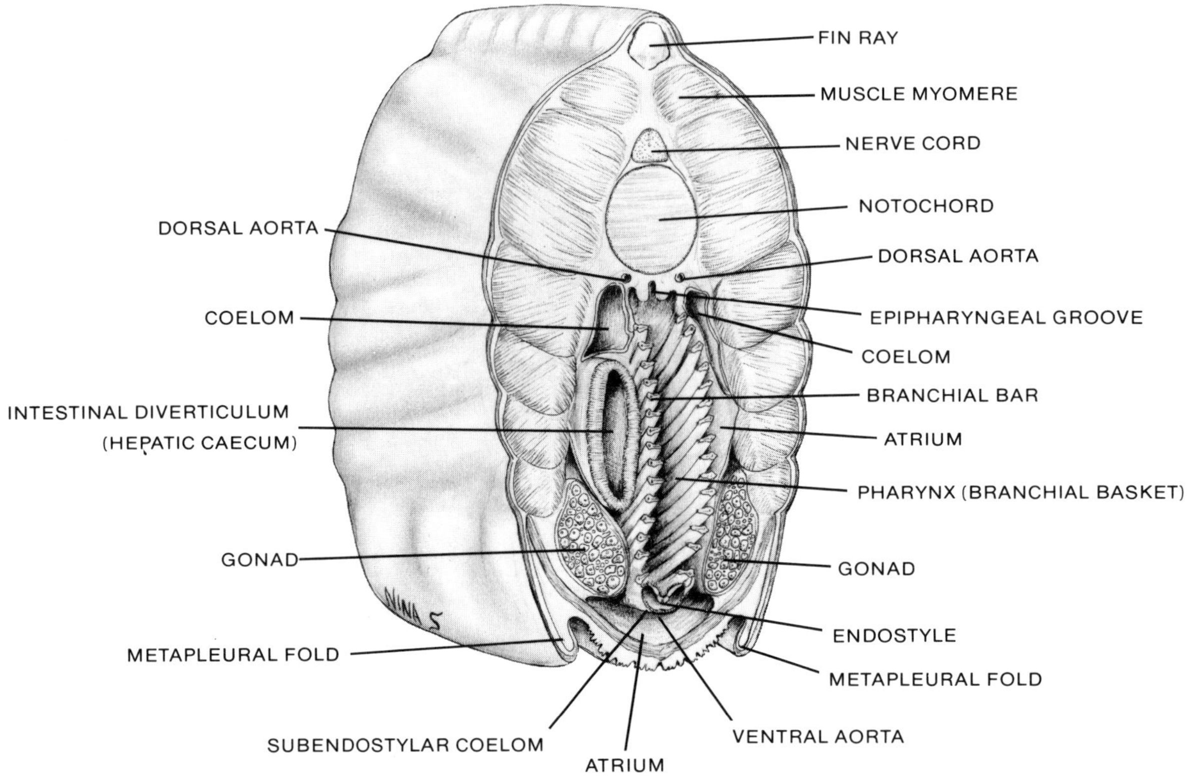

Figure 5-3. Cross section through the pharyngeal region of *Branchiostoma.*

Endostyle. At the base of the diagonal pharyngeal slits is a fold (slightly darker in whole mounts) running the length of the branchial basket. This endostyle produces abundant mucus just as in the urochordates. This mucus, carried dorsally by the ciliated gill bars, joins the mucus band started by the wheel organ at the anterior of the branchial basket. Cells in the endostyle can concentrate iodine into an organic molecule, thyroxine; this suggests homology to cells in the vertebrate thyroid gland.

Intestinal diverticulum. According to Barrington (1965), once the food chain leaves the branchial basket, the currents of cilia in the alimentary tract propel the food and sort out the food particles into the midgut and a midgut structure called the intestinal diverticulum. This structure is sometimes called the hepatic diverticulum or caecum, though it is not related to liver in function. Some scientists have not observed food passage into the diverticulum, although they agree with Barrington on the role this convoluted diverticulum plays in the production of digestive enzymes. The nature of these enzymes led to a theory that the intestinal diverticulum may be the forerunner of the vertebrate's pancreas.

Ilio-colon ring. At the junction between the midgut, or intestinal region, and the posterior alimentary tract, there is sometimes a conglomeration of food particles apparent in the cleared whole mount. At this point there is a specialized "ilio-colon ring" of cilia, which moves the food-mucus mass along the gut near the point where digestive enzymes enter the gut. The rest of the alimentary tract is a straight tube, slightly ciliated, which terminates in a highly ciliated ***anus***. It is ciliary action, not muscular contractions, that transports food through the whole gut.

FUNCTION

In addition to the production of mucus, the endostyle in *Branchiostoma* has been shown to have a definite endocrine role as a primitive thyroid gland with the ability to incorporate iodine into several compounds with thyroxine-like activity.

The amphioxus has a primitive blood vascular system which is closed in the sense of having a complex set of blood vessels. However, it is open in the tissues, so that blood fluid bathes the cells directly, in contrast to the capillary system of vertebrates. Motion of the blood is caused by regular contractions of smooth muscle cells along the walls of the larger vessels. *Branchiostoma* has no respiratory pigment, but this does not present a problem, inasmuch as the animal's general respiratory rate on a per weight basis is only about 10 percent of that measured in "cold-blooded" vertebrates.

The presence of excretory cells, the ***solenocytes***, within sacs that empty into the atrium, and the importance of osmotic balance in the estuarine habitat of *Branchiostoma*, suggest that the blood vessels of the pharyngeal region might be primarily concerned with maintenance of water and salt balance rather than respiration.

Branchiostoma has some unique sensory cells adapting it for survival in a half-buried position on a sandy sea bottom. Around the oral hood are chemoreceptors that prompt an escape response in the presence of stagnant water. Along the nerve cord in the posterior region are ***ocelli*** that are sensitive to light. When light shines on the normally buried tail, an escape response occurs. The escape response itself is initiated by a system of giant nerve fibers known as ***Rhode cells***, located near the center of the nerve cord. The axons of these giant nerves make assymetric connections to the muscle masses, causing jerky, unpredictable movements, very effective as an escape mechanism.

The reproductive system of *Branchiostoma* is quite simple. There are male and female *Branchiostoma*, but under certain temperature and salinity conditions, individuals of each sex can produce both eggs and sperm. Spawning is dictated by temperature and by seasonal factors. Larvae hatch from the fertilized eggs approximately twenty hours after spawning and spend about five months as freely mobile organisms. The larvae then metamorphose into adults.

The notochord of *Branchiostoma* also deserves comment. It is composed of a tough sheath surrounding vacuolated cells which have considerable turgor pressure; significantly, it also has contractile filament systems running transversely. These contractile systems are located segmentally and are innervated. When a wave of myomere contraction passes down a side of the body as it bends in swimming, a wave of contraction passes down the notochord at the same time. As a result that structure is tensed or made more rigid just at the point of myomere contraction. Apparently this allows the notochord to function more efficiently as a rigid rod that resists shortening of the body during myomere contraction.

What is so surprising about the presence of muscle in the notochord is that the other lower chordates and vertebrates do not have this feature. This means that a truly basic chordate organ, the notochord, has undergone a radical structural and functional modification in a single line of organisms, the Branchiostomids. The reasons are hard indeed to envision, since *Branchiostoma* is neither a more vigorous nor a more frequent swimmer than an ammocoete or a tunicate tadpole. Furthermore, the muscular notochord also suggests that the cephalochordates are not on the main line of chordate evolution leading toward the vertebrates, unless, of course, the contractile apparatus of cephalochordates was lost secondarily as the true vertebrates appeared.

MARY BETH BRYAN McNABB

6

The Larval Lamprey

Agnatha, or jawless vertebrates, include the earliest known vertebrates—ancient, bony fishes, all of which are extinct. One type of extinct agnathan, the ***osteostracans***, are believed to have given rise to an exceedingly conservative line of parasitic and scavenger fishes. These ***cyclostomes*** include hagfish and lampreys, inhabitants of the sea and some fresh waters.

The reproductive biology of lampreys has long been of interest because of the remarkable larval form, the ***ammocoete***. The ammocoete was long thought to be a member of a different genus than that of the adult lamprey. This is understandable since the larva lives for up to nine years and is very different in structure from the adult. The ammocoete phase constitutes the major part of the life cycle in some species of lamprey; thus the larva must cope with the environment in all its aspects, just as the adult must. The ammocoete is a filter-feeder like *Branchiostoma* and shares the environmental preference of those animals—a half-buried existence in sand or mud. The ammocoete has separate respiratory, alimentary, excretory, and circulatory systems. Because these systems in the ammocoete seem to be the simplest known examples of those organ systems available in living or fossil vertebrates, this organism is cited as an example of what the primitive vertebrate stock may have been like.

THE AMMOCOETE—EVOLUTION

In examining the ammocoete, we are looking at a more vertebrate-like organism than *Branchiostoma*. We believe that the modern ammocoete may be rather more like some of the ancient osteostracans than it is like modern adult lampreys. So in examining the modern cyclostomes in the laboratory, we may view the larva and the adult in different contexts: we may relate the simple organization of the larva to that of the earliest vertebrates; and we study the adult for its specializations as a parasite in addition to its more common vertebrate features.

The feature of greatest evolutionary significance in the ammocoete is its radically different method of carrying out filter feeding. The pharynx, equipped with muscles and skeletal support, periodically contracts like a force-pump, thereby driving water through the gill slits and out of the pharynx. When the muscles in the walls of the pharynx relax, the chamber expands; thereby, water and suspended food flow into the mouth. That food is filtered out during the next contraction cycle as water filters through sheets of mucus located internal to the gill slits. This mode of feeding is called ***muscular filter feeding*** to distinguish it from the ***ciliary filter feeding*** of *Branchiostoma* and urochordates.

The ammocoete has many characteristics of vertebrate organization not found in *Branchiostoma* or the tunicate larva. It has a kidney made up of tubules that collect fluid directly from the body cavity or coelom, by means of ciliated openings called nephrostomes. It has a liver, a gall bladder, a three-part brain, an inner ear, and a definite heart. For example, the ammocoete pituitary gland controls pigmentation and secondary sex changes, but perhaps little else. The pituitary in higher vertebrates controls these functions and many others. We do not know how each of these organs arose in ancestors of osteostracans and ammocoetes; nevertheless, these organs in the ammocoete do give us a "starting point" for comparison with similar organs in later vertebrates.

EXAMINATION OF THE AMMOCEOTE (Fig. 6-1)

A simple examination of a cleared whole mount of an ammocoete should reveal the similarities between this larval form and *Branchiostoma*. The notochord, the myomeres beside the dorsal nerve cord, the tentacled head, the postanal tail, and the long straight intestine should be obvious.

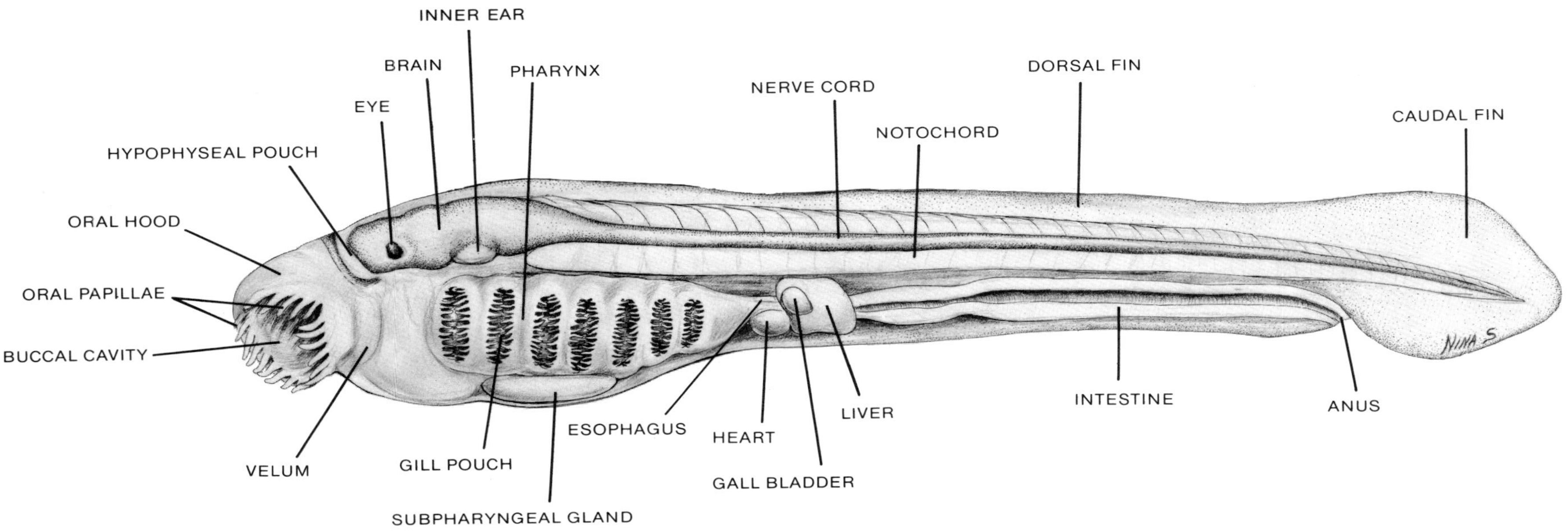

Figure 6-1. Lateral view of a cleared whole mount of ammocoete, emphasizing internal structures (scale: approximately ×25, but variable).

Eyespot. Perhaps the most striking difference from *Branchiostoma* is the large dark eyespot. In the larva the eye is not functional but represents an early developmental stage of a vertebrate eye that develops more fully during metamorphosis. The eye lies approximately over the midbrain region of the ammocoete when one is looking at the side of the animal.

Brain. The three parts of the brain—the prosencephalon, the mesencephalon, and the rhombencephalon—may or may not be distinguishable, depending on the degree to which any one specimen has been cleared.

Oral papillae. The oral papillae of the ammocoete are smaller than those of *Branchiostoma* but perform the same function, acting as a sieve to prevent large particles from entering the alimentary tract. The same kind of sensory cells occur in the oral hood of the ammocoete to test the quality of the incurrent water as are found in *Branchiostoma*.

Velum. The velum, which in *Branchiostoma* is simply a curtain of tentacles, is here composed of two muscular sheets of tissue. Along with the muscular contractions of the gill basket, the rhythmic waves of the velum promote the food-carrying current.

Gills. The gill region in the ammocoete differs from that of *Branchiostoma*. First, there are only seven pairs of gill slits compared to approximately one hundred pairs in *Branchiostoma*. Since the gill slits open directly to the outside, there is no atrial chamber or atriopore. By relying upon muscles rather than ciliary movements to propel the current of water, the ammocoete can apparently pump water more than *Branchiostoma*.

Subpharyngeal gland. Beneath the first five gill slits is a convolution of the gill basket floor, embryonically identical to the endostyle of *Branchiostoma*. In the ammocoete, this rod-shaped structure is called the subpharyngeal gland, though its endocrine function is obscure. The gland has an opening near the center of the pharyngeal chamber, where mucus for filter feeding is released. Mucus is also produced by the walls of the pharyngeal lining and the gill lamellae. The food-laden mucus strands funnel back into the alimentary canal.

Digestive tract. As the food-laden mucus strands pass from the pharynx into the esophagus, they are immediately mixed with the same sorts of digestive enzymes that are produced by pancreatic cells in higher vertebrates. Thus primitive pancreatic cells may be present in the lining of the anterior gut. Near the middle of the gut, at the beginning of the small intestine, secretions from the liver of the ammocoete enter into the tract via the gall bladder. This bile aids in the digestion of fats.

Anus. In identifying the parts of the alimentary tract, it may be easier to go from the posterior of the animal to the anterior. The anus, or termination of the tract, is an aperture at the base of the caudal fin.

Intestine. The long straight intestine extends anteriorly, eventually bulging, then sharply constricting and continuing anteriorly as the esophagus. The relatively large size of the intestine may be due to the presence of the ***typhosole*** or "spiral valve." This structure consists of a flap of intestinal wall, projects into the central lumen, and thereby provides greater surface area for absorption of digested food.

Liver and gall bladder. The posterior part of the esophagus is medial to the liver. In some specimens the gall bladder can be seen as a clear vesicle in the liver area.

Heart. Anterior to the liver and ventral to the esophagus is a clear structure with chambers. This is the ammocoete heart.

Kidney. Dorsal to the pericardial cavity, but ventral to the esophagus, is the area of the pronephric kidney tubules.

FUNCTION

The degree of complexity in body organization of the ammocoete is hardly revealed in a cleared whole mount. The endocrine glands, the circulatory system, and details of the respiratory, nervous, and excretory systems can only be appreciated by looking at sections with a microscope. The excretory system, for instance, is composed of functional units call ***nephrons*** which, as in higher vertebrates, open into a duct that in turn opens near the anus. Some of these nephrons open into the body cavity or coelom. At metamorphosis, the anterior kidney tubules degenerate, while more posterior sections of kidney tissue develop into the opisthonephric kidney of the adult lamprey (see Fig. 10-1). Interestingly, this adult kidney lacks the nephrostome openings into the coelom.

The nervous system of the ammocoete is like that of all lower vertebrates in that stimuli from the various sense organs initiate reflexive responses. The brain of the ammocoete and lamprey is divided into three sections; the same sections in sharks and mammals will be studied later.

The respiratory system of the ammocoete is more like that of higher fishes than that of the adult lamprey, for in the larva the water always enters the mouth and passes over the gills (see Chapter 7). In each of the seven gill pouches are folds of tissue bearing lamellae, which project into the lateral free space of the pharynx. The lamellae greatly increase the surface area of the gills; they are thought to be the sites of gas exchange.

The blood of the ammocoete contains a hemoglobin respiratory pigment within nucleated red blood cells. Since the ammocoete grows much larger than a *Branchiostoma* or the tunicate larva, simple diffusion is inadequate for respiratory purposes. The respiratory pigments are compatible with such large size.

With the increased size of the ammocoete, circulating the blood is also a greater task than can be handled by muscularized vessels as in *Branchiostoma*. A heart arises during embryonic development from the straight ventral vessel of the ammocoete. The parts of the heart—sinus venosus, atrium, and ventricle—are twisted in a unique manner, unlike that seen in higher vertebrates. Rather than bathing tissue cells directly, the blood is carried in capillaries through the various organs.

The endocrine system of the ammocoete is significantly different from that of most invertebrates. In addition to the pituitary gland, thyroid follicles develop in the subpharyngeal gland during metamorphosis. Some adrenal-like cells with unknown function are present, and certain "pancreatic" cells of the esophageal lining produce a hormone which controls carbohydrate metabolism.

The ammocoete is obviously very complex, even though it is a larval form. Although superficially similar to *Branchiostoma*, the complexities of the circulatory, excretory, nervous, and digestive systems show that the ammocoete is truly a vertebrate.

MARY BETH BRYAN McNABB

7

The Adult Lamprey

Much of the structure and physiology of the adult lamprey is geared to its unique parasitic existence. The lamprey feeds on the body fluids, blood, and to some degree, on the flesh of the bony fishes. The external morphology is adapted to catching up to, hooking onto, and then feeding on a passing fish. The digestive system is specialized for processing the above foodstuffs, and the respiratory system is rearranged in its pressure relationships to accomodate the specialized mode of feeding. The absence of both jaws and lateral fins (both common features of bony fish such as trout or cartilagenous fish such as sharks) from lampreys is also noteworthy.

The following description is that of a commonly dissected representative of the lampreys. In fact, the gross morphology of various lampreys may differ dramatically, so do not be confused if a brook lamprey or some other type comes to your attention yet looks very different from what we will study in the laboratory.

EXTERNAL MORPHOLOGY OF THE MARINE LAMPREY (Fig. 7-1)

The lamprey's appearance matches its eating habits! The grey cylindrical body is extremely tough-skinned, with fins being mere extensions of the body lines, accentuating the streamlined shape appropriate to swift motions.

Fins. There are two ***dorsal fins*** (anterior and posterior) and a ***caudal fin***, all of which are supported by cartilaginous fin rays. However, there are no lateral fins.

Mouth. The most striking external feature of the lamprey is the round sucker mouth, for which the cyclostomes are named. Around the perimeter of the mouth are the ***buccal cirri*** which insure a vacuum-like fit on the host fish. No jaws or true teeth of the sort found in other vertebrates are present.

Buccal Funnel. Yellow, regularly spaced, horny epidermal ***teeth*** line the inside of the buccal funnel; these grip the flesh of the host. The movable tongue is located in the center of the funnel.

Nostril. On the dorsal surface of the lamprey, quite posterior to the mouth, is the single nasal aperture, or nostil. Its posterior position is probably a result of displacement of the specialized mouth. In this position, the nostril can function when the lamprey's mouth is fixed to the side of a host fish.

Pineal eye. Immediately behind the nostril is a lighter pigmented area which overlies the pineal eye (see below).

Eyes. On either side of the head are the eyes with which the lamprey sees. (Note: the eyes of some species of lampreys are considerably smaller than those seen in the Pacific Coast lamprey.)

Lateral line system. When the specimen has dried a little, various lines of pores may be visible on the sides and the ventral surface of the head. These pores lead to open pits of the lateral line system; sensory hair cells within the pits detect changes in the water currents around the animal. (The lateral line sensory system is of great importance in all vertebrate fishes and even amphibian tadpoles. Among the specializations of the lateral line is its involvement in detecting changes in electric fields around the body of the so-called "electric fishes.") Scattered over the head and tail region are other sensory cells that react to strong light by initiating an escape reaction.

Gill slits. On the sides of the body posterior to the eyes are seven pairs of rounded gill slits.

Cloaca. The cloaca opens ventrally just anterior to the tail; within it are the anal opening (anterior) and the urogenital papilla (posterior). Slightly anterior to the cloaca are two ***genital pores*** through which the genital products escape into the surrounding water

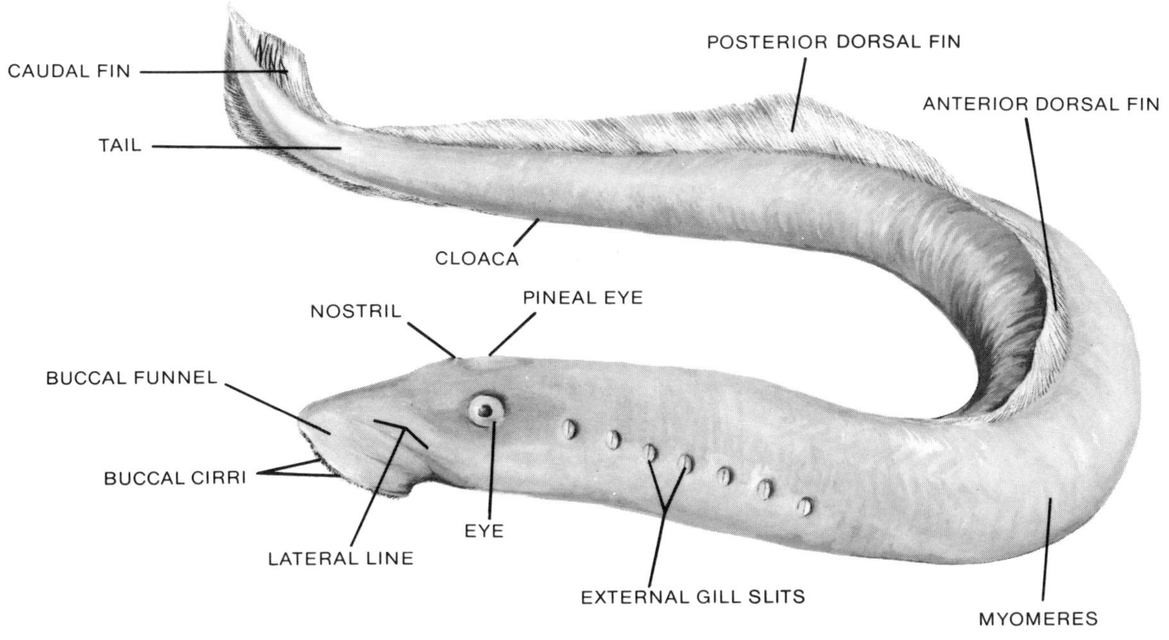

Figure 7-1. Lateral view of the Pacific coast lamprey, *Entosphenus tridentatus*, showing external morphology.

via the urogenital sinus and the urogenital papilla. Fertilization is external.

DISSECTION OF THE LAMPREY

The lamprey generally is examined in a sagittal section augmented by cross sections at various points along the body. This technique requires the use of two specimens to examine the various structures and to understand their dimensional arrangement, but it is quite an easy dissection. A more difficult dissection, which incorporates both sagittal and cross sections in one animal would require much greater skill and a knowledge of the structures beforehand.

The simple sagittal section of the lamprey consists of a single median cut extending almost the length of the animal. Usually a knife larger than a scalpel should be used, for a powerful grip is needed to cut through the cartilage of the body. Hold the lamprey in one hand, aligning it in its normal planes, as if it were swimming. Mentally sight the median line—through the nostril, the pineal, and along the mid-dorsal ridge—and with a slight sawing motion cut along the line. Preferably, the cut should also pass through the midventral line of the specimen. Extend the cut back to about an inch anterior to the cloaca. At this point, cut directly across the whole body—transversely—and separate the two sides for examination. Save the tail piece.

Cross sections are done with serial cuts at specific sites along the anterior-posterior axis of the lamprey. The ones illustrated in Figure 7.3 are found at the levels of the eye, the gill region, the posterior half of the anterior dorsal fin (the intestinal region), and the liver.

EXAMINATION OF THE SAGITTAL SECTION OF THE LAMPREY (Fig. 7-2)

The internal anatomy of the lamprey is adapted to its predatory existence. The sagittal section of the interior of the animal shows a huge amount of musculature and cartilaginous support necessary for the lamprey's feeding activities.

Buccal muscle. Around the buccal funnel is the circular buccal muscle. This muscle, also called the ***annularis***, firmly secures the lamprey to its prey once contact and suction have been established.

Annular cartilage. The annular cartilage, a ring of cartilage imbedded in the muscle, provides support for the buccal funnel.

Tongue. The tongue of the lamprey is the actual tool of the unique feeding system. It is covered by yellow horny teeth which are used to rasp away the flesh of the prey.

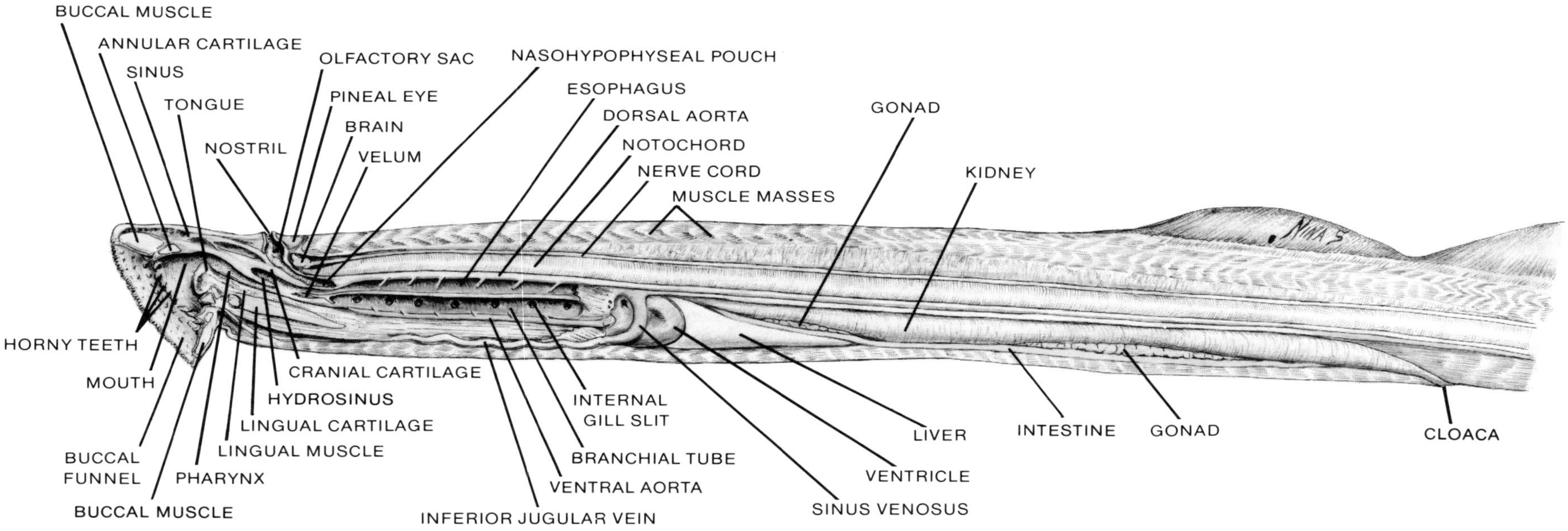

Figure 7-2. Sagittal section of the Pacific coast lamprey, from the head to the cloaca. The atrium has been removed.

Lingual cartilage. The rasping piston-like motions of the tongue are caused and supported by the lingual musculature and cartilage. The lingual cartilage rod is completely embedded in the tongue musculature, which extends posteriorly, ventral to the first gill slits, to an attachment with the pericardial cartilage.

Pharynx. Most of the cavities, valves and other structures of the head region are modified to accommodate having the mouth of the animal attached in feeding while respiration still goes on in the gill region. The ***pharynx*** is a thin cavity which connects the ***mouth*** with the ***esophagus*** and the ***branchial tube***. On the dorsal wall of the pharynx is an anteriorly projecting cavity, the ***hydrosinus***, which works with the tongue to insure a separation of feeding and respiration when the lamprey is attached to a prey. Farther down the pharynx are anteriorly projecting ***tentacles***, which are attached to the ***velum***, a flap of tissue which prevents the backflow of water from the gills (see below).

Brain and olfactory organ. The brain of the lamprey has been displaced posteriorly due to the increased muscular development around the mouth. Since there is only one nostril, there is only one olfactory sac, a dark sphere with lateral corrugations of nasal epithelium.

Nasohypophyseal pouch. Connected to the ventral side of the olfactory sac is the nasohypophyseal pouch, also called the pituitary pouch. The nasohypophyseal pouch appears to pump water back and forth over the nasal epithelium to ensure that the water sampled for smell is not stagnant. This feature may be quite important for the lamprey in detecting prey or proper spawning streams.

Pineal eye. Beneath a lightly pigmented area just posterior to the olfactory sac lies the pineal, which has a neuronal connection to the brain. The pineal organ detects diurnal light changes. This information is used by the brain to control the pituitary gland, which in turn alters the degree of pigmentation of the lamprey's skin.

Cranial cartilages and arcualia. The brain of the lamprey is ventral to the pineal eye and is partially encased in sheets of ***cranial cartilages***. Also present are the ***arcualia cartilages***, small strips of stiff cartilage dorsal to the notochord, which may be predecessors of or degenerate from true vertebrae. The presence of cranial and arcualia cartilages are among the bases for the classification of the cyclostomes as vertebrates.

Notochord. The notochord of the lamprey is the basic support structure for the body during motions of the W-shaped myomeres along the sides of the body.

Dorsal aorta. Just ventral to the notochord lies the dorsal aorta. This blood vessel carries oxygenated blood from the gills to the rest of the body. Anteriorly, it splits into two ***carotid arteries*** which feed the brain, eyes, and dorsal buccal musculature. If you are lucky in making your sagittal section, the ***inferior jugular vein*** may be bisected. This vein, immediately dorsal to the branchial basket cartilages which support the gill musculature, returns venous blood from the ventral head region to the heart.

Heart. The heart of the lamprey rests in the ***pericardial cavity*** and is surrounded by thick noncompressible ***pericardial cartilage*** that is very important in helping blood to return to the heart (see below). The lamprey heart pumps venous blood brought to it by several veins—two pairs of ***cardinal veins*** (anterior and posterior) which join on each side and enter the only common cardinal, the right; the ***hepatic vein*** draining the liver; and the ***inferior jugular***. All these vessels flow into the ***sinus venosus***, a thin-walled chamber which lies between the ***atrium*** and the ***ventricle***. The blood then passes through the sino-atrial aperture to the thin-walled atrium, which is often filled with clotted blood in preserved specimens. From the atrium, the blood passes through the atrio-ventricular aperture to be pumped out by the ventricle. The ventricle is the most muscular part of the heart and is on the right side of the animal. The fourth chamber of the heart found in higher vertebrates, the conus arteriosus, is really only a ring of tissue around the one exit vessel from the heart, the ***ventral aorta***.

The ventral aorta is an artery, since it carries blood away from the heart. It is an atypical artery, however, in that it carries venous blood from the heart to the gill regions for oxygenation.

Small intestine. Passing along the dorsal surface of the heart, parallel to the cardinal veins, is the small intestine. This yellow tube has a fold in its inner surface called the ***typhosole***, which increases the surface area of the intestine. There is no true stomach in the lamprey, but the transition from esophagus to intestine is marked by a valve.

Liver. Arising embryonically from the intestine, the lamprey's liver is a yellowish organ immediately posterior to the heart. The ***hepatic portal system*** routes the blood which absorbs food in the small intestine to the liver so that nutrients can be removed for processing or storage.

Gonad. Though the ammocoete shows evidence of paired immature gonads, only one is present in the adult. The single gonad occupies most of the visceral cavity. The gonad, producing either eggs or sperm, is supported by a ***mesentery***, a sheet of tissue extending downward from the median dorsal line. Gonadal products are released directly into the body cavity and escape the body of the lamprey through genital pores opening to the outside via the urogenital papilla. The testis is extensively lobulated, while the ovary is distinguished by the presence of many small eggs.

Kidneys. The pronephric kidney of the ammocoete, which would be dorsal to the liver in the adult lamprey, degenerates during metamorphosis. More posterior segments of kidney tissue develop; the adult lamprey has an opisthonephric kidney along each side of the dorsal visceral wall. The tubules pass their wastes into the ***archinephric ducts***, which may be seen extending posteriorly in the sheet of tissue separating the kidney from the visceral cavity. This kidney is more complex than the ammocoete's pronephric kidney in several ways: the glomeruli, or capillary tufts, are more complex, and there are no connections between the tubules and the body cavity. The archinephric ducts unite and form a ***urogenital sinus***, which empties into the cloaca through the urogenital papilla.

EXAMINATION OF CROSS SECTIONS OF THE LAMPREY (Fig. 7-3)

Most of the structures found in the representative cross sections of the lamprey were discussed in the examination of the sagittal section and should merely be examined with reference to their position. Two structures were not described in the sagittal section, the ***eyes*** and the ***pharyngeal glands***.

Pharyngeal glands. Ventral to the pharynx, in the head of the lamprey, lie a pair of pharyngeal or buccal glands, which produce a blood anticoagulant. Ducts carry this chemical to pores under the tongue, where it is released during feeding to combat the prey's blood-clotting mechanisms. The relationship of these glands to the ammocoete's subpharyngeal gland is unclear.

Eye. Each lateral eye of the lamprey is a typical vertebrate eye with some modifications. It has a clear round lens common to most fishes, the usual neural and pigmented retinas, and the six optic muscles that will be discussed in the shark. The lamprey eye has a peculiar manner of focusing an image. The lens, which lacks a well-developed musculature, sits close to the inner surface of the cornea. A set of cornealis muscles moves the lens closer or farther away from the neural retina as it pulls to varying degrees on the cornea of the eye. In this way, an image is focused on

Figure 7-3. Representative cross sections of the Pacific coast lamprey.

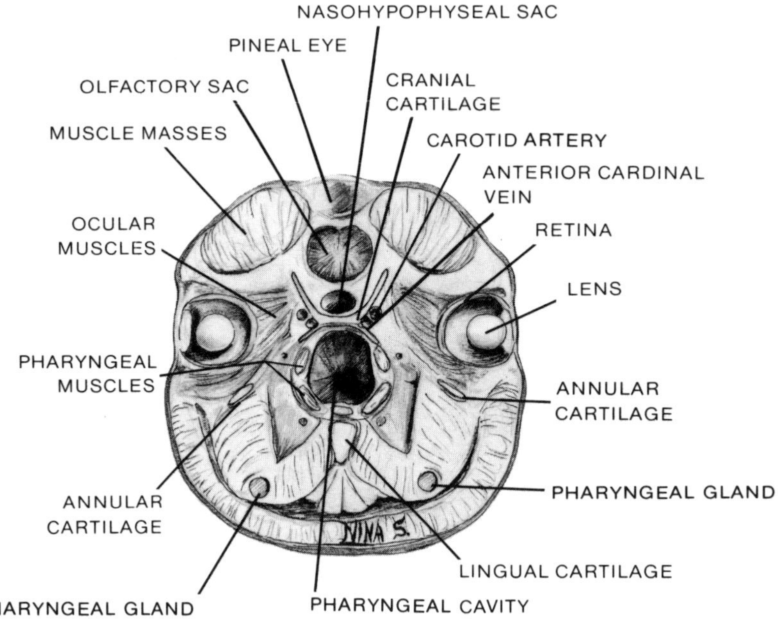

a. SECTION THROUGH THE EYES.

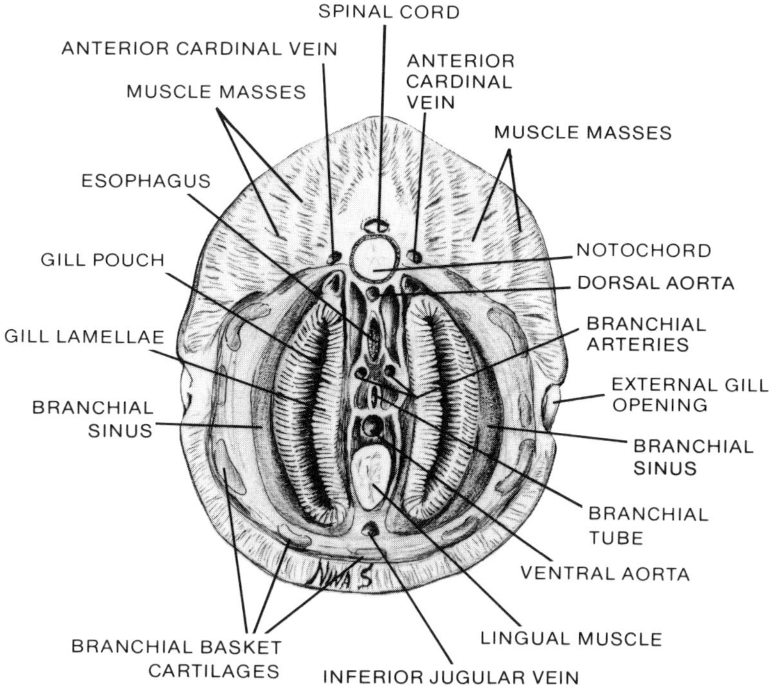

b. SECTION THROUGH THE BRANCHIAL TUBE.

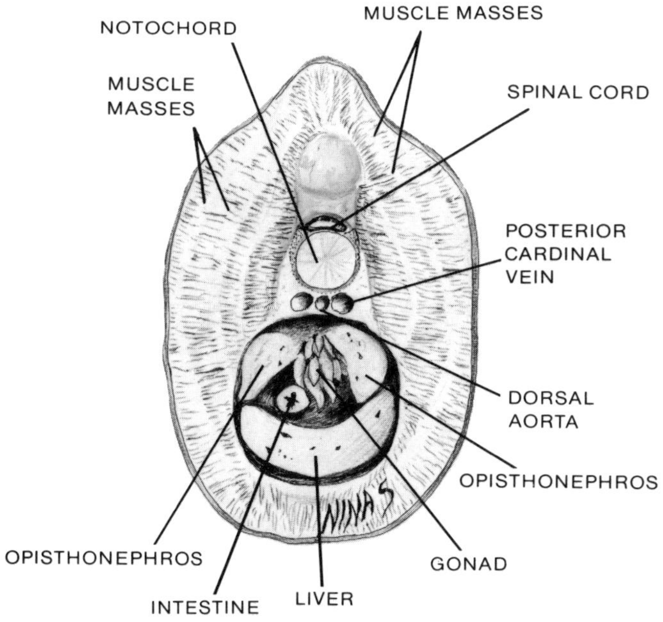

c. SECTION THROUGH THE LIVER.

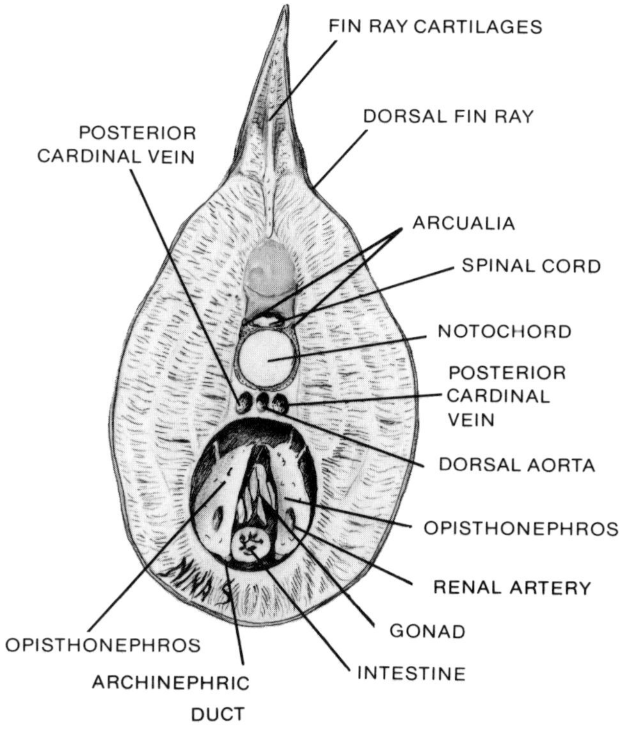

d. SECTION THROUGH THE POSTERIOR TRUNK.

the light-sensitive cells. This unique means of achieving focus is not seen in other vertebrates.

FUNCTION

The feeding process of the lamprey accounts for the functioning of structures of the head. The lamprey approaches a fish and presses the buccal funnel against its side, squeezing the water which was between the lamprey and fish into the hydrosinus. The tongue and the annular muscles pull back from the side like a suction cup, sealing off the pharynx and creating a very low pressure in the funnel. The normal water pressure in the pharynx is then changed by the contraction of the muscles around the hydrosinus. The water in the hydrosinus and the pharynx is forced back between the tentacles and the velum and out the gills. The hydrosinus expands, and the velum is sucked into place, preventing the anterior flow of water into the pharyngeal region. A low pressure area is created in the pharynx. The tongue then rasps the side of the prey so that it bleeds. Due to the suction created, there is little likelihood of detachment from the host.

For this system to work, the esophagus must branch off the pharynx posterior to the velum, and it does. The esophagus has a very small opening just dorsal to the tentacles, a site quite different from the more posterior connection of esophagus and pharynx in the ammocoete and hagfish. In the lamprey, the esophagus extends from its anterior origin straight back past the gill area to the small intestine.

The gill region is quite separate from the feeding areas in the adult lamprey and is primarily devoted to respiration. During feeding the branchial tube is completely closed off by the velum at one end and the pericardial cavity at the other. The tube functions only when the organism is not feeding. Water for respiration then enters through the mouth and passes through the velum and through the branchial tube to the internal gill slits, the openings to the actual gill filaments. The seven pairs of external gill pouches of the lamprey are unique among fishes. Due to the animal's need to respire even while the mouth is closed during feeding, the gill pouches and the musculature around them are adapted to water flow in two directions. This feature yields a different arrangement of gill filaments than that seen in higher fishes, for in the constantly changing currents within the gill pouches, countercurrent optimization for respiratory exchange is impractical.

The lamprey's circulatory system uses the stiffened pericardial cavity as a solid chamber, except for the incoming blood vessels. As the ventricle contracts, pushing the blood within it forward, a larger unoccupied space is present within the pericardial cavity because its outer walls remain rigid due to the cartilage. To fill this low pressure region, venous blood is sucked into the heart, filling the sinus venosus and the atrium. When the ventricle's muscles relax and expand again, they push against the thin-walled atrium, squeezing blood out of the atrium and into the ventricle. Another contraction causes the atrium to be filled again. Blood pressure in this system is very low, so even though the blood is pushed through the gill regions by the ventricle, the system is dependent upon the "suction pump" effect as well.

The great number of nephrons in the kidney of the lamprey, along with their glomeruli, raise questions as to the maintenance of osmotic balance by the lamprey. The large surface area in a glomerulus allows a great amount of fluid to pass from the blood into the excretory system. In fresh water, where the salt-free outside medium is, in a sense, "diluting" the lamprey, a great deal of water excretion is required. Thus, the glomeruli and nephrons perform a vital function. In the sea, however, where the marine lamprey lives for several years, such excretion would seriously dehydrate the animal. The lamprey relies on a number of adaptations to prevent dehydration: its tough skin minimizes absorption of salt by diffusion; salt-secreting cells in the gills get rid of any salt absorbed by diffusion; and circulatory controls on the kidney prevent excessive water loss. There appear to be actual valves on the blood vessels leading into the glomeruli which constrict in times of dehydration to lower the amount of fluid being filtered. These adaptations, along with a comparably watery diet and a fairly high concentration of salts in the body fluids, evidently save the lamprey from dehydration.

In summary, we see that the lamprey, though a relatively simple fish, is a conglomeration of specialized and primitive features. Those features can only be understood by studying the full functional biology of living specimens.

The Dogfish Shark

NEIL J. BAKER

8

The Dogfish Shark: External Morphology and Musculoskeletal System

In the early agnathans, the mouth consisted of a simple anterior opening which conducted a current of water to the pharynx for filter-feeding. Eventually, a more efficient method of obtaining food evolved: the skeleton and muscles of the anterior gill region were modified to form the jaws and jaw musculature.

The appearance of jaws was extremely important in the evolution of vertebrates. Filter-feeding was appropriate only for certain food sources and modes of existence. Jaws, however, allowed vertebrates to evolve into a tremendous number of new niches. For example, new predator-prey relationships emerged, since vertebrates could now eat other vertebrates as well as previously unavailable invertebrates or large plant materials. This well may have contributed to selection pressures for the development of more efficient locomotion. Greater maneuverability was made possible by the appearance of paired fins for fine control of vertical or horizontal changes in direction. Also, more complex nervous and sensory systems facilitated pursuit or escape. Equally significant modifications developed in the digestive system, including the appearance of the stomach, where the larger morsels of food could be stored and digested.

The first jawed vertebrates, the acanthodians, possessed jaws and paired fins on each side of the body. The descendents of the early jawed fishes eventually branched into two classes—the *Chondrichthyes* or cartilaginous fishes, including such animals as sharks and rays (the elasmobranchs), and the *Osteichthyes*, or bony fishes.

EVOLUTION

Thoughout their three hundred million years of history, sharks have retained some body features characteristic of ancient vertebrates, such as the series of gill slits with separate openings to the exterior. At the same time, they have highly specialized reproductive structures and a relatively well developed nervous system.

There are more than three hundred species of living sharks. They range in size from the 6- to 8-inch midwater shark to the 45-foot whale shark; in temperament they range from the lethargic basking shark, which has reverted to filter feeding, to the ferocious human-eating great white shark. Despite this great diversity, most species of sharks have nearly the same streamlined body shape. In the dogfish, for example, the head is flattened dorsoventrally, while the trunk is flattened laterally and tapers off toward the tail.

Such body streamlining minimizes the energy the shark must expend in swimming. This is essential because, lacking the swim bladder present in bony fishes, sharks may not always achieve neutral buoyancy. Sharks do contain certain oils with specific gravities lower than water, which are concentrated in the liver and lighten the body. Sharks are fairly large carnivores. They move constantly at slow or moderate speeds and depend on short powerful bursts of speed for the capture of prey.

Perhaps to lighten the body for the shark's constantly roving predatory existence, the bony plates which protected their ancestors are reduced in sharks to microscopic scales embedded in the skin. The skeleton itself is lightened by its entirely cartilaginous composition. In most other jawed vertebrates, a completely cartilaginous skeleton is a feature only of embryos or the very young.

THE DOGFISH SHARK

According to most academic sources, the dogfish sharks are named both for their tendency to hunt in packs and for their side-to-side "sniffing" movements reminiscent of dogs. Old seafarers, however, contend that the dogfish is named for its unusual symbiotic relationship with the unique "fire-hydrant fish." Despite intensive research, this relationship has not yet been documented.

Sharks have long been thought of as offensive creatures. Dogfishes are a nuisance to fishermen, whose catches and nets they often destroy. As long ago as the sixteenth century, Rabelais found the flesh of a shark foul smelling, loathsome, and hard to digest. In fact, shark steaks are so delicious an alternative to tuna, that there is now grave fear that Atlantic and Pacific shark populations will be decimated. Unlike bony fish (such as tuna), sharks produce very small numbers of young after long gestations. Boycott shark meat!

Dogfishes are graceful and beautiful creatures when alive. They are classified in the order of "spiny dogfishes"—***Squaliformes***, named for the stiff spine in front of each dorsal fin. Those which inhabit the Pacific are the *Squalus suckleyi*, while those in the Atlantic are the *Squalus acanthias*. *Squalus* species are generally 2 to 3 feet in length; they migrate north in spring and south in the fall in schools of thousands. They have few predators except for the species *Vertebratus labstudentias*!

EXTERNAL ANATOMY (Fig. 8-1)

Body divisions. The body may be roughly divided into the ***head***, which includes the gill region; the ***trunk***, extending from the gills to the cloaca; and the posterior ***tail***.

Snout. The area of the head anterior to the eyes is designated the snout.

Eyes. The eyes are laterally situated on the head. Note the folds of skin which cover the outer margins of the eyeball.

Mouth. The ventral position of the mouth facilitates seizure of prey swimming below the shark. Pockets of tissue at the corners of the mouth are called the ***labial pouches***. The furrows called ***labial grooves*** are the anterior extensions of each pouch.

Spiracle. The spiracle is a modified gill slit on each side of the head located just posterior to the eyes. A fold of tissue closes the spiracle to prevent water from leaving during part of the respiratory cycle. Attached to the spiracle's internal wall is a small strip of vestigal gill lamella called the ***pseudobranch***.

Endolymphatic pores. Two holes located mid-dorsally between the spiracles are the endolymphatic pores. The endolymphatic ducts of the inner ear open to the surface through these pores.

External nares. These are openings in the ventral snout, divided into incurrent (lateral) and excurrent (medial) apertures for water entering and leaving the olfactory sacs (Fig. 8-11). They do not connect with the mouth cavity.

Ampullae of Lorenzini. The pores of these sensory organs are located on many parts of the head, including the dorsal and ventral snout.

Figure 8-1. Lateral view of an adult dogfish shark, showing external morphology.

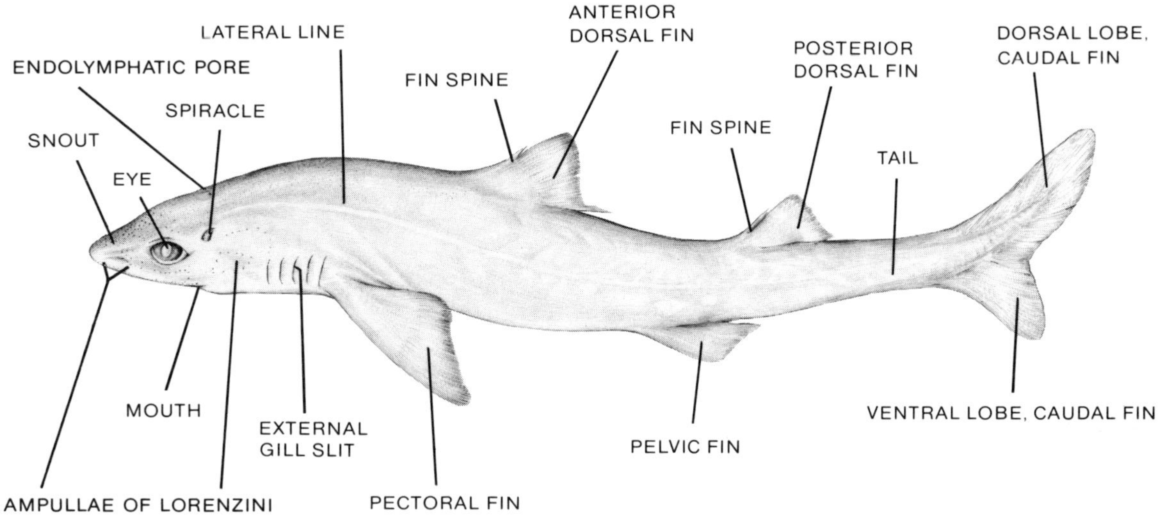

Lateral line. The main lateral line is visible as a white line extending along the trunk. Other smaller branches extend into the head. (See Chapter 14 for a discussion of the lateral line system.)

External gill slits. These are openings from the five gill pouches to the exterior. Insert a probe into one of the slits and note the oblique angle of the gill chamber with respect to the vertical axis of the animal.

Pectoral fins. The pectoral fins are paired anterior appendages which control changes in direction during swimming, particularly in the dorso-ventral plane.

Dorsal fins. The dorsal fins are median dorsal fins which serve as stabilizers.

Fin rays. (Ceratotrichia) The fin rays are fibrous rods supporting the fins. These are visible only in a prepared skeleton, or through careful dissection.

Denticles. These are microscopic scales embedded in the skin. Rub your hand anteriorly across the skin to detect the caudally projecting spines.

Spines. Enlarged denticles called spines are found anterior to each dorsal fin. They may help to hold the fin rigid.

Pelvic fins. These are paired appendages which serve as stabilizers. They mark the posterior end of the trunk.

Claspers. The claspers are extensions of the pelvic fins in males which serve as the copulatory organs.

Cloaca. This is a chamber between the pelvic fins through which digestive and excretory wastes and genital products are released.

Abdominal pores. These are small openings on each side of the cloaca which connect the body cavity (coelom) to the exterior. Excess coelomic fluids may be vented here. However, they frequently are not open, and their function is unknown.

Caudal fin. In the caudal fin the vertebral column curves upward into the dorsal lobe. The ventral lobe of the fin is smaller than the dorsal lobe; thus, the tail is classified as a "positive heterocercal" one.

FUNCTION

The function of many of the structures listed above will be discussed in subsequent chapters. Here we will deal only with those specifically relevant to external anatomy.

Skin and Its Derivatives

The skin consists of an outer layer of epithelium (the epidermis) and a much thicker inner tissue (the dermis), composed of densely packed fibrous connective tissue. The skin contains no fat deposits and, as a result, conforms very tightly to the streamlined body.

As with many aquatic organisms, the dorsal and ventral surfaces of the dogfish have contrasting dark and light colorations. With such shading, the animal blends in with the depths when viewed from above and with the lighter surface waters when viewed from below. The dark pigment granules are contained within ***melanophores*** in the upper dermis close to the epidermis. Changes in shading occur when the granules either "wander" or aggregate close together, probably as the result of hormones released from the pituitary gland.

The ***denticles*** or ***placoid scales*** are thought to be evolutionary remnants of the bony armor of ancestral placoderms. Each one consists of a basal plate anchored in the dermis and a spine which projects out of the epidermis, pointing caudally. Both the plate and the spine have an internal pulp cavity containing blood vessels which supply nutrients to the thick outer layer of dentine. The surface of the denticles is covered with a hard material variously identified as vitrodentine or as enamel.

The teeth of the shark are somewhat similar to the denticles with regard to their mode of formation and dentinal composition. In fact, these two structures are homologous, since the mouth lining from which the teeth develop is derived from inturned skin. All the teeth are adapted for cutting and shearing and are more or less of the same shape; near the angle of the jaws, however, they decrease in size. (In some sharks, teeth on the upper jaw are narrow and pointed for penetration and holding; those on the lower jaw are broader-based and used for cutting out pieces of seized prey.) The teeth are arranged in several rows, the outermost of which are functional. As outer teeth are worn away, rows of teeth produced in the mouth tissues gradually grow and move outward to the edge of the jaws in a constant replacement process.

Fins and Locomotion

Sharks retain the primitive mode of generating thrust by the serial contraction of segmental muscles. The alternating waves of contraction which proceed down each side give the appearance of a sine wave

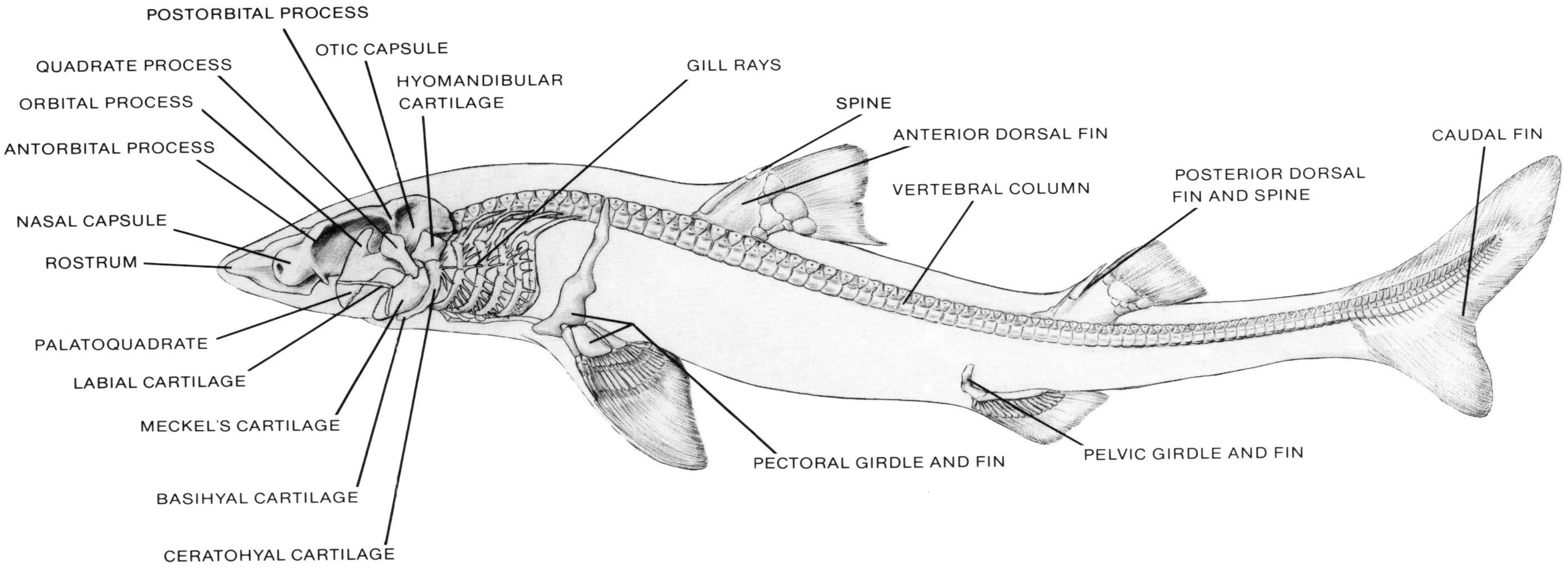

Figure 8-2. Lateral view of an adult dogfish shark, including full skeleton.

traveling along the body. These "waves" push laterally and posteriorly against the water to provide forward thrust. The main thrust is produced by the tail, which is moved back and forth with the largest force and through a greater distance than any portion of the trunk.

In fact, when a shark (or bony fish) cruises along at a slow speed, a portion of each myotomal muscle mass contracts actively. This is "red muscle," which may appear dark in fixed or cooked sharks or bony fish and which is biochemically suited for sustained activity. The rest of each myotome is "white muscle"—tissue adapted for brief, intense contractile activity. Thus, the great bulk of the white muscle "goes along for the ride" without working during normal swimming. White muscle is called into play when a prey item is spotted or a predator must be avoided.

In the tail, the vertebral column and spinal cord curve upward; the dorsal lobe of the caudal fin is larger than the ventral lobe. When such a tail is moved laterally, the water is forced at an angle downward and backward. In accordance with Newton's Second Law, the tail is then forced in the opposite direction—forward and upward (e.g., lift is generated at the tail). The head would then tend to angle downward as the animal swims. The combined use of the waves of myotome contractions and the dorsal and ventral lobes of the caudal fin may be considerably more complicated than we describe here; accordingly, some sharks can "fine tune" their propulsion and lift-generating systems.

The tendency for the heterocercal tail to rise and the head to sink is counteracted by the ventrally flattened snout and the pectoral fins. Since the snout is flattened on its ventral side, it forms a plane which pushes against the water as the animal swims and thus forces the head upward.

To change its vertical direction, the animal simply alters the angle at which it holds the pectoral fins; when horizontal they give no net lift, but when tilted up at the front edge lift is produced. In order to turn in a horizontal direction, a shark may change the angle of only one pectoral fin. At the same time, varying numbers of myomeres along the same side of the body may be contracted simultaneously to curve the body to the right or left.

The dorsal fins serve as stabilizers to prevent rolling and excess lateral turning because of the side-to-side movements of the tail. During swimming, the pelvic fins undulate and probably help stabilize the animal.

Note that none of the fins except the caudal are involved in generating propulsive forces. On the other hand, in skates and rays the pectoral fins are expanded into large, flat, "wing-like" structures. Sequential dorso-ventral waves pass posteriorly along such fins, push backwards on the water, and thereby generate thrust forward.

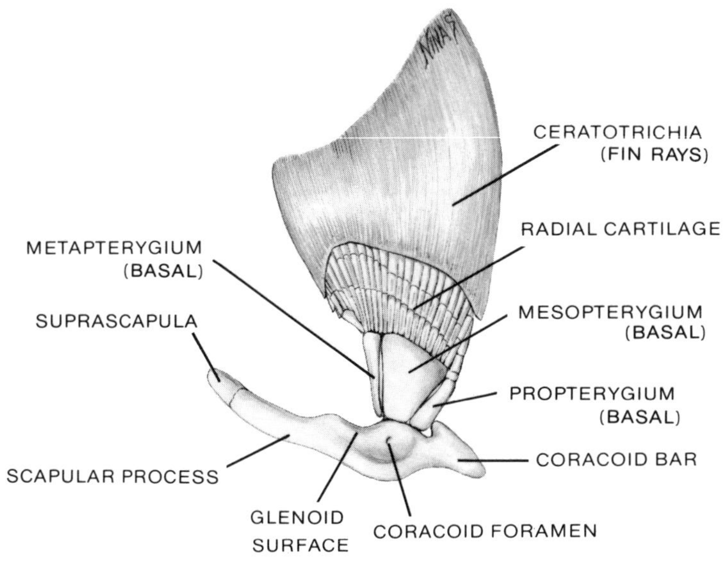

Figure 8-3. Left lateral view of the pectoral girdle and fin.

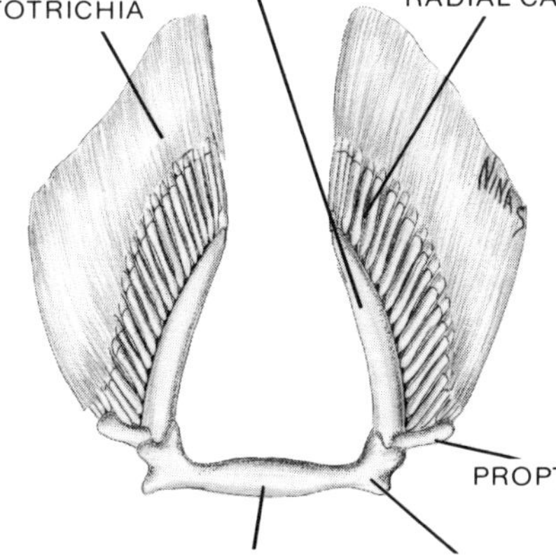

Figure 8-4. Dorsal view of the pelvic girdle and fins of a female shark.

THE MUSCULOSKELETAL SYSTEM

Because of the cost of prepared skeletons, the skeleton and muscles of the dogfish may prove inconvenient to examine in some labs. However, a general familiarity with these systems is essential in understanding various aspects of shark anatomy which will be studied in later chapters.

The musculoskeletal system may be divided functionally into two divisions. Locomotion is a function of the ***somatic division***, while the ***visceral division*** is associated with feeding, digestion, and respiration. (The terms ***somatic*** and ***visceral*** usually have much broader definitions; at this point, however, only their applications to the musculoskeletal system will be considered.)

The skeleton of each of these divisions is entirely cartilaginous. In larger sharks, some calcification does take place—in the vertebrae, for example, for strengthening. This calcification of cartilage occurs in a completely different way from bone formation, or ossification.

Terminology concerning the skeleton and muscles associated with the visceral arches may seem confusing; however, the following Greek derivatives may be helpful:

Branchial. From the Greek *branchia*, meaning "gills."

Epibranchial. (*Epi*-, "above") Above the gills.

Hypobranchial. (*Hypo*-, "below") Below the gills.

Basibranchial. (*Basi*-, "base") At the base of the gills.

SOMATIC DIVISION: AXIAL SKELETON AND MUSCLES (Fig. 8-2 through 8-8 and 8-10)

The axial skeleton consists of the braincase or ***chondrocranium*** and the vertebral column.

Chondrocranium. Unlike the skull of mammals which consists of several bones, the chondrocranium of sharks is essentially a single block of cartilage. It houses the brain and the major sensory organs of the head, including the olfactory sacs, eyes, and inner ears.

The brain is located within the ***cranial cavity***, which is enclosed by the medial portions of the chondrocranium. The cavity extends as far forward as the posterior edge of the ***rostrum*** (Fig. 8-5), an oval cartilage supporting the snout. Ventrally, the floor of the cavity is very narrow except caudally where it widens to form the ***basal plate*** (Fig. 8-6).

The olfactory and auditory organs are enclosed within cartilage, whereas the eyes are suspended by the ocular muscles. The ***olfactory sacs*** are enclosed in spherical shells of cartilage called the ***nasal capsules*** located on each side of the rostrum. In Fig. 8-6, the openings for the ***external nares*** may be seen in the ventral surface of the capsules.

Orbit. Each eye is situated within a recess in the side of the chondrocranium known as the ***orbit***. The anterior wall of the orbit is continuous with the

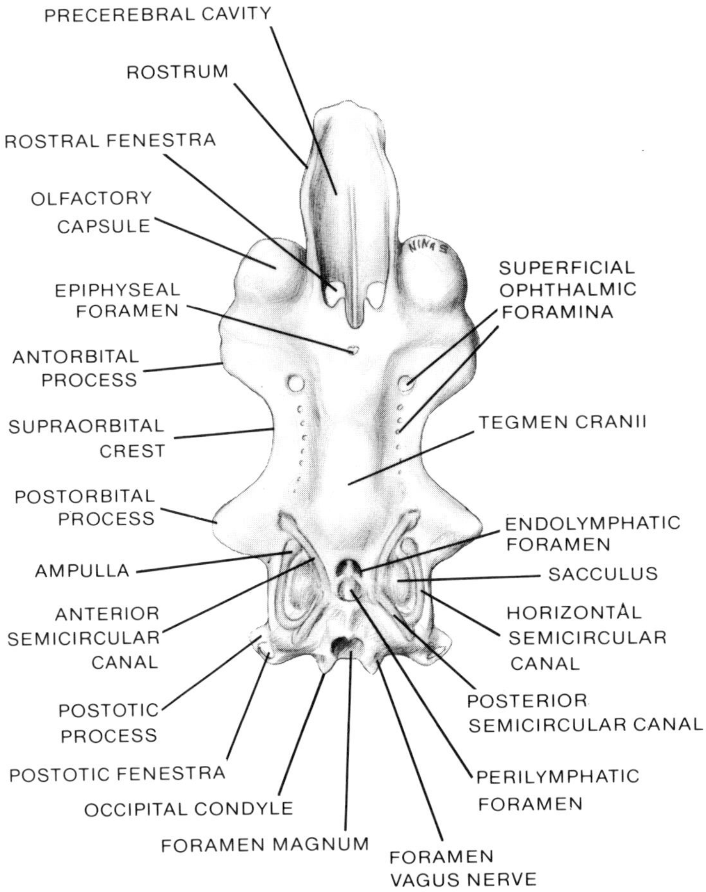

Figure 8-5. Dorsal view of a dogfish chondrocranium and associated ear structures.

posterior side of the nasal capsule, and the posterior wall bears a triangular, lateral projection of cartilage, the ***postorbital process***.

Supraorbital crest. Dorsally, a shelf of cartilage called the supraorbital crest extends over a portion of the eye.

Otic capsules. The inner ears are embedded in box-shaped otic capsules just posterior to each orbit.

Other structures and the foramina labeled in these drawings will be referred to in later chapters.

Vertebral column. **(Fig. 8-7)** This is the main structural element on which the myomeres act to whip the body and tail back and forth in swimming. Both trunk and tail vertebrae have a dorsal ***neural canal*** which contains the spinal cord. The sides of the canal are formed by the ***neural arch*** which extends dorsally as a narrow "spine." Accessory plates of cartilage fill out the spaces between the spines in the roof of the canal. Ventrally, the vertebral column consists of ***centra***, biconcave discs of calcified cartilage. Tiny remnants of the notochord are found in the central portion of each centrum, whereas the notochord appears to be expanded in the intervertebral spaces—the result is to give the notochord a "beaded" appearance.

Tail vertebrae may be distinguished from those of the trunk by the presence of a ventral ***hemal arch*** surrounding the ***hemal canal***, which contains the caudal artery and vein (Fig. 8-8).

Axial muscles. The axial musculature consists chiefly of the myomeres involved in swimming and therefore forms the bulk of shark muscle. As in amphioxus and the lamprey, the myomeres have a zig-zag pattern which extends their action over a longer portion of the body; this produces smoother waves of contractions along the body.

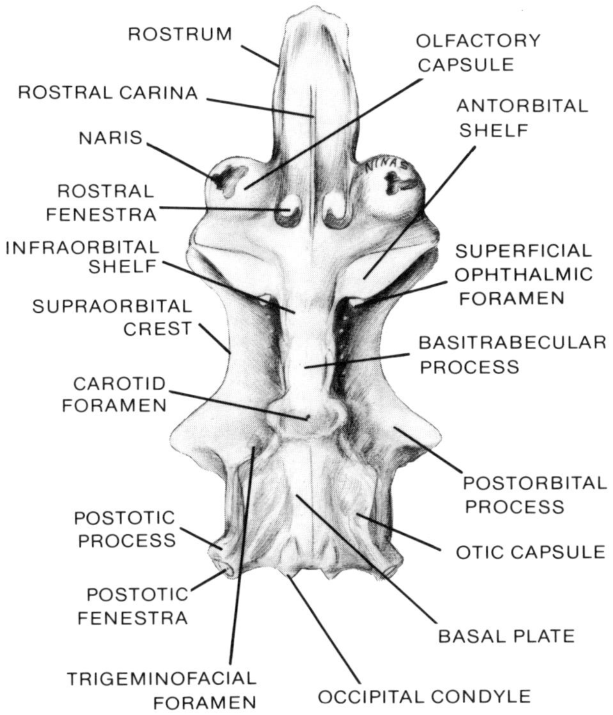

Figure 8-6. Ventral view of a dog fish chondrocranium.

(Note: It may be helpful to steam or gently boil one dogfish or piece of dogfish tail containing several myomeres. After such treatment, individual trunk myomeres may be separated from each other so that their true shapes and relationships may be appreciated.)

Epaxial and hypaxial. A horizontal septum of connective tissue extends from the vertebral column to the body surface (Fig. 8-9) at the level of the lateral line. This divides the myomeres into dorsal and ventral sections called ***epaxial*** and ***hypaxial*** musculature, respectively. In the trunk region, the epaxial muscles are bulkier. (Most of each epaxial mass is "white muscle," and is not used for routine swimming. Your dissection may reveal a dark stripe of "red muscle" extending down the side of the body; these are the muscle fibers that usually propel the shark.)

The hypaxial musculature of the trunk forms a sheath for the body cavity or coelom. Since the coelom does not extend into the tail, the hypaxial muscles are much bulkier there than in the trunk.

Epibranchial. (Fig. 8-10) The gill region disrupts the myomeres along the sides of the body. Therefore, the epaxial musculature is limited to thin strips of myomeres above the gills called ***epibranchial muscles***. These extend forward to cover the posterior half of the otic capsules.

Hypobranchial. Beneath the gills, the myomeres are modified extensively to form the ***hypobranchial musculature*** which aids in opening the mouth, feeding, and swallowing. This musculature also strengthens the floor of the mouth and the walls of the pericardial cavity.

Except for the ***common coracoarcuals***, the hypobranchial muscles are concealed by the ventral constrictors of the gill arches (see Fig. 8-11). The coracoarcuals form a triangular mass just ventral to the pericardial cavity; they are attached posteriorly to the pectoral girdle.

Figure 8-7. Craniolateral view of a typical series of trunk vertebrae of a dogfish shark.

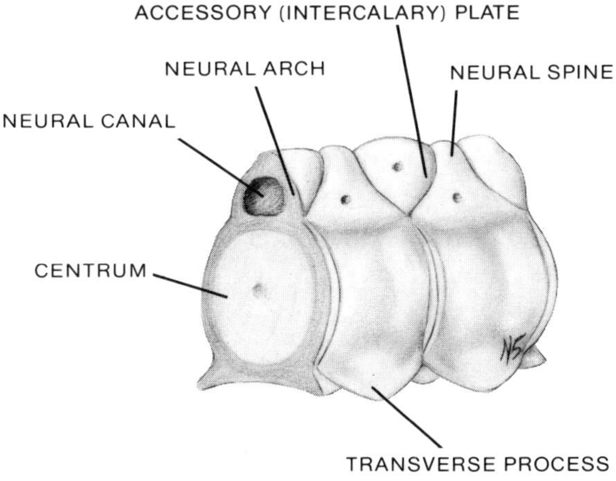

Figure 8-8. Craniolateral view of a typical series of caudal vertebrae of a dogfish shark.

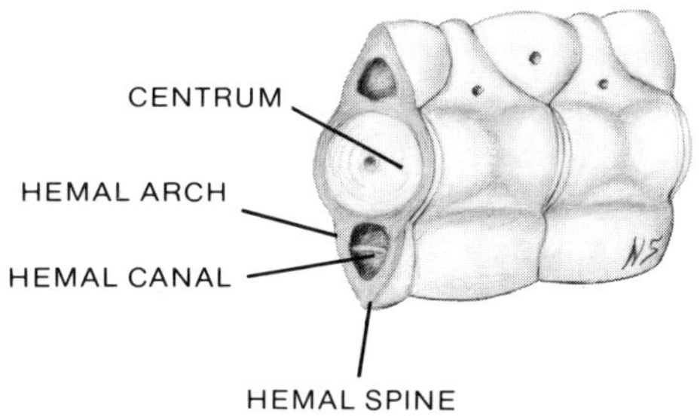

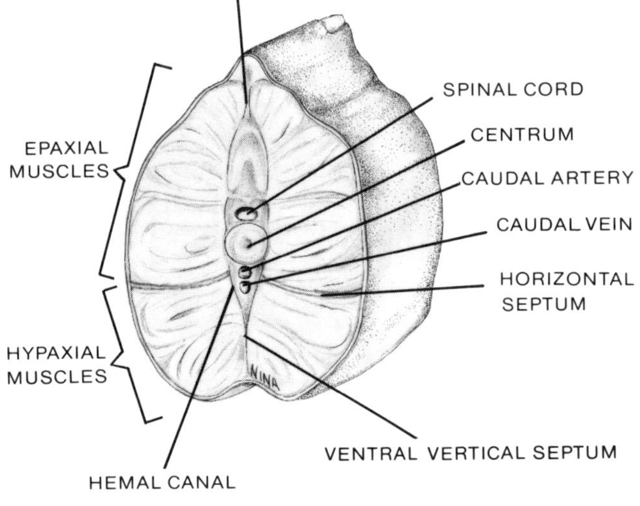

Figure 8-9. Craniolateral view of a cross section of tail musculature of a dogfish shark.

SOMATIC DIVISION: APPENDICULAR SKELETON AND MUSCLES (Fig. 8-2)

The appendicular skeleton of the shark includes cartilages in the paired fins and supporting structures in the body called girdles. Both the ***pectoral*** and ***pelvic girdles*** (Fig. 8-2) are embedded in surrounding tissue and do not articulate with any part of the axial skeleton.

Pectoral girdle. **(Fig. 8-3)** The pectoral girdle consists of a curved bar of cartilage immediately posterior to the last gill slit; the girdle articulates with the three large ***basal cartilages*** of each pectoral fin. More laterally in the fins, a series of numerous smaller, ***radial cartilages*** fan out from the basals and attach to the ***fin rays.***

Pelvic girdle. **(Fig. 8-4)** The pelvic girdle consists only of a ventral ***pubioishiac bar***. Each pelvic fin articulates at the ***acetabular surfaces*** on the bar by two basal cartilages. In males, several ***radial cartilages*** extend from the basal into the fin. These cartilages are lengthened and modified to form the skeleton of the claspers.

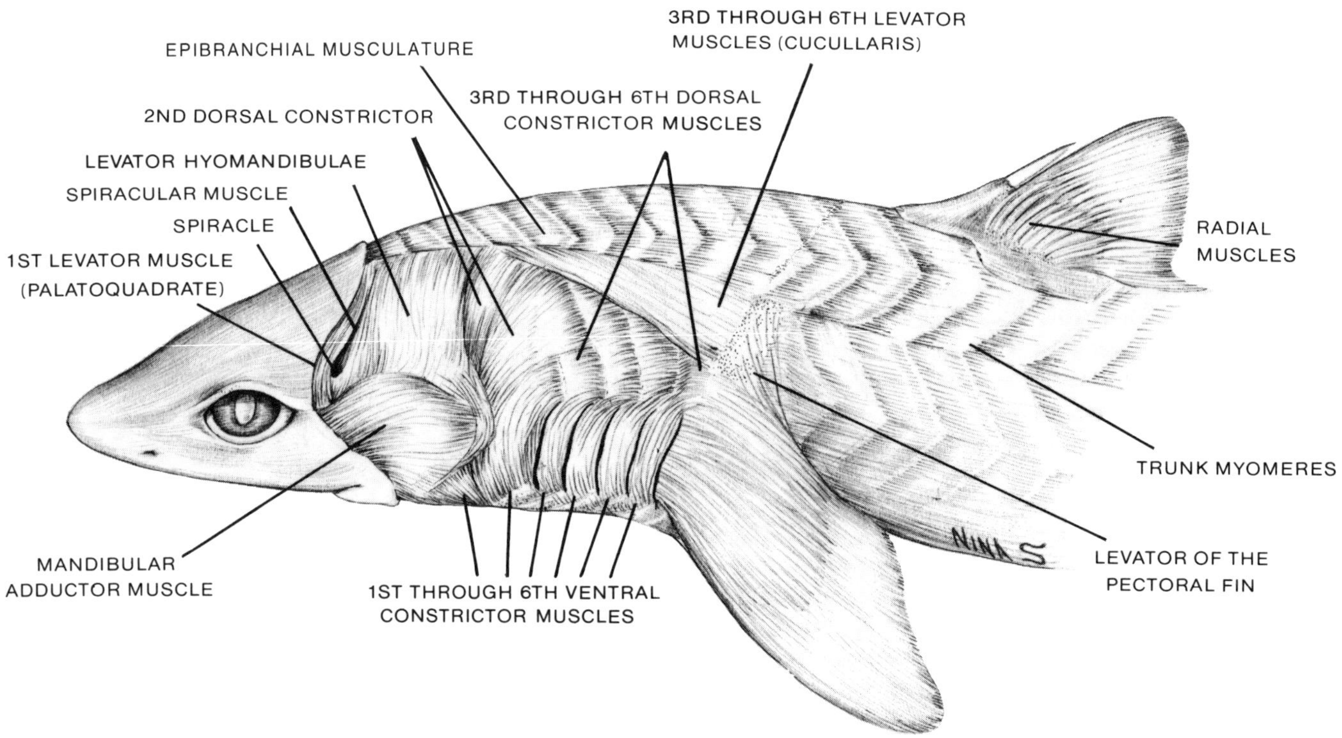

Figure 8-10. Lateral view of the head and anterior trunk of a dogfish, with the superficial skin removed, showing appendicular and branchiomeric musculature.

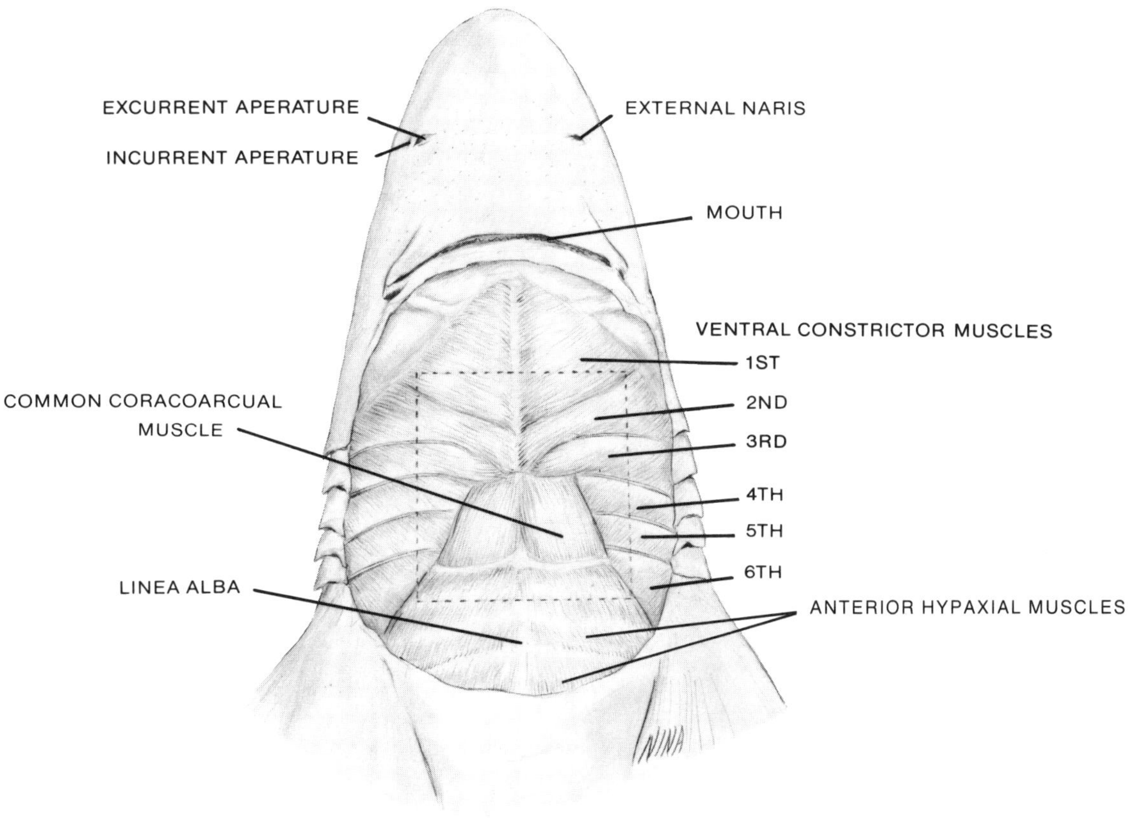

Figure 8-11. Ventral view of the head and anterior trunk of a dogfish, with the skin removed. The superficial hypobranchial musculature is indicated.

The appendicular muscles consist of small dorsal and ventral masses that elevate and depress each fin (see Fig. 8-10).

VISCERAL SKELETON OR SPLANCHNOCRANIUM (Figs. 8-12, 8-13, 8-14)

The visceral skeleton is composed of a series of seven cartilaginous structures called ***visceral arches***. In the jawless vertebrates ancestral to sharks, the visceral arches all supported structures and tissues of seven hypothetical gill pouches in the branchial region. It is thought that the first two arches were modified to form the jaws and jaw supports, while the five remaining visceral arches were retained as gill supports. These five are known as the ***branchial arches*** or ***branchial basket***. Therefore, the first *branchial* arch in sharks is also the third *visceral* arch (See Fig. 8-13).

The term *branchial arch* may refer not only to skeletal structures but also to the tissue between gill slits, including muscles, nerves, blood vessels, and lamellae. We will restrict usage of "branchial arch" to the cartilaginous support. ***Pretrematic*** tissues are those located in the anterior wall of a gill slit; those structures in the posterior wall are ***posttrematic***.

Since the first two visceral arches, the ***mandibular*** and ***hyoid arches***, are derivatives of branchial-type arches, they will be discussed after the branchial arches are considered.

Branchial Arches (Visceral Arches 3-7)

The branchial arches (Fig. 8-12) are a series of five horseshoe-shaped structures which curve around the branchial region, passing between the gill slits. They do not articulate with any part of the somatic skeleton. Each side of each arch is formed by a series of three cartilages. Ventrally, the arches are united by three pairs of ***hypobranchial*** cartilages which articulate medially with a small anterior and a large posterior ***basibranchial*** cartilage (Fig. 8-14).

Along the inner margin of the cartilages forming the sides of each arch are a series of small projections

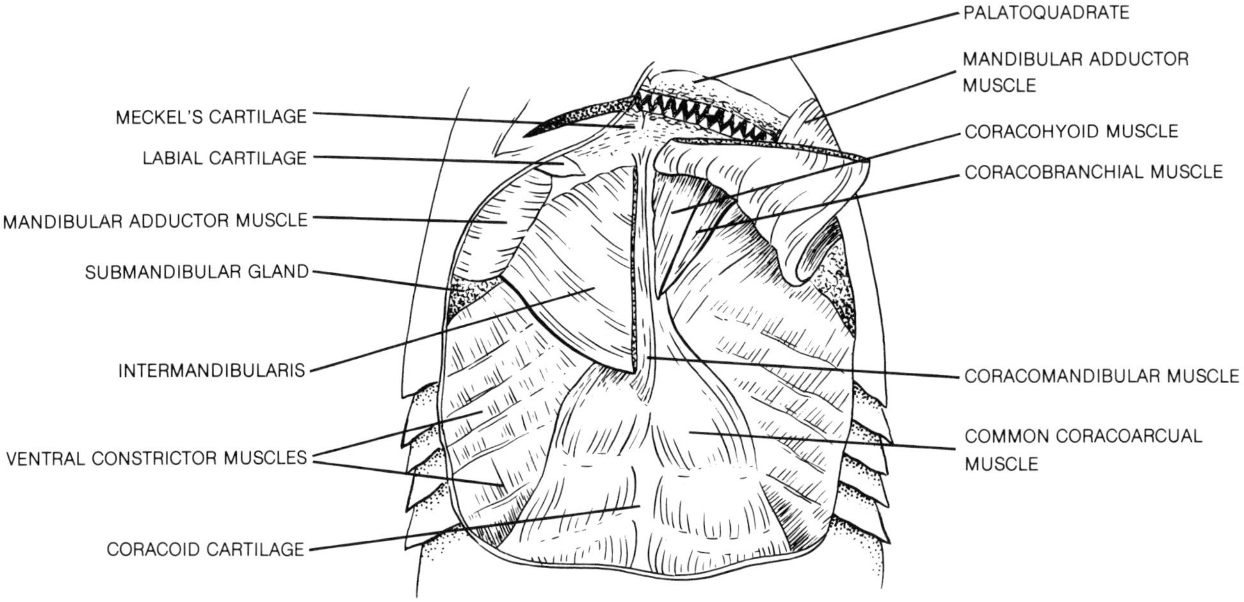

Figure 8-12. Ventral view of the jaws and hypobranchial musculature of a dogfish, with deep muscles exposed.

called ***gill rakers***. These point forward into the cavity of the branchial region (the pharynx) and prevent food from entering the gill slits (Fig. 8-15). Attached to the outer lateral margins of these same cartilages are the ***gill rays***, a series of longer projections which help to support the gill structures. The rays also support the sheets of skin that separate consecutive gill slits and extend outward to form flaps covering each succeeding gill slit.

On each side of the septum containing the gill rays are numerous strips of tissue called ***gill lamellae*** (Fig. 8-16). These structures form the actual respiratory surfaces (see Chapter 12). Only the last arch embedded in the posterior wall of the last gill slit possesses no gill rays or lamellae.

Branchial muscles. **(Fig. 8-10)** Contraction of the branchial musculature compresses the gill chambers, thereby forcing water across the lamellae and out the gill slits. Re-expansion of the branchial basket occurs by elastic recoil rather than by muscular action. Three groups of muscles—the ***dorsal constrictors***, ***ventral constrictors***, and ***levators***—are associated with each of the branchial arches except the fifth.

Constrictors. Since the fibers of the dorsal and ventral constrictors of each arch run together, the two muscles are not easily distinguished. The constrictors of each arch form a sheet of muscle which overlaps each succeeding constrictor by half its width. They do not insert on the cartilagenous skeleton but adhere to vertical sheets of connective tissue between the consecutive muscles.

Levators. The ***levators*** of each arch are united to form a single muscle mass called the ***cucullaris***, above the series of constrictors, which inserts on the pectoral girdle and last branchial arch. Beneath these superficial muscles are small muscles which extend between the gill arch cartilages and pull them together to assist in compression of the branchial basket.

Hypobranchial muscles. **(Fig. 8-12)** These muscles consist of the common ***coracoarcuals***, the ***coracomandibular***, the ***coracobranchials***, and the ***coracohyoids***. These muscles play an important role in opening the mouth and provide support for the pharynx. To expose the deeper ones, make a midventral incision beginning just anterior to the coracoarcuals and extend it to the lower jaw. The coracomandibular is most anterior and attaches to the jaw. Make an incision into the posterior part of this muscle and reflect it anteriorly to expose the underlying thyroid.

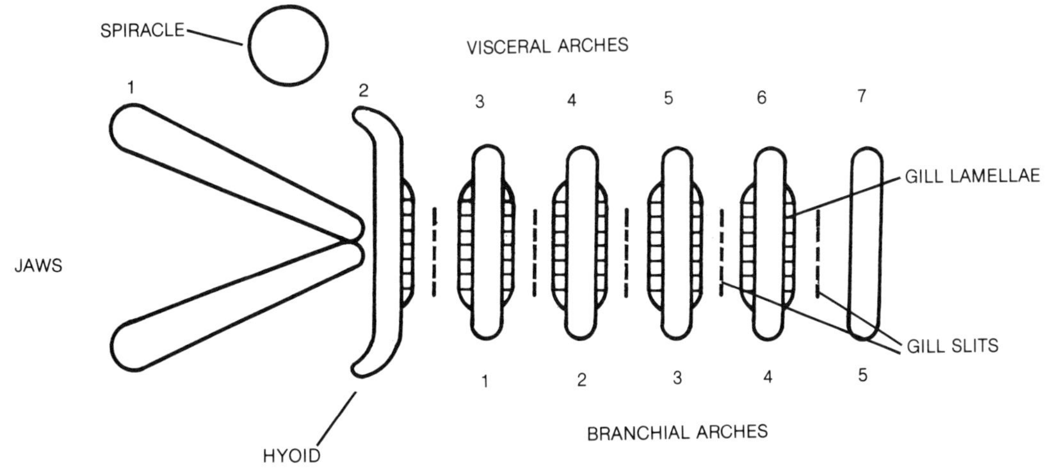

Figure 8-13. Diagram of the visceral arches 1–7 and the branchial arches (1–5) of a dogfish.

Figure 8-14. Ventral view of the chondrocranium and splanchnocranium of a dogfish.

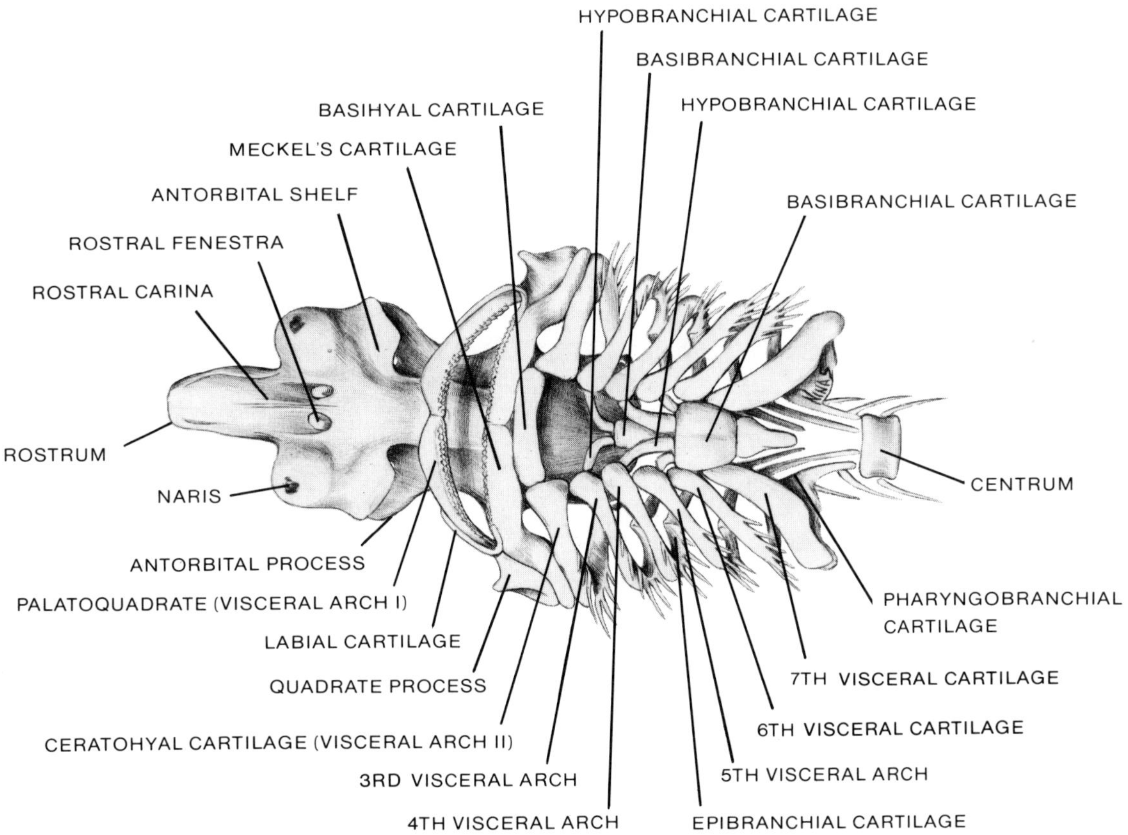

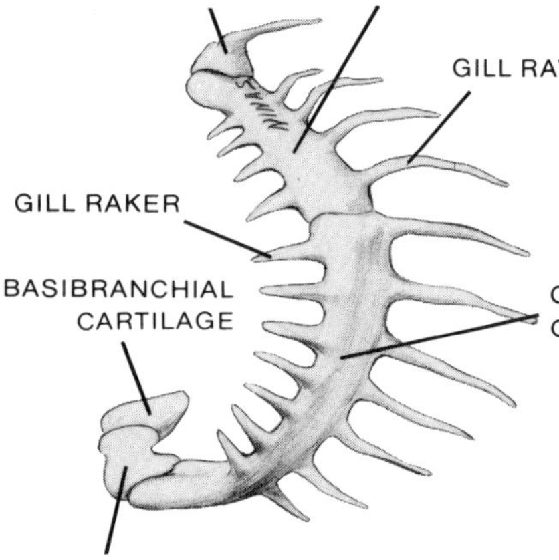

Figure 8-15. Craniolateral view of an intact half of a branchial arch showing component cartilagenous structures.

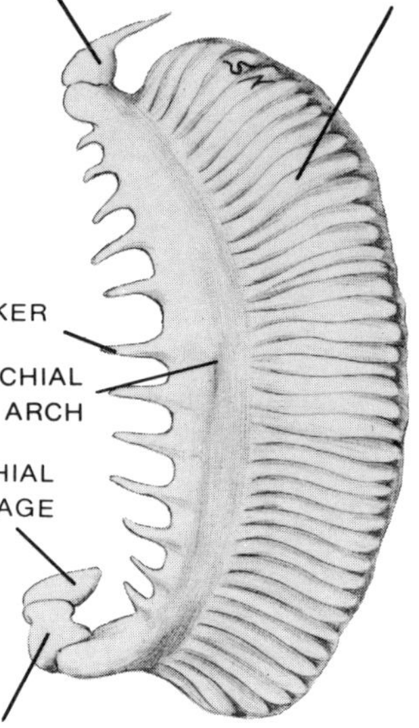

Figure 8-16. Craniolateral view of an intact half of a branchial arch with a full complement of lamellar gas exchange tissue.

Mandibular and Hyoid Arches (First and Second Visceral Arches) (Figs. 8-12, 8-13)

As the mouth openings of gnathans were enlarged during the course of evolution to incorporate the anterior gill slits, the branchial cartilages of the first visceral arch were apparently bent forward to form the jaw cartilages or ***mandibular arch***. The upper jaw is called the ***palatoquadrate***, and the lower jaw is known as ***Meckel's cartilage***. (See Fig. 8-12.)

The hyomandibular, ceratohyal, and basihyal cartilages make up the ***hyoid*** or second arch. This arch supports the jaws by means of ligaments which attach the hyoid to the mandibular arch. In addition, ligaments brace the dorsal hyomandibular cartilage against the otic capsule. The ***spiracle***, which was previously mentioned under external anatomy, is a vestige of the gill slit originally located between the first and second visceral arches.

Unlike the upper jaws in mammals such as cats or humans (which are fused to the skull), the palatoquadrate is supported only at its posterior end by the hyoid arch. At its anterior end it has a loose ligamentous articulation with the braincase so that the anterior end of the jaw may actually protrude somewhat during biting. In addition, the posterior end of the upper jaw is quite mobile, permitting a wide gape when the jaws are opened.

Branchial muscles of the first and second visceral arches. Constrictor and levator muscles of the first two visceral arches are modified according to new skeletal functions (Fig. 8-10). For the mandibular arch, the constrictors are divided into several muscles, the most important of which is the powerful ***mandibular adductor*** which closes the jaws. The "second dorsal" or "hyoid" constrictor is much broader than other constrictors and aids in compressing the gill chambers. In bony fishes, the hyoid constrictor is expanded to move the operculum, a series of bony plates which covers the gills in the place of separate external gill slits.

FUNCTION

Differences between the somatic and visceral muscle types may be correlated with the functions of their respective divisions. The ***striated*** somatic musculature is well suited for the vigorous contractions associated with locomotion. Visceral muscle is classified as either ***smooth*** or ***branchial*** muscle. Smooth muscle, primarily associated with the digestive tract, has a different embryonic origin and histological structure from striated muscle; it provides the slow, constricting movements characteristic of peristalsis of the gut. ***Branchial muscle*** is, in fact, striated. It is associated with the jaws (feeding) and the branchial or gill region (respiration). This unique muscle is similar in embryonic origin to the smooth muscle of the gut, but similar in structure and mode of action to skeletal muscle. The striated nature of this visceral muscle is essential for the vigorous movements required in feeding and respiration (see Chapters 9 and 12).

NATHAN C. SCHAFER

9

The Dogfish Shark: Digestive System

The digestive system, which processes food so it can be absorbed and metabolized by the cells of the body, is made up of the "gut" and its derivatives. The acquisition of moveable jaws and teeth in vertebrates allowed large pieces of food to be bitten off and swallowed. But with that ability came the need to store and process the larger food morsels. In sharks the three main changes from the digestive system studied in the lamprey are the presence of a stomach, a well-developed "spiral valve" in the ileum, and a more definitive pancreas.

DISSECTION INSTRUCTIONS

Locate the small midventral incision—made during the injection and preservation process—in the abdomen of your specimen. With a scalpel, extend the cranial edge of the incision to the pectoral girdle cartilage. As you approach the pectoral girdle, lift the skin and observe the falciform ligament between the liver and the ventral body wall. At the pectoral girdle, cut a semicircular arc laterally to the caudal edge of each fin to facilitate reflection (folding back) of the anterior body wall flap.

In female specimens, which lack claspers, extend the caudal edge of the original incision through the pelvic girdle cartilage to the cloaca. At the cloaca, cut around the rectum to isolate the rectum-cloaca as a single integral unit apart from the ventral body wall. Do not disturb the urogenital structures found dorsal to the rectum. At the pelvic girdle, cut the body wall laterally to free the body walls for easy reflection.

In male sharks (which possess claspers), in order to preserve part of the male reproductive system, extend the original incision caudally only to a point approximately 3 cm anterior to the pelvic fins; then cut the body wall laterally at this site.

To examine the internal structures and contents of the stomach, make a longitudinal incision in its ventral wall. The coveted Golden Scalpel Award should be presented with appropriate fanfare to the student whose shark's stomach has the strangest contents. A longitudinal incision in the ventral wall of the ileum, or valvular intestine, will expose the spiral valve. Wash out the spiral valve region with a water spray to reveal the internal structure.

IDENTIFICATION (Figs. 9-1, 9-2, 11-4)

Teeth. The jaws are lined with teeth which are more or less alike. These teeth are rootless, anchored only to the tissues of the jaws. The teeth serve more for ripping and tearing than they do for chewing, for the shark swallows its food nearly whole. As the teeth of one row wear down, those behind replace them.

Buccal cavity. **(Oral cavity)** The buccal cavity of the shark is the chamber of the mouth anterior to the points of entry of the gill slits. The oral cavity is characterized by transverse folds of the mucosa or lining tissues; in this respect it differs from the pharynx.

Pharynx. This portion of the "mouth" cavity lies between the oral cavity proper and the esophagus. The pharynx is common to both the digestive and respiratory tracts, inasmuch as food passes here on its way to the stomach, and water passes outward through the pharyngeal slits so that respiratory exchange can occur. The pharynx is characterized by longitudinal mucosal folds and by the presence of the gill slits. The pharynx is discussed more fully in Chapter 12 as part of the respiratory system.

Tongue. The tongue of the shark is a virtually immobile structure found in the floor of the mouth. It

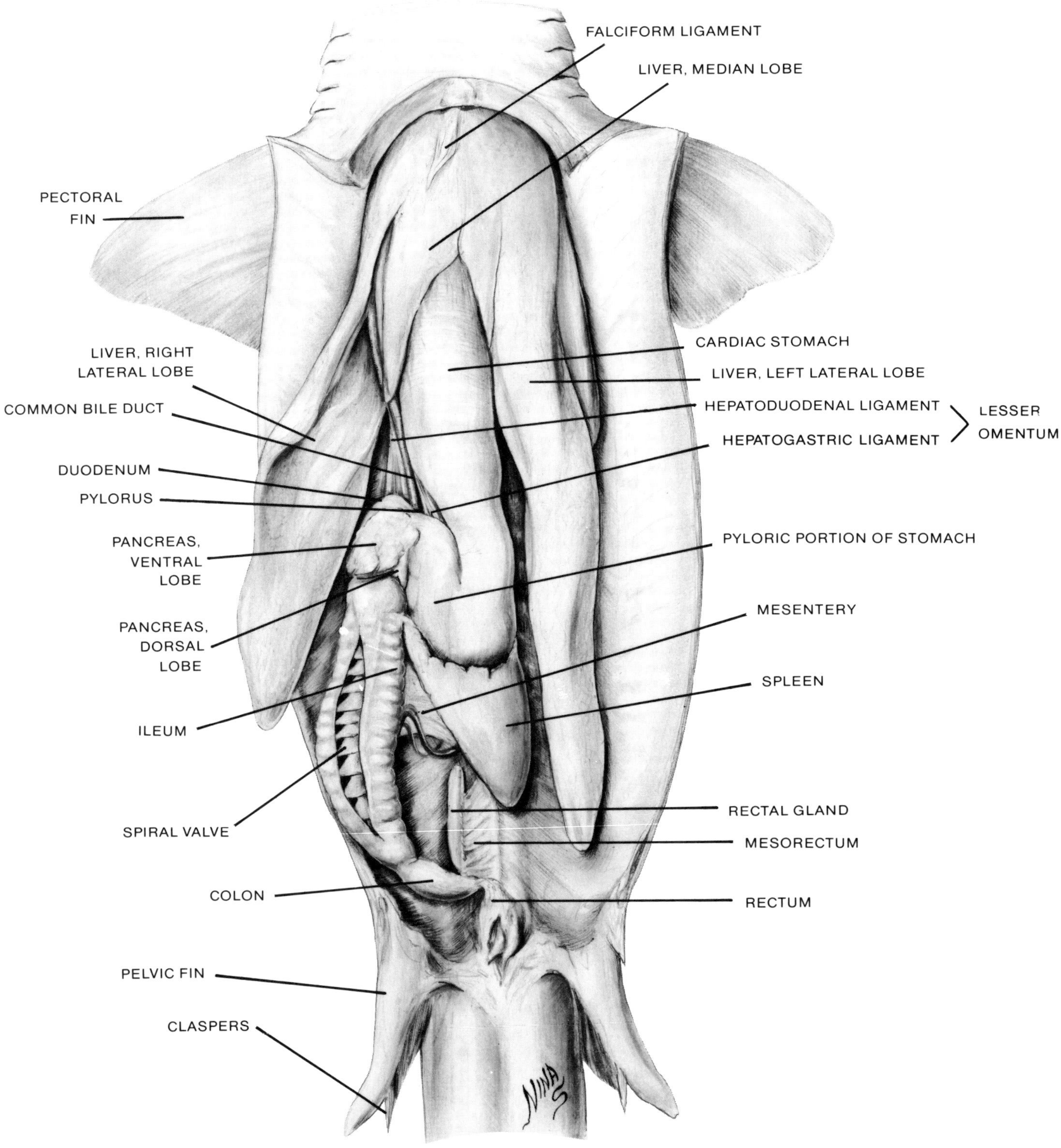

Figure 9-1. Ventral view of the abdominal cavity of a dogfish after midventral incision, including partial disclosure of internal structures of the ileum.

has no intrinsic muscles and is often referred to as a "primary tongue."

Esophagus. The esophagus is the continuation of the pharynx which leads into the stomach. Its walls are characterized by numerous conical projections called ***papillae***.

Stomach. The stomach is a J-shaped organ divided into three portions: the anterior, ***cardiac*** portion; the central portion, or ***body***; and the posterior, ***pyloric*** portion. There is no distinctive external demarcation between the esophagus and the cardiac stomach. However, a distinguishing internal feature of the cardiac portion is the presence of rugae.

Rugae. The rugae are longitudinal folds found on the internal surface of the cardiac stomach; they serve to increase the surface area utilized during digestion.

Pyloric sphincter. The pyloric sphincter, made up of circularly arrayed muscle fibers, is found surrounding the ***pylorus***, the caudal opening of the stomach. The pyloric sphincter controls the admission of food to the anterior portion of the small intestine, the duodenum, and prevents the return of intestinal contents to the stomach.

Small intestine. The small intestine begins at the pyloric sphincter. It consists of two parts, the ***duodenum*** and the ***ileum***.

Duodenum. The duodenum is the short segment of the intestine immediately following the pyloric sphincter.

Ileum. The ileum is the longer, thicker segment of the small intestine which follows the duodenum. Its internal surface is characterized by the presence of the spiral valve.

Spiral valve. The spiral valve consists of interconnected transverse sheets of tissue on the inner wall of the ileum. The sheets provide extra surface area for absorption of nutrients.

Colon. The colon is the segment of the digestive tract after the ileum. It terminates at the point where the rectal gland enters the gut.

Rectal gland. The rectal gland is a digit-shaped gland with an apparently excretory function. It is likely that the rectal gland removes excessive monovalent salts from the shark's blood and excretes them into the rectum. Thus the gland helps to keep the shark in osmotic equilibrium with the sea water. For this reason, the rectal gland might be thought of as complementing the function of the kidney where

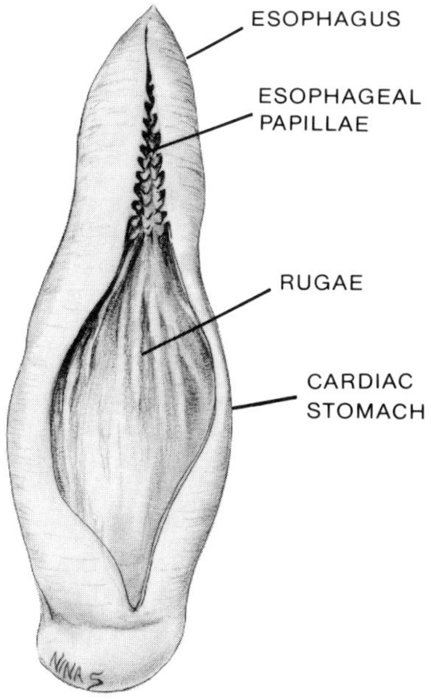

Figure 9-2. Internal features of the esophagus and stomach of a dogfish after midventral incision of digestive tract wall.

divalent ions such as magnesium and sulfate are excreted.

Rectum. The rectum is the continuation of the colon, from the rectal gland to the cloaca.

Cloaca. The cloaca is the common chamber through which intestinal, rectal gland, and urinary wastes leave the body. In the male, sperm also passes through this chamber. In the female, the cloaca serves as the birth canal. In higher vertebrates, such as the cat, these ducts tend to be separated so that waste products and gametes leave the body by separate openings.

Accessory Structures: Digestive Glands

Liver. The liver is a very large, grey organ of great importance to the shark, partly because its light, oily contents are a source of buoyancy. It consists of ***right*** and ***left lobes***, both of which extend much of the length of the pleuroperitoneal cavity, and a smaller, ventral ***median lobe***. Bile, which is produced in the liver, leaves it through very small ***hepatic ducts*** which drain toward and terminate in the gall bladder. Squalene, the light oil, is stored in large quantities in the liver.

Gall bladder. The gall bladder, a small organ on the right side of the median lobe of the liver, is the storage site for the green-colored bile. Bile leaves the gall bladder by means of the common bile duct.

Common bile duct. The common bile duct connects the anterior end of the gall bladder with the small intestine. This duct may be colorless in your specimen, but it can be found running parallel to a large blood vessel, the hepatic portal vein.

Pancreas. The pancreas is an important digestive gland consisting of two lobes. The more flattened ***ventral lobe*** may be found along the ventral surface of the duodenum. The ***dorsal lobe*** is longer and narrower; it is located in the curve between the pyloric stomach and the duodenum.

Spleen. The spleen is a dark triangular organ attached to the posterior end of the pylorus at the point where the "J" of the stomach turns upward. It is actually an organ of the circulatory system, since it is involved in the production, storage, and elimination of blood cells.

Mesenteries

Mesenteries are membranous structures which extend from the body wall to the viscera of the pleuroperitoneal cavity. In the shark the pleuroperitoneal and pericardial cavities make up the coelom, or body cavity (see Chapter 20 for further discussion of body cavities). Sheets of tissue covered on each side by epithelium and containing nerves, blood vessels, fat cells, and connective tissue cells between these epithelia make up the mesenteries. Embryologically, a dorsal mesentery suspends viscera from the dorsal body wall, and a ventral mesentery connects the primitive gut with the ventral body wall. As development proceeds, most of the ventral mesentery degenerates. The greater omentum, mesentery proper of the small intestine, mesorectum, and gastrosplenic ligament are all derived from the dorsal mesentery; the falciform ligament and the lesser omentum (hepatogastric ligament and hepatoduodenal ligament) are derived from the ventral mesentery.

Falciform ligament. The falciform ligament is a mesentery which connects the anterior portion of the liver with the midventral body wall.

Lesser omentum. (Gastrohepatoduodenal ligament) The lesser omentum is the mesentery which extends from the liver to the digestive tract. Specifically, it consists of two parts, the ***hepatogastric ligament*** and the ***hepatoduodenal ligament***.

Hepatogastric ligament. The hepatogastric ligament, made up of two thin sheets, stretches from the liver to the curved portion of the "J" formed by the stomach and the pylorus.

Hepatoduodenal ligament. This ligament stretches from the liver to the duodenum. The common bile duct runs along this ligament on its course from the gall bladder to the duodenum.

Greater omentum. (Mesogaster) The greater omentum is the mesentery which extends from the dorsal stomach and esophagus to the middorsal line of the body wall. (It is not shown in Fig. 9-1.)

Gastrosplenic ligament. The gastrosplenic ligament is the mesentery which connects the stomach to the spleen.

Mesentery proper. This is the part of the dorsal mesentery passing from the middorsal line of the body wall to the anterior portion of the small intestine.

Mesorectum. The mesorectum connects the rectum and the rectal gland to the middorsal line of the body wall.

FUNCTION

Anatomical features such as the well-developed spiral valve and even the stomach itself are important adaptations, essential to a shark's way of life. Both of these structures permit foods such as an entire squid or small fish to be ingested and digested. The sheets of tissue which make up the ***spiral valve*** serve to increase the absorptive surface area of the intestine. Food moves through the gut of sharks because of smooth muscle activity in the gut wall. Large pieces of food can be pushed along. All of this is in distinct contrast to chordates such as *Branchiostoma* or the tunicates, where cilia are the sole means of moving minute food particles along in the gut. One result of the ingestion, movement, and digestion of large pieces of food is the acquisition of the amount of nutrients needed to sustain the shark's large body.

The ***liver*** of the shark usually contains large quantities of squalene. This hydrocarbon has a specific gravity less than 1, meaning it is lighter than water. As a result it serves as an important source of buoyancy for the shark. This adaptation differs from that of many bony fishes, in which the swim bladder is employed as a source of buoyancy. On the other hand, some bony fishes also store light oils in their bodies, thus increasing buoyancy; this storage may occur in the bones or head—not just the liver.

The liver also carries out numerous metabolic functions. It processes nutrients into other substances—amino acids into blood proteins, for example. It stores nutrients, for instance, glucose in the form of glycogen. Liver cells also produce and excrete bile. Major components of bile are organic waste products such as hemoglobin catabolism (bile pigments). Bile salts act as emulsifying agents to aid in fat digestion and catalyze certain enzyme actions on fats.

Two kinds of cells are normally found in the vertebrate ***pancreas***. The predominant gland cells are the so-called "exocrine" cells responsible for producing the pancreatic juice secreted into the duodenum. This juice contains digestive enzymes which act on proteins, carbohydrates, fats, and nucleic acids. The other type of cells found in the pancreas are the endocrine cells of A and B type (which produce the hormones insulin and glucagon). In sharks these cells are ***not*** gathered together into discrete ***Islets of Langerhans***, as in the cat.

NEIL J. BAKER

10

The Dogfish Shark: Urogenital System

Although the excretory and reproductive systems have entirely different functions, they have acquired close structural associations in all vertebrates. They are, therefore, studied together as the ***urogenital system***. However, before considering morphological aspects, evolutionary developments in the structure and function of the kidney tubules will be examined.

EVOLUTION (Fig. 10-1)

The most primitive vertebrate kidney, the ***holonephros***, is believed to have consisted of a series of tubules arranged segmentally along the entire length of the trunk, drained by a single ***archinephric duct*** (also called ***Wolffian duct*** or ***mesonephric duct***) on each side (Fig. 10-1A). At present, such a kidney is found in hagfish larvae and certain amphibian larvae and adults.

In adult cyclostomes, the most anterior tubules (the pronephros) have disappeared; the remaining more posterior portions, called the ***opisthonephric kidney***, are drained by the archinephric duct. Although lampreys may have several tubules per segment, their kidney is essentially similar to that in Figure 10-1B.

Figure 10-1. Diagram showing three stages in the evolution of the primitive male vertebrate kidney: *A*, postulated holonephros; *B*, adult cyclostomes; *C*, adult sharks. (Modified from Romer.)

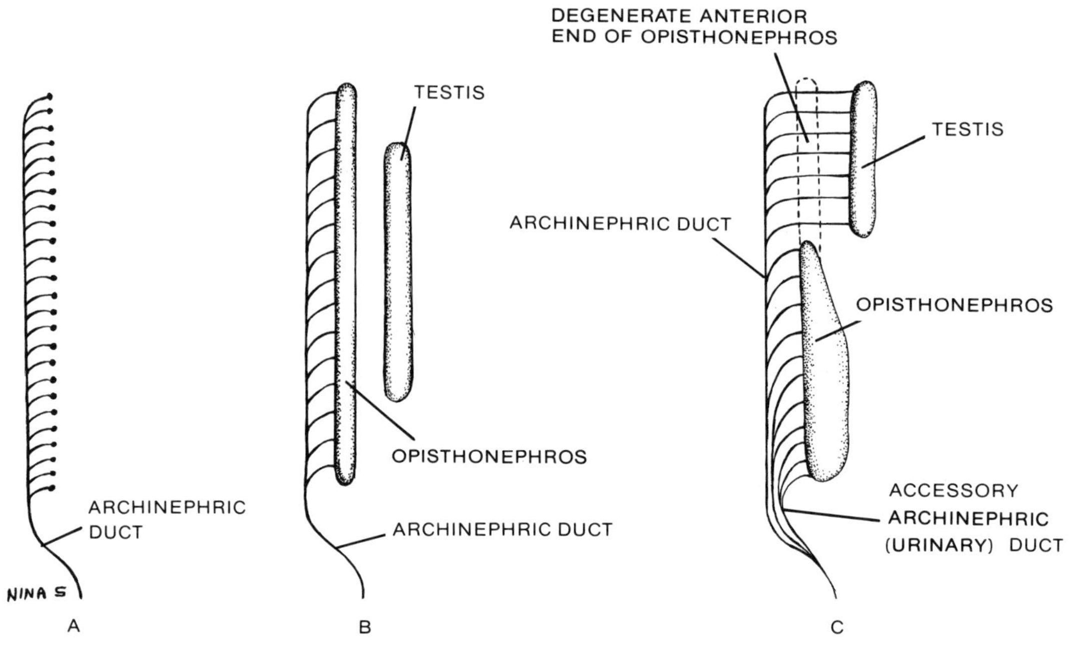

The gonads have no system of ducts for transporting reproductive products. The eggs and sperm are simply released into the body cavity and reach the exterior through genital pores.

In adult male sharks, the anterior kidney tubules and the archinephric duct are modified for sperm transport (Fig. 10-1C). This portion of the kidney and the anterior part of the archinephric duct are homologous to the mammalian epididymis (a storage organ for sperm). With this new use of the anterior kidney system, the kidney tubules elsewhere in a shark lose their segmentation and are massed in large numbers in the posterior portion of the opisthonephros. The increase in tubule number provides greater surface area for filtration of the blood.

A series of collecting ducts drains the tubules, eventually uniting to form a single ***accessory archinephric duct***, or urinary duct. The archinephric duct itself drains only a small portion of the tubules—or none in some sharks—and is chiefly used for sperm transport in males. In females, no accessory ducts are present; the unmodified archinephric ducts drain the unchanged kidneys.

The reproductive system of female sharks differs from that in males and does not incorporate structures of the excretory system. The origin of the female genital ducts may be understood by examining the embryonic development of sharks in general. Before the embryo has begun overt development as either a male or a female, two ducts run posteriorly on each side of the body; they are the archinephric duct and the Mullerian duct. At this point, the embryo is at the ***indifferent stage*** in which the gonads have not differentiated into ovaries or testes and in which the primordia of both the male and female ducts are present. If the gonads develop into testes and produce male sex hormones, the Mullerian ducts degenerate and the male ducts and accessory sex structures, such as claspers, develop. If the gonads develop into ovaries, however, the Mullerian ducts form oviducts which become functional.

Thus the general trends in the evolution of the urogenital system include a gradual posterior concentration of the kidney and the gradual formation of completely separate urinary and genital tracts.

GENERAL DISSECTION INSTRUCTIONS

Manually displace the main visceral structures without removing them. Examine the pair of thin grey kidneys located on each side of the dorsal surface of the peritoneal cavity just lateral to the dorsal aorta. With a scalpel, separate a kidney from the peritoneum along one lateral edge. Lift the kidney slightly and observe its compact tubular composition.

The prominence of the paired archinephric ducts is variable. In the female, the ducts are barely visible, at best, as thin transparent tubes on the ventral surface of each kidney. In mature males, they are visible as light beige, coiled tubes running the length of each kidney.

Abdominal pores open into the pleuroperitoneal cavity along the posterior-lateral margins of the cloaca. These are most evident in mature females.

Male (Figs. 10-2, 10-3)

In the male, beginning one inch cranial to the pelvic fins and progressing caudally, remove the skin of the pelvic fins and their associated claspers. Directly beneath the skin, medial to the fins are a pair of thin, greenish siphon sacs. Each siphon sac gives rise to a duct that leads caudally to hidden dorsomedial folds in each clasper.

With a blunt probe, gently separate the coiled archinephric duct from one kidney. Caudally it widens to form the seminal vesicle. Take care to differentiate the seminal vesicle from the light, almond-shaped sperm sac located cranial and slightly lateral to the cloaca. Where the archinephric duct bulges laterally to form the seminal vesicle, notice the accessory archinephric duct coursing along the ventral kidney medial to the seminal vesicle.

In the male shark, the following structures can be identified.

Caudal ligament. The caudal ligament is a thick white strip of connective tissue between the kidneys. This should not be mistaken for urinary or genital ducts, which are tubular in structure.

Testes. The testes are paired, elongated oval gonads, located at the anterior end of the body cavity. They may be concealed by the lobes of the liver.

Mesorchium. This is an epithelial and connective tissue sheet extending from the dorsal body wall to support the testes.

Efferent ductules. These are modified kidney tubules suspended within the mesorchium. They are used for transporting sperm from the testes to the archinephric duct.

Archinephric duct. The archinephric duct, also called the ***ductus deferens*** or sperm duct, adheres to the ventral surface of each kidney. In immature males, the convolutions and the duct itself may not be so prominent.

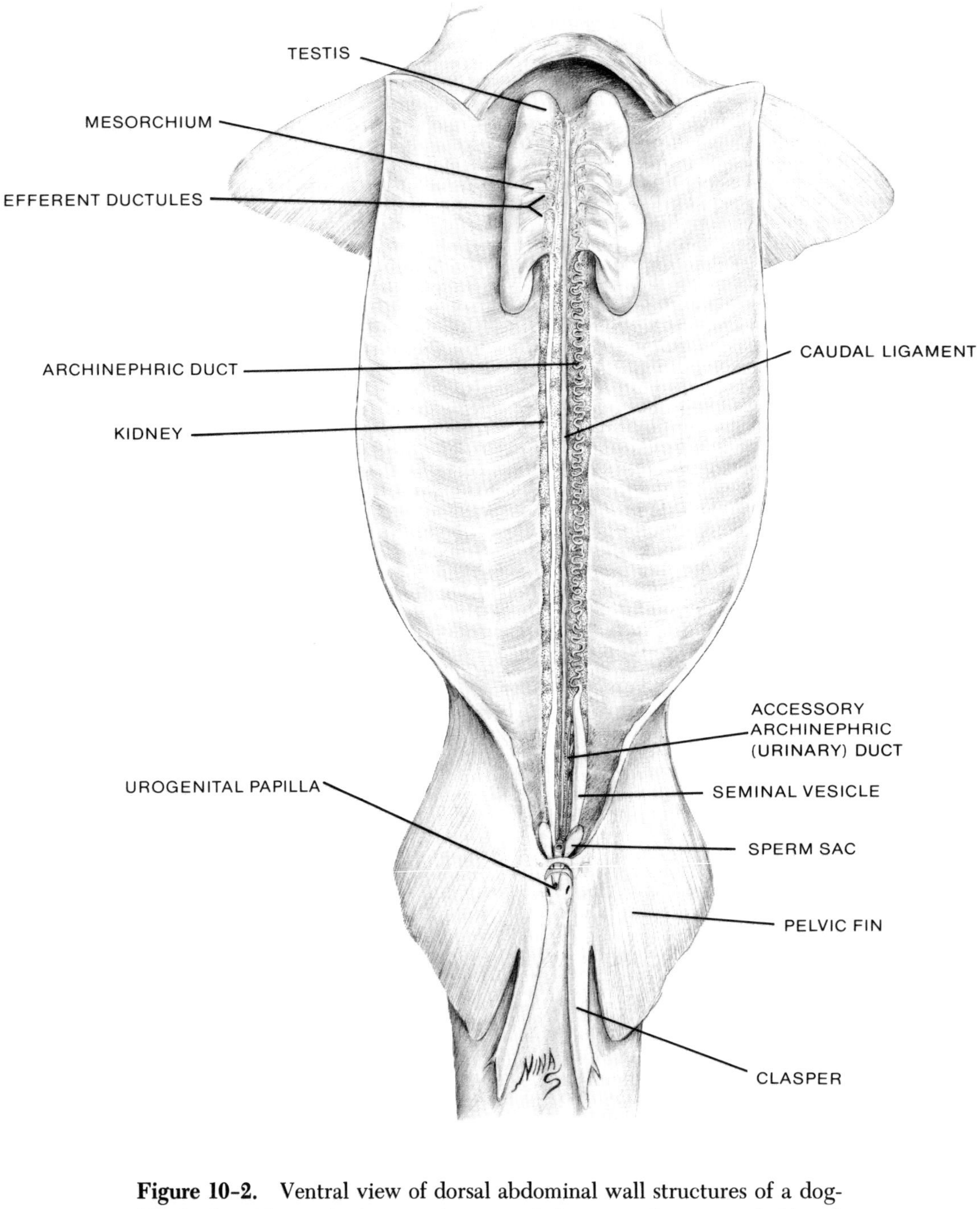

Figure 10-2. Ventral view of dorsal abdominal wall structures of a dogfish shark, with emphasis on male urogenital system: immature (left) and mature (right).

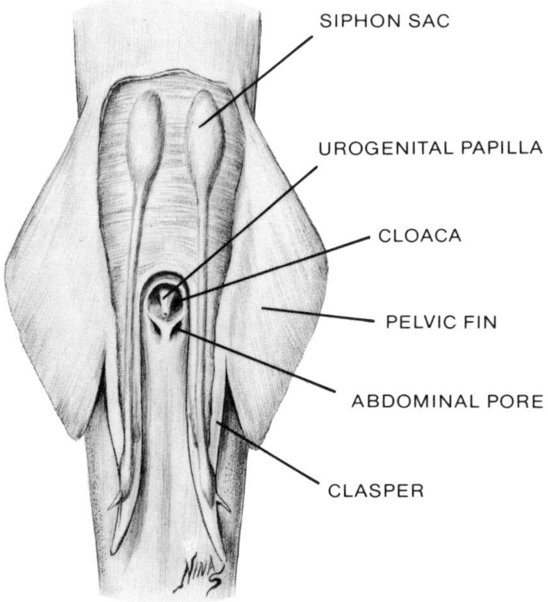

Figure 10-3. Ventral view of pelvic fin region of a male shark. Skin has been partially removed to show underlying structures.

Seminal vesicle. The seminal vesicles are the enlarged posterior portions of each archinephric duct, which open into the sperm sac. The seminal vesicles contribute secretions to the seminal fluid in which the sperm are transported. In immature males they are poorly developed.

Sperm sac. A sperm sac receives secretions from each seminal vesicle. Each sperm sac empties into the urogenital sinus, after making further contributions to the seminal fluid.

Urogenital sinus. The sperm sac and seminal vesicle join to form the urogenital sinus. The two sinuses unite in a cavity within the urogenital papilla.

Urogenital papilla. This is a conical structure located dorsally within the cloaca. Urine and sperm are released into the cloaca through the posterior urogenital pore.

Opisthonephric kidneys. These are dark, elongated organs near the middorsal line. the posterior portion is the functional excretory portion.

Accessory archinephric ducts. Formed by the union of several enlarged collecting tubules from the opisthonephros; they enter the urogenital sinus posterior to the seminal vesicle. The accessory archinephric ducts drain most of the urine from the kidneys.

Cloaca. The cloaca receives feces from the intestine and urine and sperm from the papilla for elimination from the body.

Claspers. These modified portions of the pelvic fins are the male copulatory organs. Note the medial groove along which sperm are transported and the opening of the siphon sacs at its anterior end.

Siphon sacs. These are muscular-walled sacs which secrete a lubricating fluid to aid sperm transport. Muscular action of the walls aids in propelling the sperm suspension.

Female (Figs. 10-4, 10-5, 10-6)

In the female, examine the bilateral pair of oviducts, checking for the caudally located enlargement, the uterus. The uterus is greatly enlarged in pregnant specimens, dominating a large portion of the caudolateral peritoneal cavity. In immature specimens, the uterus appears as a minor widening of the oviduct just anterior to the cloaca.

If your specimen is carrying young, cut through the uterus longitudinally and expose the embryo for observation by cleaning out extra contents of the cavity with a wash bottle. Free the embryo and yolksac from the uterine wall. If desired, the embryo can be stored in a preservative solution.

Ovaries. The ovaries are oval gonads at the anterior end of the body cavity, concealed by the lobes of the liver.

Mesovarium. This is an epithelial and connective tissue sheet which suspends the ovaries from the kidneys in mature females. In immature females, the ducts adhere to the surface of the kidneys; the mesotubarium, the mesentery which supports the oviduct of the mature female, is not evident.

Oviducts. Posteriorly, the oviducts open into the cloaca on each side of the urinary papilla. Anteriorly, they curve around the anterior end of the liver and enter the falciform ligament where they unite to form the ostium (ostium tubae).

Ostium. The ostium is the anterior opening of the oviduct and receives ovulated eggs. It consists of a slit within the falciform ligament and may be difficult to find without seriously damaging your specimen. The ostium still may be closed in immature females but is apparenτ in large adult females.

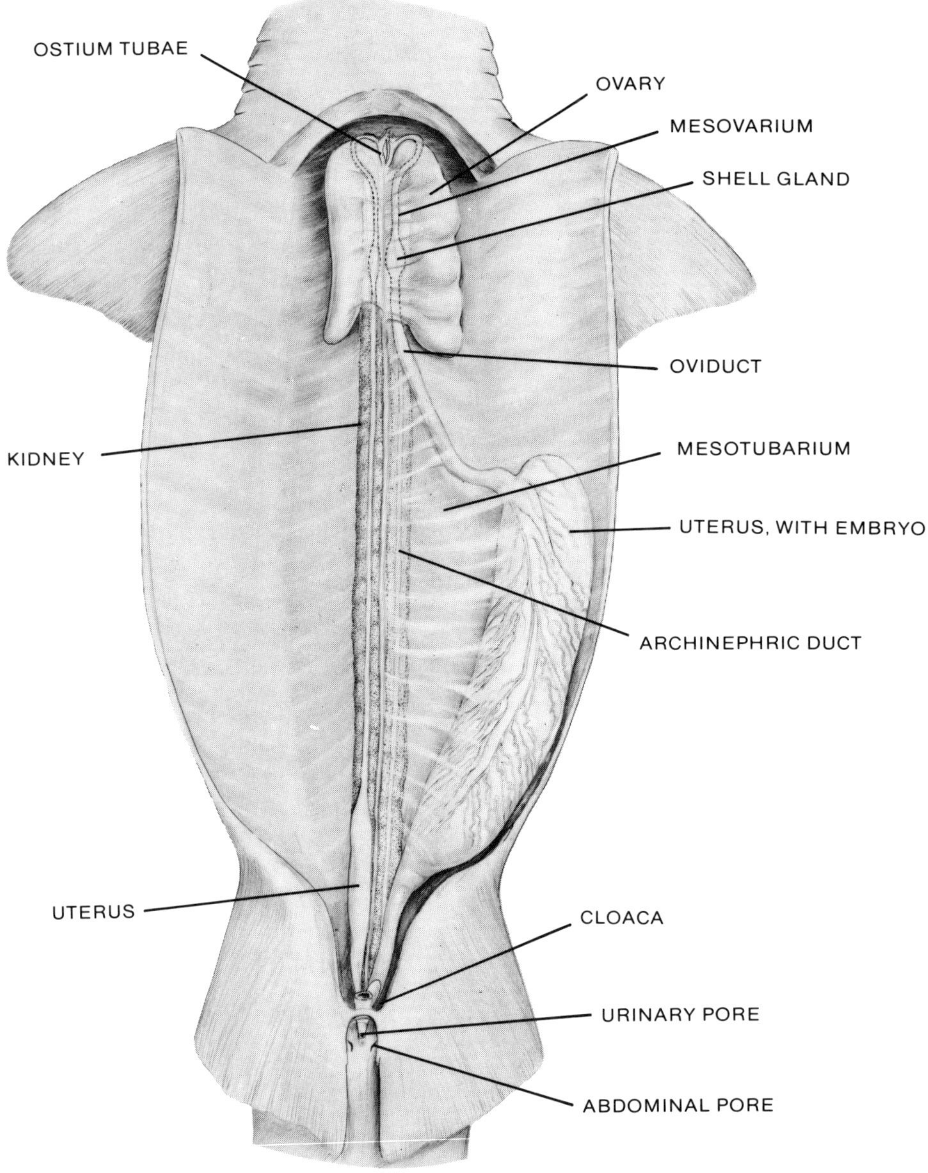

Figure 10–4. Ventral view of dorsal abdominal wall structures of a dogfish shark, with emphasis on female urogenital structures: immature (left); mature and pregnant (right).

Nidamental or shell gland. This is an expanded oval portion of the anterior oviduct which secretes a horny shell around the eggs.

Uterus. The uterus is the enlarged posterior portion of the oviducts (seen in mature females) in which embryos develop. Frequently the walls of the oviduct and uterus appear to be heavily vascularized if embryos are present; this suggests that some gas exchange or other exchanges may occur at such sites.

Archinephric duct. The duct is not convoluted in the female and is less prominent than in the male. It drains the urine from the kidney.

Urinary papilla. This is a conical structure located dorsally within the cloaca. Its ***urinary sinus*** receives urine, which is released through the urinary pore.

Cloaca. The cloaca receives the feces and urine and serves as a birth canal.

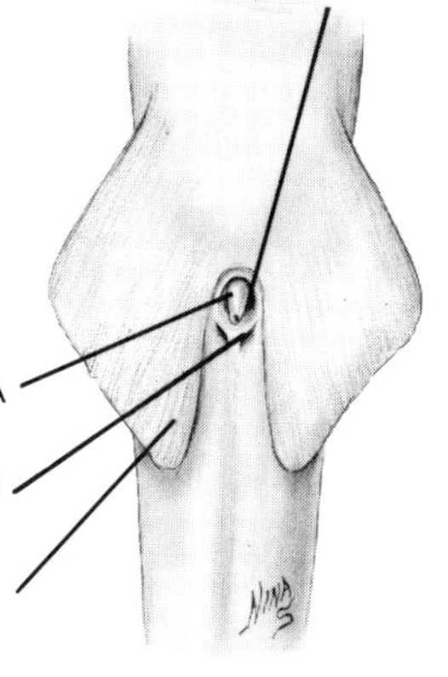

Figure 10-5. Ventral view of pelvic fin region of a female dogfish shark.

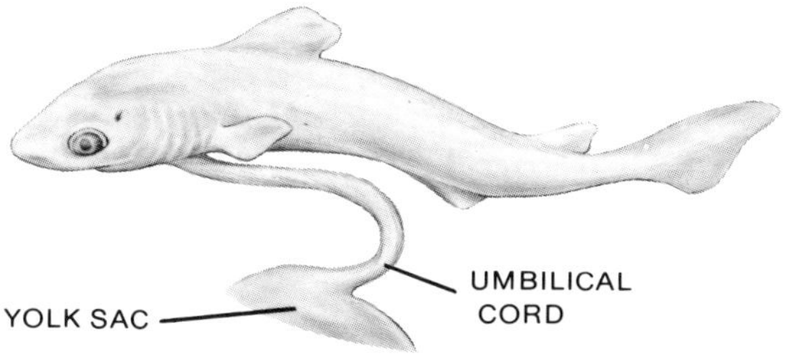

Figure 10-6. Intact shark embryo removed from the uterus.

FUNCTION: URINARY SYSTEM

The ***nephrons*** are the basic functional units of the kidney. They filter various substances out of the blood in order to maintain the proper composition of the internal environment and eliminate waste products. Such filtering takes place in the portion of the nephron called the ***renal corpuscle***; this consists of a spherical mass of capillaries called the ***glomerulus*** surrounded by a hollow, cup-shaped structure called ***Bowman's capsule***. The filtrate is produced as fluids are forced from the capillaries into the space within the capsule. However, substances are filtered indiscriminately and in large quantities; they must be selectively ***reabsorbed*** as they pass through the ***tubule***. The filtrate remaining within the tubule consists mainly of water, with varying concentrations of salts and nitrogenous wastes such as urea. This filtrate is excreted as urine through systems of collecting ducts.

Nephrons vary in structure according to the osmotic relationship of the organism to its environment. For example, in fresh-water fishes the body tissues and fluids are ***hyperosmotic*** to the surrounding medium. Water continually enters the body, particularly through thin, vascularized surfaces such as the gill lamellae. In the kidney tubules, large renal corpuscles provide a voluminous filtrate to eliminate excess water and thereby prevent dilution of the internal environment.

Vertebrates living in salt water face entirely different problems. The body tissues and fluids are ***hypoosmotic*** to the surrounding medium, so that water tends to diffuse out of the body. Two distinct strategies are used to avoid dehydration. In many modern bony fishes, excess salt entering the body is discharged by salt-secreting cells in the gills. In addition, renal corpuscles are reduced or absent, so only a small volume of filtrate and urine is produced. The nitrogenous end-product of protein metabolism is ammonia, a compound formed in the gills and excreted there into the sea water.

The second strategy, which is used by sharks, includes retention of renal corpuscles and resultant generation of significant volumes of filtrate. However, large volumes of water are not lost because urea is reabsorbed from the kidney tubules into the body fluids. Consequently, the body tissues and fluids contain such a high concentration of urea that the body is ***isoosmotic*** or even slightly hyperosmotic to sea water. The rectal gland (see Chapter 9) which excretes excess monovalent salts also helps to maintain osmotic balance.

FUNCTION: REPRODUCTIVE SYSTEM

Both internal fertilization and intrauterine development of the embryo are features of reproduction in many sharks. Internal fertilization is possible because of modification of pelvic fins in the male to form ***claspers*** (Fig. 10-3). In smaller species the male lies twisted around the female during copulation (Fig. 10-7).

Muscles bend the claspers forward to be inserted either together or separately (reports conflict) into the cloaca of the female. Sperm reach the female along the medial groove of the clasper, aided by muscular action in the walls of the siphon sacs and by the lubricating fluid the sacs secrete.

In females, the eggs develop in the ovaries; at maturity they are about one inch in diameter. At

ovulation, the eggs burst through the walls of the ovary into the coelom fairly close to the ostium. Unlike the paired ostia in other vertebrates, the openings of the shark oviducts are fused into this large funnel-shaped structure in order to receive the huge eggs. Since the oviducts are highly elastic, they stretch greatly as the egg descends to the shell gland, where it receives a thin, horny covering. This "shell" later breaks down and disappears. The egg must, of course, be fertilized before reaching the shell gland.

Gestation periods in some sharks are among the longest of all vertebrates. In ***Squalus***, embryos develop in the uterus for approximately two years; at birth they are 12 to 20 cm long. Embryos may develop in both uteri simultaneously and may range from two to fourteen in number, depending on the species. Much of the yolk is suspended from the ventral side of the embryo in a vascularized ***external yolk sac***. There is also a smaller ***internal yolk sac*** within the body cavity which still contains a small amount of nutrients at birth.

Vascularized villi project from the uterine wall and come in contact chiefly with the extenal yolk sac. Dogfish embryos receive water and, possibly, a few nutrients through this so-called placenta. Because of their minimal dependence on the mother, dogfish sharks are called ***ovoviviparous***. Those sharks which depend a great deal on the mother for nutrients are ***viviparous***. Finally, many primitive ***oviparous*** sharks actually lay eggs with a large amount of yolk encased in a horny shell.

Figure 10-7. Copulatory position of dogfish sharks. (After Budker.)

HENRY B. KISTLER, JR.

11

The Dogfish Shark: Circulatory System

The circulatory system of the dogfish shark is a primitive closed system. The heart pumps only blood with relatively low oxygen content from the systemic veins to the gills. The heart consists of four chambers arranged in a dorsoventral S-shaped structure. Blood passes from the dorsal receptacle chamber, the ***sinus venosus***, to the ***atrium***. The blood then passes ventrally into the muscular ***ventricle***, the contraction of which forces it through the valvular ***conus arteriosus*** to the ***ventral aorta*** and on toward the ***gills***.

The arrangement of ***aortic arches*** between the ***ventral aorta*** and the ***dorsal aorta*** shows little variation from more primitive vertebrate forms. The first aortic arch is present in the embryo but disappears in the adult. Gas exchange with sea water takes place in the gills from arches two through six.

The venous system shows major alterations from more primitive chordates, particularly the full separation of the ***renal portal system***. Primitive vertebrates and many embryos possess paired posterior cardinal veins that bypass the kidney tissue laterally as they carry blood to the heart. In embryonic elasmobranchs (sharks, skates, and rays), ***subcardinal veins*** arise along the medial surface of the kidneys and course cranially to fuse with the posterior cardinals. Subsequently, the embryonic posterior cardinals degenerate between their point of fusion with the subcardinals and the cranial edge of the kidneys. In the adult, blood flows from the tail capillary beds through the caudal vein, which splits to form the ***renal portal veins*** (part of the embryonic posterior cardinals). Blood then passes medially in the body, first through the kidney capillaries, and then via renal veins. It is collected by the subcardinal extension of the posterior cardinal vein and transported to the heart.

Due to the multiple capillary arrangement, the overall blood pressure in the shark is very low, except in the ventral aorta. Blood pressure is reduced by at least 25 percent as it passes through the gill capillaries alone. Blood returning to the heart may even have a slightly negative pressure. Because of this extremely low pressure, thin-walled sinuses have evolved to collect and store large amounts of blood.

DISSECTION AND IDENTIFICATION: HEART (Figs. 11-1, 11-2, 11-3)

Cut through the skin along a midventral line from the mouth to the pectoral girdle. Palpate the position of the coracoid bar of the pectoral girdle and the mandibular cartilage before you being to cut. Then cut the skin at a right angle to the first incision to form two lateral incisions parallel to the coracoid bar and running toward the fifth gill slit on each side (Fig. 11-2). Reflect and then remove the skin flaps; you have exposed the ventral pharyngeal region. The superficial layer consists of muscles involved in ventilating the gills. The sheets of muscle with fibers oriented more transversely are the ventral portions of the constrictor muscles (branchial group; see Chapter 8). More posteriorly there are two large muscles with fibers more sagittally; these are the coracoarcual muscles introduced in Chapter 8. They represent "fused" portions of separate slips going to each of the pharyngeal arches. (You will see the separate slips in the next dissection.)

Remove the superficial fascia around the coracoarcuals and coracoid bar, then palpate the anterior margin of the coracoid bar where the coracoarcual fibers insert. Slide your scalpel along the anterior margin of the coracoid and sever the muscle attach-

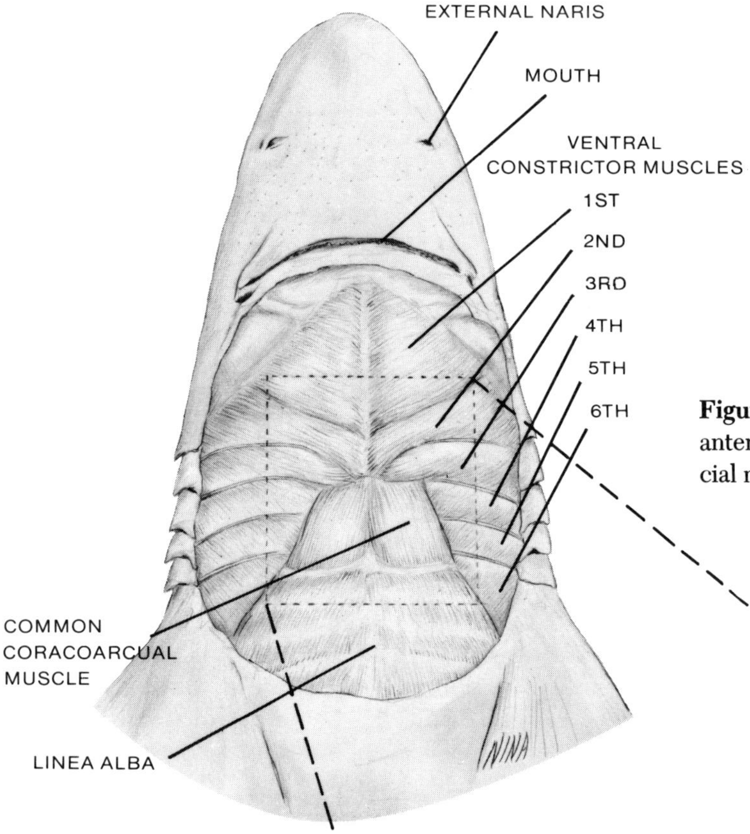

Figure 11-1. Ventral view of a dogfish head and anterior trunk, after skin is removed, showing superficial musculature.

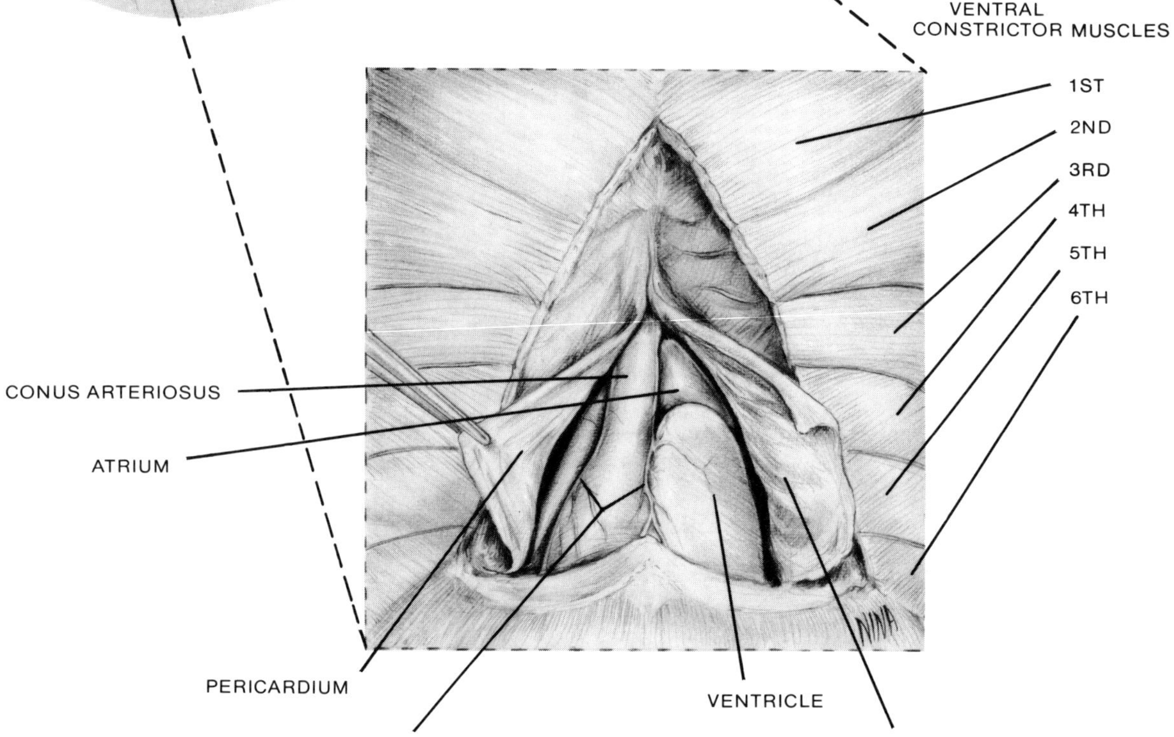

Figure 11-2. Ventral view of the hypobranchial region after removal of musculature to expose the pericardium and the ventral surface of the heart.

ment. Begin with an incision of a few millimeters and repeat the process several times to control the depth of the cut. Avoid cutting into the pericardium, a tough leathery sheet lying just a bit deeper in the body than the coracoarcuals. Separate the two coracoarcual muscles and remove the caudal portions that overlie the pericardium.

Cut through the coracoid bar on one side just lateral to the pericardium and pry it up slightly. Separate the coracoid from the underlying structures, particularly the thin-walled, sac-like sinus venosus and Ducts of Cuvier (refer to Figs. 11-3 and 11-6). Then sever the coracoid bar on the other side and remove it. You have now exposed the ventral surface of the pericardium; examine its relationships before opening it (see following text).

Open the pericardial cavity with a midventral and a transverse incision. The heart is folded into an "S"-shape, and you are viewing it from the bottom of the "S." This bottom part of the "S" is formed by the conus arteriosus. Use this as a landmark and follow the "S" through the ventricle and atrium to the sinus venosus, which lies dorsally and posteriorly in the chamber.

Pericardium. The pericardial membrane has two layers, an outer ***parietal pericardium*** and an inner ***visceral pericardium***. The outer membrane is rigid because of numerous connections with adjacent cartilage and branchial musculature. Its caudal base is the fibrous pericardio-peritoneal septum. The pericardial cavity is bounded on its ventral and lateral caudal surfaces by the cartilage of the pectoral girdle. The dorsal roof of the cavity is bounded by the basibranchial cartilages of the pharyngeal skeleton. Rostral limits are provided by paired branchial muscles. Small quantities of lymph fluid exude from the parietal pericardium into the pericardial cavity. This fluid lubricates the heart as it moves within the rigid membrane.

Conus arteriosus. This is a barrel-shaped chamber situated anterior to the ventricle. It extends rostrally to the ventral aorta through the cranial surface of the pericardial wall. It is made up of cardiac muscle that contracts in sequential series with the other chambers of the heart. Along the inner wall are four longitudinal rows of semilunar valves shaped to prevent backflow of blood. These valves are necessary

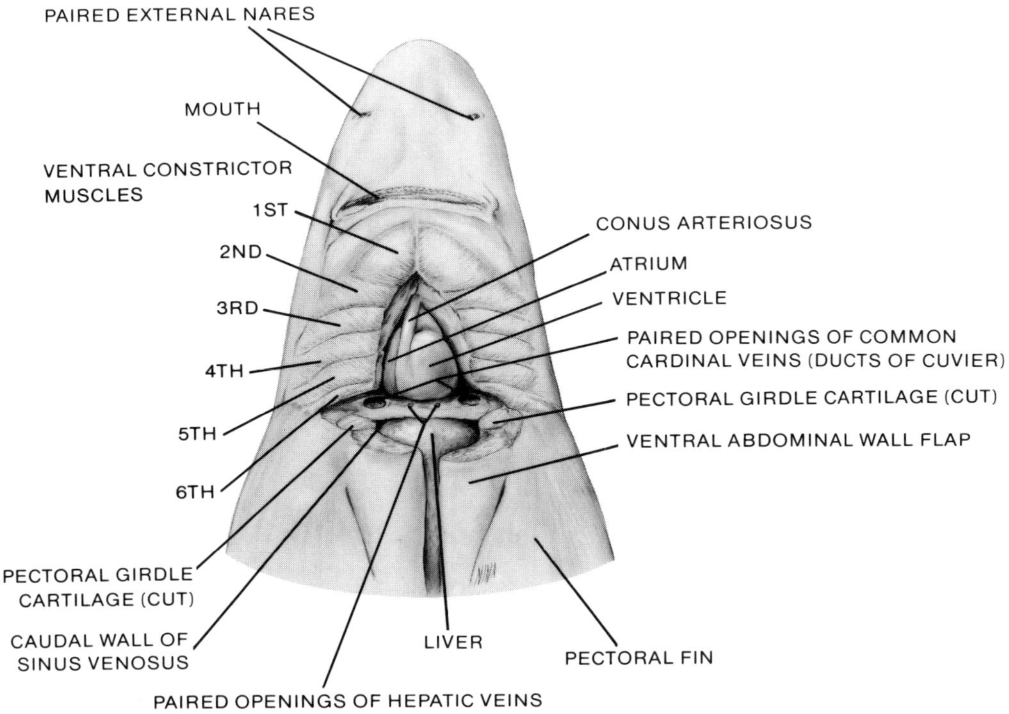

Figure 11-3. Ventral view of the hypobranchial region after removal of musculature to expose the pericardium and the ventral surface of the heart, and after removal of the medial pectoral girdle to expose the sinus venosus.

because of high blood pressure created by the powerful ventricular contractions and resistance to flow by the gill capillary beds.

Ventricle. The ventricle is a thickly muscular chamber lying ventral to the atrium. Its caudal wall is against the pericardioperitoneal septum and is partially obscured ventrally by the pectoral girdle. Its walls have two layers, an outer smooth cortical layer and an inner spongy fibrous layer. Ventricular contractions force blood cranially into the conus arteriosus.

Atrium. This single large, dark-colored chamber is lined with two fans of muscles. When these muscles contract, they draw the dorsal atrial roof ventrally to force blood through the atrioventricular valve. The valve itself consists of two flaps projecting ventrally into the ventricle.

Sinus venosus. This large, triangular, sac-like compartment can be exposed by lifting up the ventricle. The sinus venosus lies caudal to the atrium and just dorsal to the pectoral girdle. Its very thin walls contain delicate cardiac muscle. Its caudal base lies adjacent to and may be fused with the transverse abdominal septum. Blood enters from the systemic venous system through two sets of paired apertures and flows cranially through the sinus venosus and through the sinoatrial aperture into the atrium.

DISSECTION AND IDENTIFICATION: ARTERIAL SYSTEM (Fig. 11-4)

Using heavy scissors, cut through the right corner of the shark's mouth. With a large scalpel, cut through the jaw cartilage and along the ventral margin of the external gill slits on this side until the

Figure 11-4. Ventral view of the pharynx and gill circulation after lateral incision through the gills and reflection of lower jaw structures. (Refer, also, to plate of the same number in the color section at the end of the book.)

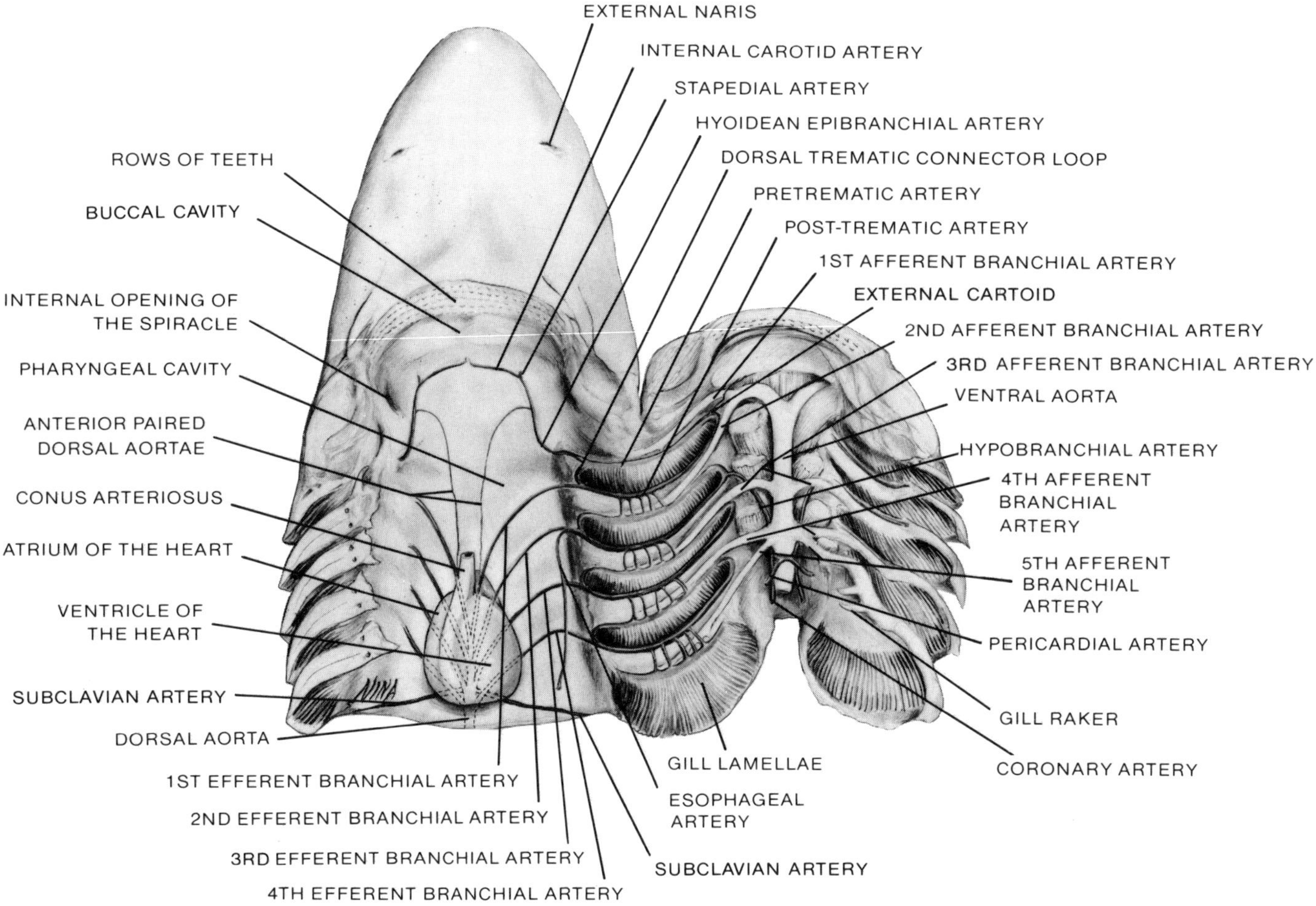

pectoral girdle is reached. Cut through the conus arteriosus as close to its origin at the ventricle as possible. Then lift the heart and cut medially through the pharyngeal skeleton that lies dorsal to the heart. Cut toward the region just posterior to the left fifth gill slit. ***Make no cuts in the gill region on the left side of the shark.*** *Fold back the ventral portion of the jaw toward the left (Fig. 11-4). Cut through the mouth and jaw cartilage on the left to relieve the tension and allow the ventral portion to lie flat.*

Remove the mucous membrane and cartilage covering the ventral aorta, being careful not to damage underlying blood vessels. This cartilage can be cut away in small pieces or pulled off after making two initial cuts lateral to the aorta cutting in a cranial direction on each side of the ventral aorta. To remove the tissue obscuring the gill circulation, first either pull off the mucous membrane with forceps or cut it away carefully. The gill cartilages may then be removed by grasping each cartilage with forceps and pulling it free.

Completely remove the mucous membrane and connective tissue in the roof of the mouth to reveal the efferent branchial arteries, the dorsal aorta, and additional rostral arteries.

Trace the branches of the ventral aorta into the interbranchial septum. Note that they begin to give off small branches to supply unoxyenated blood to the gill lamellae. Trace one afferent branch until it enters the gill arch. The pretrematic and posttrematic arteries lie on either side of the afferent branch. Trace the loops formed by four pairs and note that each loop leads into an efferent branchial artery.

Branchial Arteries (Fig. 11-4)

Ventral aorta. Unoxygenated blood leaving the conus arteriosus flows cranially through this single, large uninjected artery. This artery courses ventrally along the basibranchial cartilages and gives rise anteriad to two primary pairs of branches before bifurcating cranially near the mouth.

Afferent branchial arteries. Unoxygenated blood from the ventral aorta is carried dorsally to the gills through five pairs of uninjected arteries. The first, most caudal branch from the ventral aorta bifurcates immediately to supply the fifth and fourth arch lamellar systems. The second branch from the ventral aorta proceeds directly, without splitting, to the lamellae of the third arch. Finally, each branch formed by the terminal bifurcation splits to feed the first and second arch gills. The branchial arteries are numbered from one to five in a craniocaudal direction.

Efferent branchial arteries. The system returning blood from the gill capillaries begins with a series of trematic or "collector" loops, each associated with one gill opening. Each injected collector loop arises from a ***pretrematic artery*** that collects blood from the cranial surface of the gill opening and a ***posttrematic artery*** that collects blood from the caudal surface of the same gill opening. ***Pretrematic*** and ***posttrematic*** arteries unite at the dorsal and ventral limits of each gill opening, creating a ***trematic loop*** that surrounds the opening. In addition, several cross branches connect adjacent trematic loops. Each of the four trematic loops gives rise to an ***efferent branchial artery***. These branchial arteries angle caudomedially toward the midline of the roof of the pharynx. They ultimately fuse to form the dorsal aorta. (For an understanding of relationships between the circulatory and respiratory systems in this region, see Chapter 12.)

Rostral Arteries (Fig. 11-4)

Several pairs of rostral arteries arise from trematic collector loops, efferent branchial arteries, and the dorsal aorta.

External carotid artery. This small artery arises from the ventral turn of the first trematic loop. It extends ventromedially to vascularize the lower jaw region.

Hyoidean epibranchial. (Efferent hyoidean) This is a substantial superficial artery starting at the dorsal turn of the first trematic loop. It courses craniomedially along the roof of the mouth until it branches to give off the stapedial artery.

Stapedial artery. This is the main branch off the hyoidean epibranchial. It extends dorsally into the stapedial foramen of the cartilagenous chondrocranium. Once through the chondrocranium, it proceeds to the medial surface of the eye and to the snout region.

Internal carotid artery. This artery appears as the continuation of the hyoidean epibranchial artery. As a bilateral pair, the internal carotids meet and fuse at the carotid foramen of the chondrocranium. Once through the chondrocranium, they separate laterally to supply the brain.

Anterior or dorsal aortae. (Radix aortae) These L-shaped arteries originate near the midpoint of each first efferent branchial artery. They pass superficially in a cranial direction, then turn laterally to fuse with the hyoidean epibranchial artery caudal to the stapedial foramen.

Vertebral arteries. This pair of arteries is seldom seen because of inefficiency in the latex injection process. They proceed medial and parallel to the anterior paired dorsal aortae along the ventral surface of the vertebral column.

Hypobranchial arteries. These paired arteries originate on each side of the animal at the ventral turn in the second trematic collector loop. They run caudomedially, just ventral to the afferent branchial arteries and the ventral aorta. As they approach the heart, they trifurcate. One pair of branches, the ***pericardial arteries***, course ventrally along the pericardial wall. A second pair of branches, the ***coronary arteries***, extend to the heart along the lateral walls of the conus arteriosus. The third pair of branches proceed medially and fuse dorsally across the anterior end of the conus arteriosus.

Esophagial arteries. These bilaterally paired arteries extend caudally from the second efferent branchial arteries. They extend superficially along the sides of the pharynx, continuing caudally to the walls of the esophagus.

Subclavian arteries. These paired arteries originate from the dorsal aorta, cranial to its union with the fourth efferent branchial arteries. The subclavian arteries extend laterally to the pectoral fins. Near each fin they branch several times.

Dorsal Aorta and Its Branches (Fig. 11-5)

Reflect the lateral edges of the midventral abdominal incision and enter the peritoneal cavity. Reflect the lateral lobes of the liver cranially. Locate the dorsal aorta and its branches. Arterial branches often run adjacent to veins and can be identified by the organ or region they supply (see Table 11-2).

Dorsal aorta. This large artery is formed by the convergence of four efferent branchial arteries. Once formed, it courses caudally through the transverse septum and into the pleuroperitoneal cavity. It runs caudally down the middorsal line of the pleuroperitoneal cavity. Posterior to the paired iliac arterial branches near the pelvic fins, it is referred to as the ***caudal artery***.

Celiac artery. (Coeliac artery) This major unpaired artery branches off the dorsal aorta caudal to the transverse septum and dorsal to the cranial edge of the liver. It courses caudally along the right side of the cardiac stomach in close proximity to the bile duct and the hepatic portal vein. It bifurcates at the lesser omental mesentery.

Gastrohepatic artery. This is a short branch formed by the bifurcation of the celiac artery. It subsequently splits to form the ***gastric artery***, which supplies the cardiac stomach walls, and the ***hepatic artery***, which proceeds cranially into the right lateral lobe of the liver.

Pancreaticomesentric artery. (Intestinopyloric artery) This artery appears as the continuation of the celiac artery caudal to the gastrohepatic branch. It in turn splits to form four or more branches, two of which are readily accessible. The ***anterior intestinal artery*** runs along the lateral edge of the ileum and gives rise to a series of parallel ***annular arteries***. These traverse the valvular intestine and delineate the fusion of the spiral valve with the intestinal wall. The ***intraintestinal artery*** is concealed as it courses through the interior of the ileum, providing blood to the spiral valve. The ***duodenal artery*** is a short internal artery branching to the interior duodenum. The ***pyloric artery*** extends into the pylorus, where it gives rise to three minor branches.

Posterior intestinal artery. This is a short unpaired artery arising from the dorsal aorta adjacent to the spleen. It extends laterally to the medial surface of the ileum where it splits and gives rise to a series of transverse ***annular arteries*** (which fuse with the parallel annular arteries).

Gastrosplenic artery. (Lienogastric artery) This unpaired artery arises from the aorta just caudal to the posterior intestinal artery. It extends craniolaterally to enter the spleen and then continues on to the cardiac stomach.

Note: In some specimens, the posterior intestinal artery and the gastrosplenic artery branch as one from the dorsal aorta and then subsequently split. In these cases, the single branch is referred to as the ***anterior mesenteric artery***.

Posterior mesenteric artery. This short (sometimes double) artery branches from the dorsal aorta just posterior to the gastrosplenic artery. It passes directly to the rectal gland.

Several sets of small paired arteries also branch from the dorsal aorta. These smaller branches are described below.

Genital arteries. This pair, often called the ***testicular arteries*** in males and ***ovarian arteries*** in females, arises from the dorsal aorta, the celiac artery, or both. The genital arteries are small and difficult to locate without unduly disrupting the integrity of abdominal structures.

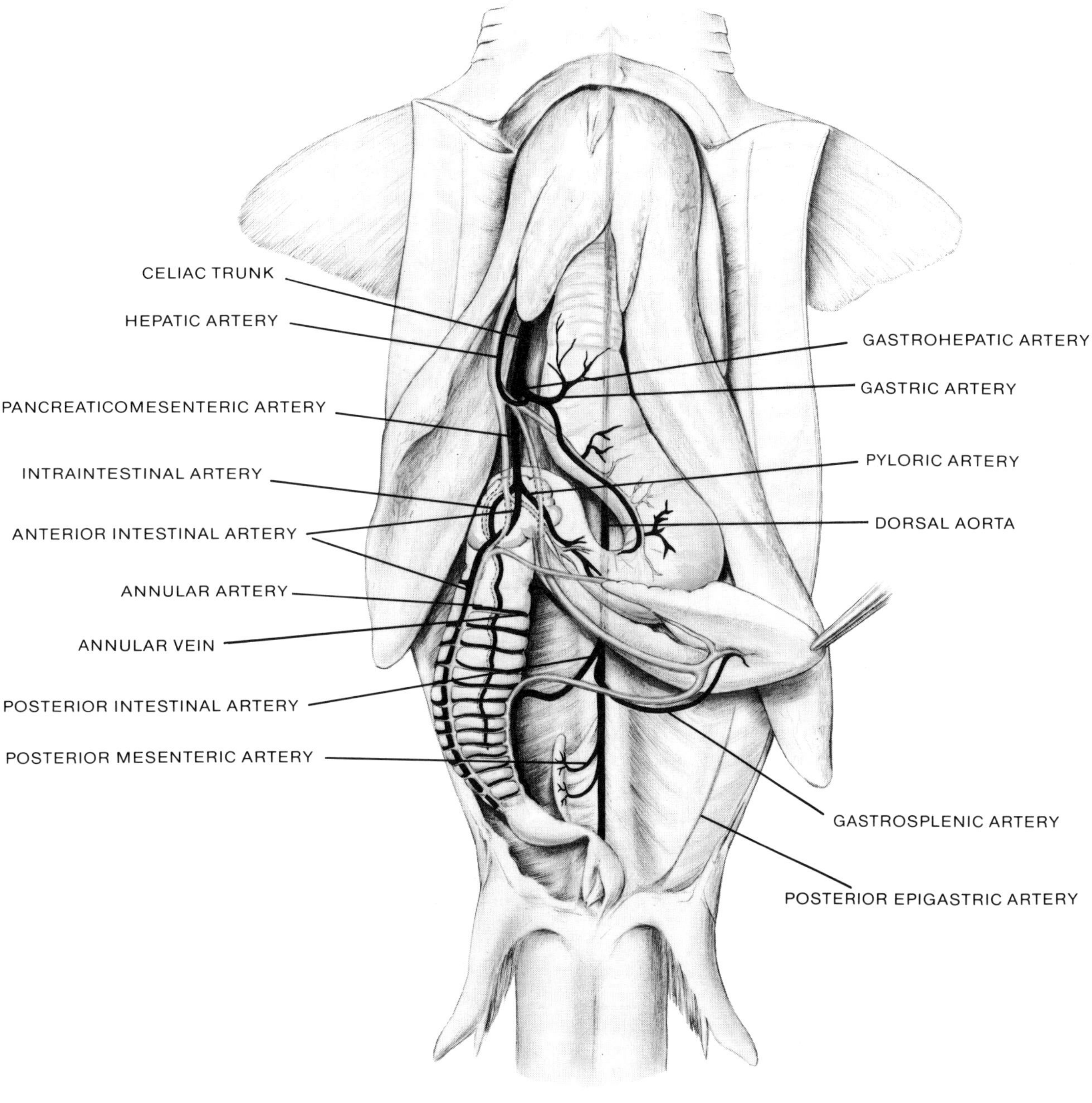

Figure 11-5. Ventral view of the abdomen after midventral incision, with emphasis on the dorsal aorta and its main branches. (See also color plate of the same number.)

Segmental arteries. These are a series of tiny arteries associated with individual muscle segments. Each artery branches into three separate arteries (***vertebromuscular***, ***renal***, and ***parietal arteries***) that in turn supply the spinal cord and epaxial musculature, the kidneys, and the muscle segments of the body wall.

Renal arteries. These are found between the kidneys and the dorsal body wall.

Iliac arteries. (Not shown in Fig. 11-5) This pair of small arteries originates caudal to the posterior mesenteric artery. The iliac arteries pass caudolaterally into the pelvic fin region. Before leaving the pleuroperitoneal cavity, they branch to form the ***posterior epigastric arteries***. The posterior epigastric arteries course cranially, embedded in the body wall. The latter vessels fuse with the ***anterior epigastric arteries***, branches of the subclavian arteries that extend caudally along the lateral body wall from the fin area.

Femoral arteries. This pair of arteries is the continuation of the iliac arteries caudal to the posterior epigastric branches. They extend deep into the pelvic fin areas.

DISSECTION AND IDENTIFICATION: VENOUS SYSTEM

Reenter the pericardial cavity (Fig. 11-3) and observe the cranial surface of the sinus venosus. Remove the ventral wall of this sinus, then clear away its contents with water spray from a wash bottle. Observe and identify the paired apertures in the sinus wall. Removal of more of the central portion of the pectoral girdle may be necessary for an unobstructed view. Cut dorsally with a scalpel through the pectoral cartilage near each pectoral fin, remove any connective tissue, and then remove the segment of cartilage.

Systemic Veins (Figs. 11-3, 11-6)

Sinus venosus. This thin-walled, membranous sac was previously discussed as the initial reception chamber of the heart. As such, it is the ultimate destination for all venous blood. ***Two*** main ***pairs*** of veins enter the sinus—the ***hepatic veins*** and the ***common cardinal veins.***

Hepatic veins. This large pair of veins enters the sinus venosus near the midline of its caudal surface. The hepatic veins drain large ***hepatic sinuses*** located inside the liver, which in turn collect blood from ***sinusoids***—vessels of varying diameter—throughout the liver tissue.

Common cardinal veins. (Ducts of Cuvier) This major pair of veins enters the lateral corners of the triangular sinus venosus. Each vein appears as a short, stubby trunk, formed mainly by the union of the ***anterior cardinal vein*** and the ***posterior cardinal vein*** on each side of the body. The ***inferior jugular veins*** and the ***subclavian veins*** also contribute venous blood to the ducts of Cuvier.

Anterior cardinal veins. These thin veins are usually uninjected as they enter the ducts of Cuvier on each side of the animal. These veins drain the bilateral ***anterior cardinal sinuses***, which are situated just medial and dorsal to the gill region. Several minor veins from the brain, the upper jaw, the rostral snout area, and the spiracle contribute blood to the anterior cardinal sinuses.

Subclavian veins. This pair of veins enters the lateral corner of each common cardinal vein. The subclavian veins are formed by the fusion of two tributaries, the ***brachial veins*** and the ***lateral abdominal veins***. The brachial veins proceed medially from the extremity of each pectoral fin. Each lateral abdominal vein, embedded in the lateral body wall, is formed posteriorly by the fusion of the very short ***cloacal vein*** and the ***iliac***, a continuation of the ***femoral***, vein from the lateral pelvic fin region. A relatively large vein, the ***subscapular*** (not shown in Fig. 11-6), enters the brachial medially.

Inferior jugular veins. These paired veins run caudally to enter the ducts of Cuvier along their cranial surfaces. They drain the lower jaw, pericardium, and pharyngeal regions; they are frequently found extending in an anterior-posterior direction beneath the afferent branches of the ventral aorta.

Posterior cardinal veins. This pair of veins is located medial to the kidneys along the dorsal wall of the pleuroperitoneal cavity. Cranially, these veins expand into two fully injected blue pouches, the ***posterior cardinal sinuses***. These sinuses fuse directly with the anterior cardinal veins to form the common cardinals. Tributary veins to the posterior cardinals are a series of short efferent ***renal veins*** (discussed below) and a series of segmentally arranged ***intersegmental*** or ***parietal veins*** that drain the lateral body walls.

Genital sinuses. (Spermatic sinuses in males and ovarian sinuses in females) Blood from the gonads is collected in large pouch-like sacs that encircle the gonads. These sinuses are seldom seen due to inefficiency in the injection process. Each sinus drains medially into the adjacent posterior cardinal sinus via numerous channels (not shown in Fig. 11-6).

PORTAL SYSTEMS (Figs. 11-7 and 11-8)

A portal system is simply a set of veins that transports blood from one capillary bed to a second capillary bed, from which the blood is subsequently returned to the heart. The hepatic portal system transports nutrient-rich blood from capillary beds in the digestive tract to the liver, where the nutrients can be extracted and in some cases stored. The renal

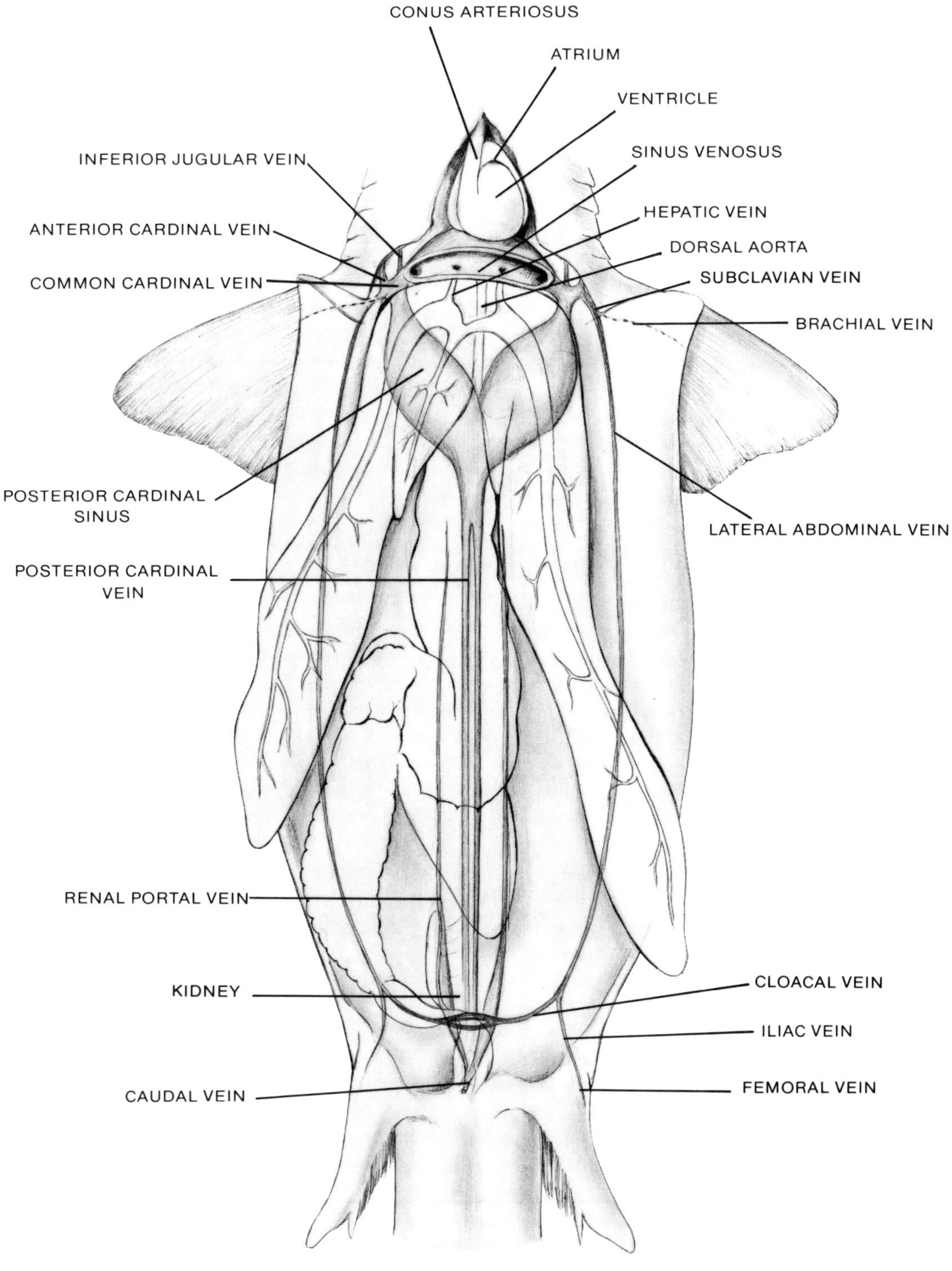

Figure 11-6. Ventral view of the abdominal cavity, with special attention to the systemic veins.

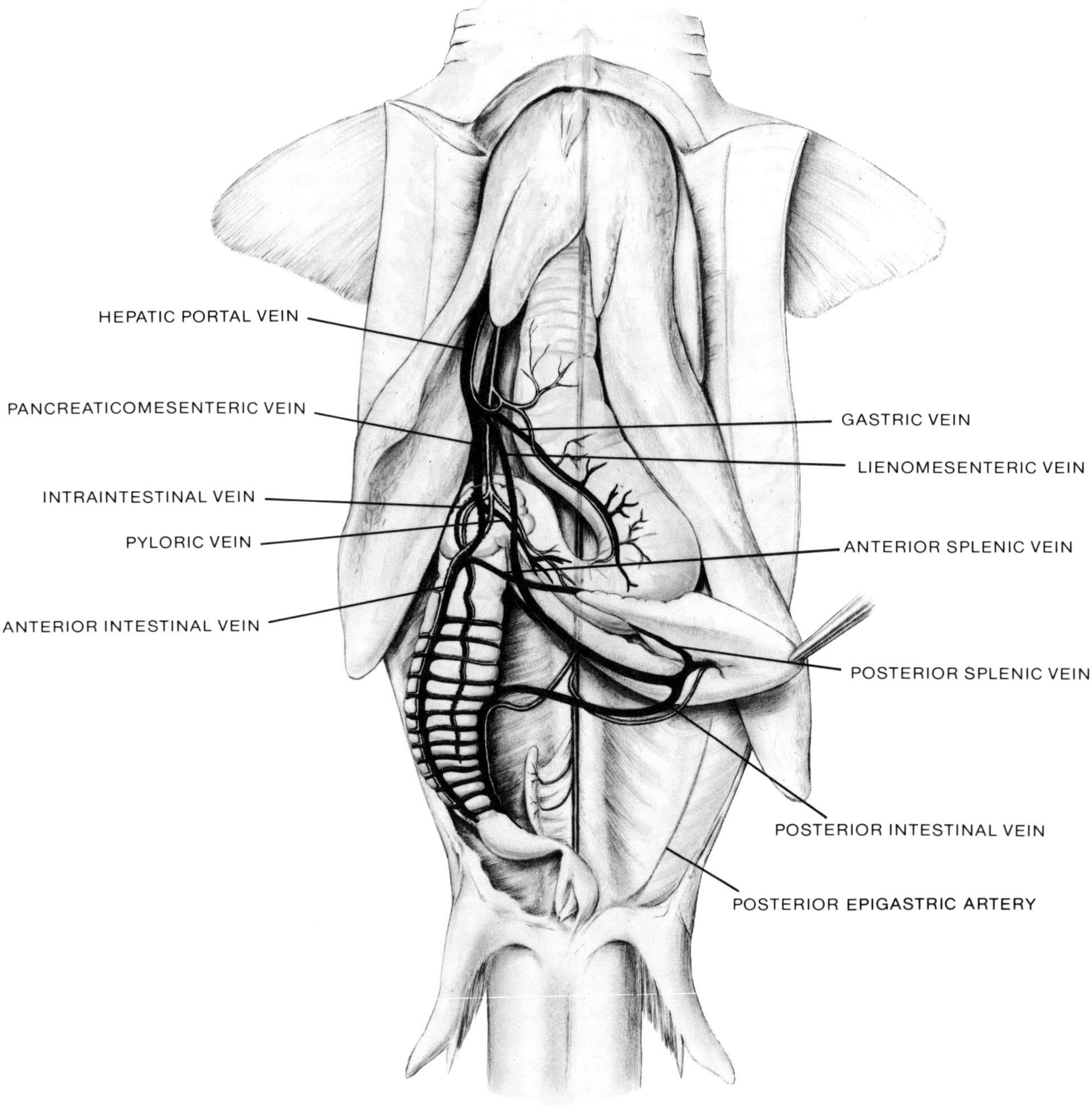

Figure 11-7. Ventral view of the abdominal cavity, with special attention to the branches of the hepatic portal system. (Refer, also, to the associated color plate.)

portal system channels the blood flow returning from the muscular tail and posterior trunk region through the capillaries around the kidney tubules.

Hepatic Portal System (Fig. 11-7)

Again, enter the peritoneal cavity and locate the ***hepatic portal vein****. It runs laterally adjacent to the celiac artery within the lesser omental mesentery. Follow the hepatic portal vein caudally, observing the veins which drain the digestive organs as they converge to join it. Each tributary vein can best be identified by tracing it to the organ or region it drains.*

Hepatic portal vein. This is a large unpaired vein collecting blood from three main tributaries. The three branches can be seen as they converge at the lesser omentum. They are the ***gastric vein***, the ***lienomesenteric vein***, and the ***pancreaticomesenteric vein***.

Note: Many biological supply houses will, upon request, inject the hepatic portal system separately with a different color. Yellow latex is standard for such injections. Because the colors may vary, no mention of color will be made in this identification section.

Gastric vein. This is the most cranial branch leading to the hepatic portal vein. It is formed by the union of dorsal and ventral branches which drain the walls of the cardiac stomach.

Lienomesenteric vein. **(Gastrosplenic or lienogastric vein)** In conjunction with the gastric vein, this unpaired branch joins the hepatic portal vein. The lienomesenteric, in turn, has three main tributaries: the ***posterior splenic vein*** (or ***posterior lienogastric***), draining the dorsal portion of the spleen; the ***posterior intestinal vein***, coursing laterally to drain the medial surface of the ileum and the annular venous system; and the ***pancreatic veins***, draining portions of the ventral lobe of the pancreas. These branches are small and seldom seen.

Pancreaticomesenteric vein. **(Gastrointestinal vein)** The third major tributary to the hepatic portal vein, this large unpaired vein appears to be an extension of the hepatic portal vein. It is partially visible where it enters the pancreas. Several minor veins converge to contribute blood to this vessel. The ***pyloric vein*** courses within the stomach wall from the pylorus after draining that area. The ***intra-intestinal vein*** joins the pancreaticomesenteric vein near the point of entry of the pyloric branch and extends caudally into the central portion of the valvular intestine, draining the spiral valve. The ***anterior intestinal vein*** runs superficially along the lateral surface of the ileum, collecting blood from the series

Figure 11-8. A series of three cross sections of the posterior trunk, showing the caudal vein, the renal portal veins, and the posterior cardinal veins.

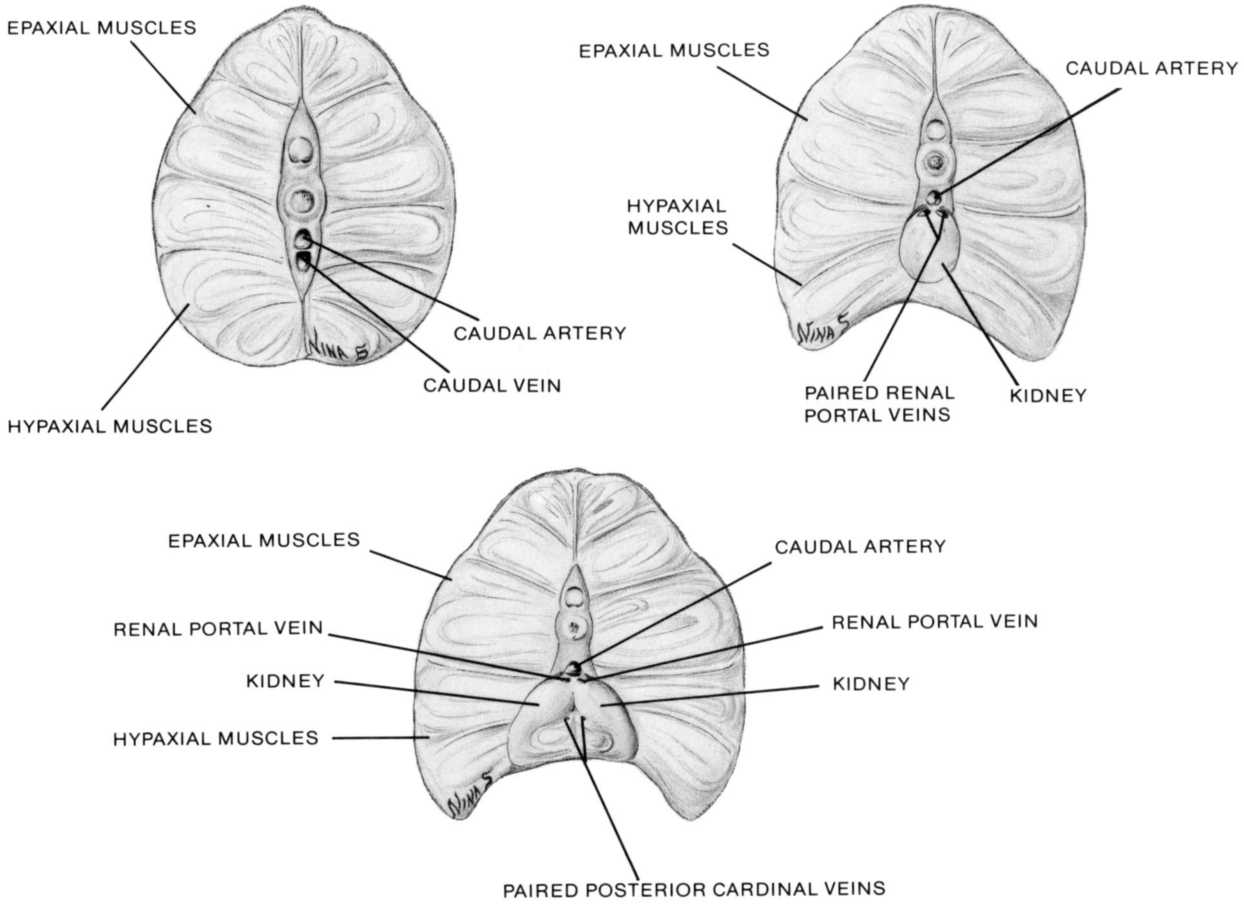

of annular veins associated with the junction of the spiral valve, and the external wall of the ileum. The ***anterior splenic vein*** joins the pancreaticomesenteric caudal to the pyloric. It runs along the caudal surface of the dorsal lobe of the pancreas to the anterior portion of the spleen.

In addition to those listed above, there are several secondary veins that enter the hepatic portal vein, including the ***choledochal*** veins that drain the bile duct. These veins are seldomly seen because of their small size and of inefficiency in the injection process.

Renal Portal System (Fig. 11-8)

Starting approximately four cm caudal to the cloacal aperture, begin making a series of transverse sections. Cut from the dorsal surface of the dogfish ventrally until both the caudal vein and artery have been severed. The initial section reveals a single caudal artery dorsal to a single caudal vein. Subsequent anterior sections reveal the bifurcation of the caudal vein and formation of the paired renal portal veins. More anteriorly in the series of sections, observe the appearance of the posterior cardinal veins. These veins are located ventral and medial to the renal portal veins. Note that the right posterior cardinal extends more caudally than the left. Close inspection of transverse sections through the kidney may reveal afferent and efferent renal veins as they course through the kidney tissue.

Posterior cardinal veins. This fully injected pair of veins, discussed previously, collects blood from both kidneys and transports it to the sinus venosus.

Efferent renal veins. These tiny veins course medially from the capillary beds of the kidney to the posterior cardinal veins. They are seen in transverse section as one approaches the posterior cardinal veins.

Afferent renal veins. These veins enter the lateral surface of the kidney from the renal portal veins that parallel the kidneys. They are not visible superficially but can also be seen in the series of transverse sections.

Renal portal veins. This large pair of veins courses along the lateral edge of each kidney. The pair is formed by the bifurcation of the caudal vein just posterior to the cloacal aperture. As they proceed cranially, they diminish in size until they disappear at the cranial limit of the kidneys. They can be seen more easily in transverse section.

Caudal vein. This is the unpaired main vein of the tail region. It courses cranially from the tip of the tail within the hemal arch just ventral to the spinal cord. It bifurcates just caudal to the cloacal aperture.

FUNCTION

The elasmobranch cardiac cycle relies on a dual pump mechanism within the pericardial cavity, Because of rigidity of the pericardial wall, relative volume changes of the cardiac chambers result in pressure changes within the cavity.

Expulsion of blood from the heart by ventricular muscle contraction momentarily reduces the total fluid volume within the pericardial cavity. Reduction of fluid volume reduces cavity pressure, creating a suction force. Blood is drawn from the sinus venosus into the atrium by this suction pump. Contraction of the atrium forces the blood ventrally into the ventricle. Thus the atrium and the ventricle reciprocally fill each other. A slight lag between ventricular expulsion of blood and refilling of the atrial chamber leaves a transient negative pressure in the pericardial cavity.

Heartbeat in elasmobranchs and other vertebrates is myogenic—that is, the beat originates within the heart itself. Heartbeat is generally initiated at the sinus venosus or its derivative in land vertebrates, the sino-atrial node. However, pacemaker properties have been demonstrated at various locations. From the point of initiation, an electrical wave of excitation spreads over the heart. Fast and slow conducting pathways temporally distribute excitation and thus control the contraction sequence. The rate of pacemaker discharge is an intrinsic property. Nevertheless, changes in heart rate result from acetylcholine released by the vagus nerve and from changes in body temperature.

Table 11-1. Organs and regions served by arterial and venous flow in the dogfish shark.

Region or organ	*Artery*	*Vein*
bile duct	choledochal	choledochal
brain	internal carotid	anterior cardinal sinus
duodenum	duodenal	duodenal
esophagus	esophogeal (small branch of celiac)	
eye	opthalmic (via stapedial)	orbital sinus
gonads	genital	genital
heart	coronary	coronary
ileum	annular, anterior intestinal, posterior intestinal	annular, anterior intestinal, posterior intestinal
jaw, lower	external carotid	inferior jugular
jaw, upper	hyoidean epibranchial	anterior cardinal sinus
kidney	renal	renal, posterior cardinal
liver	hepatic	hepatic, hepatic portal
myomeres	epigastric, segmental	parietal (intersegmental)
pancreas, dorsal	branch of pyloric	lienomesenteric
pancreas, ventral	pancreaticomesenteric	pancreaticomesenteric
pectoral fin	brachial, subclavian	brachial
pelvic fin	femoral	femoral (via iliac)
pericardium	pericardial	inferior jugular
pharynx	esophageal	inferior jugular
pylorus	pancreaticomesenteric, pyloric	anterior splenic, pyloric, pancreaticomesenteric
rectal gland	posterior mesenteric	posterior intestinal
snout	stapedial	anterior cardinal sinus
spiracle	afferent spiracular	anterior cardinal sinus
spiral valve	intraintestinal	intraintestinal, pancreaticomesenteric
spleen	lienogastric	anterior splenic, posterior splenic
stomach	lienogastric, celiac, gastric	gastric, anterior splenic
tail	caudal	caudal

Table 11-2. Major arteries and veins of the dogfish shark.

1. *Arteries*

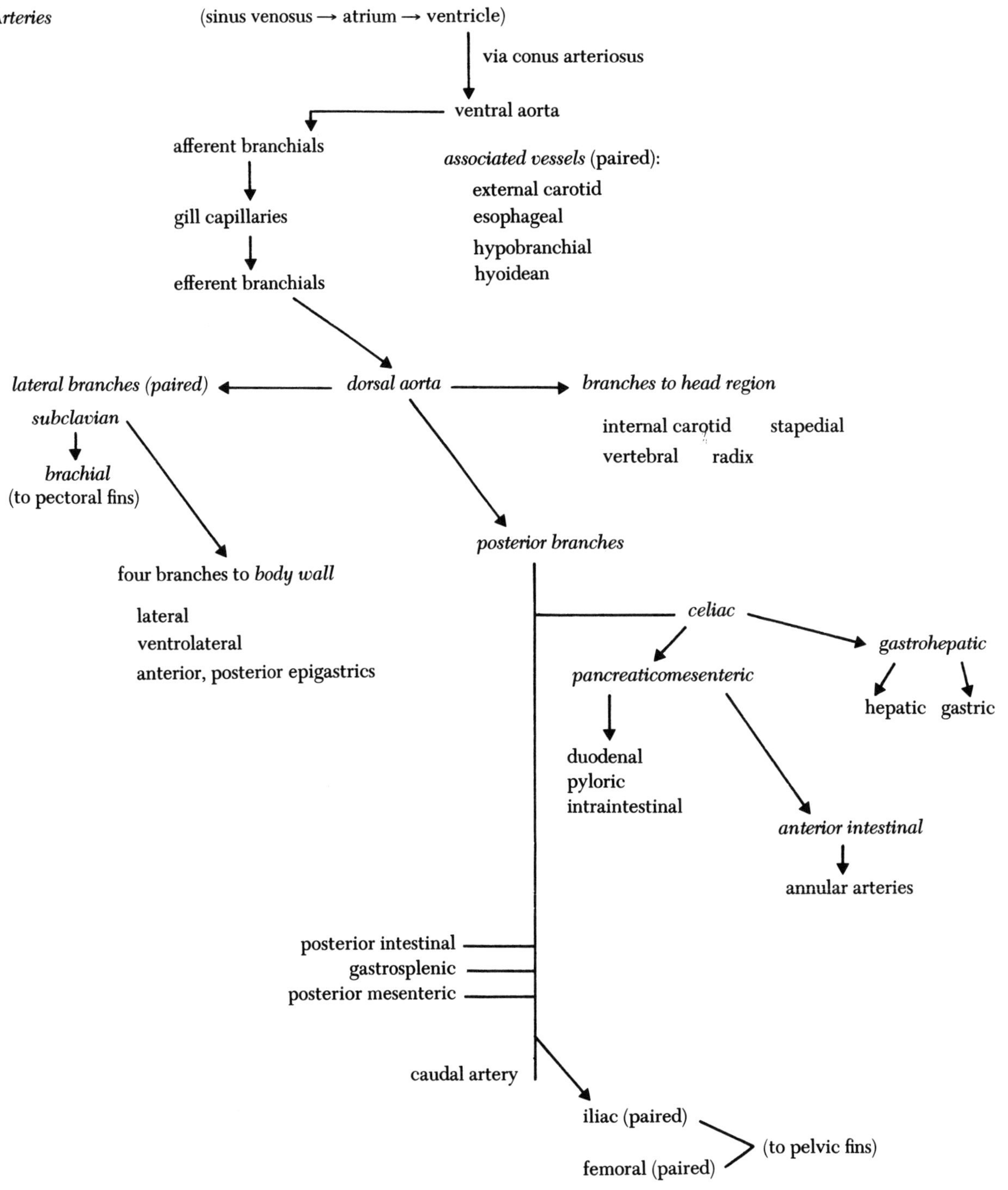

2. *Veins*

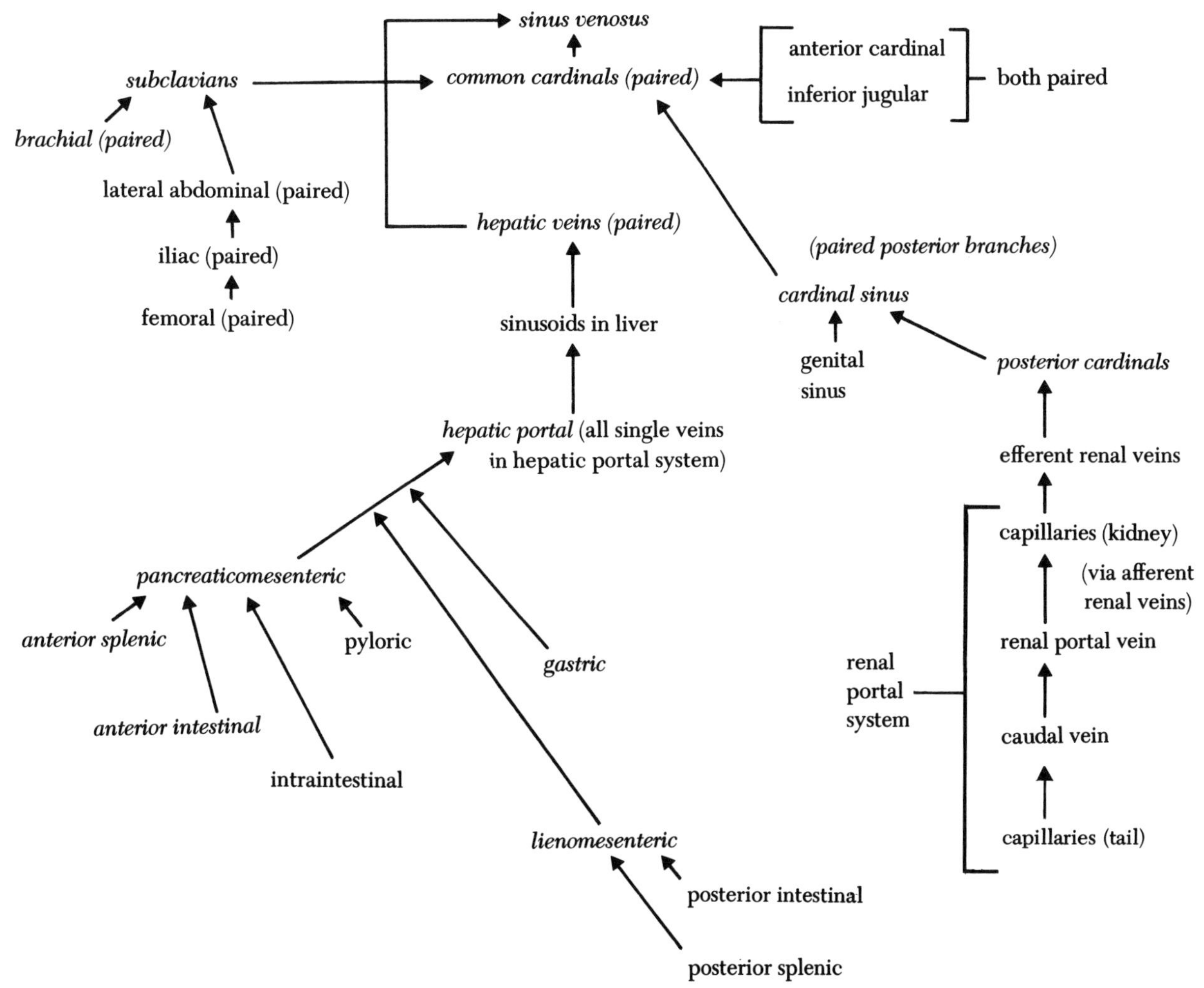
sinus venosus
common cardinals (paired)
anterior cardinal
inferior jugular
both paired
subclavians
brachial (paired)
lateral abdominal (paired)
iliac (paired)
femoral (paired)
hepatic veins (paired)
sinusoids in liver
hepatic portal (all single veins in hepatic portal system)
(paired posterior branches)
cardinal sinus
genital sinus
posterior cardinals
efferent renal veins
capillaries (kidney)
(via afferent renal veins)
renal portal vein
caudal vein
capillaries (tail)
renal portal system
pancreaticomesenteric
anterior splenic
pyloric
gastric
anterior intestinal
intraintestinal
lienomesenteric
posterior intestinal
posterior splenic

HENRY B. KISTLER, JR.

12

The Dogfish Shark: Respiratory System

The respiratory mechanism of elasmobranchs is a modification of the primitive system seen in urochordates and cephalochordates. Water ingested via both mouth and spiracle is forced laterally through five pairs of internal gills. (A few elasmobranchs have other numbers of gill slits; for instance, the saw shark, *Pliotrema*, has six pairs.) The gill arrangement differs anatomically from that of the bony fishes, such as trout or goldfish. In the shark, water leaves each gill via an individual external gill slit instead of being first collected in an external chamber before being discharged through a single opercular opening.

DISSECTION INSTRUCTIONS

On the dogfish shark's left (intact) side, separate each of the gill units by cutting dorsally and ventrally from the corners of each external gill slit. Observe the structures of the intact gills on this side of the body; then observe the gills in cross section on the opposite side. Examine the secondary lamellar structures with a hand lens or a dissecting microscope after cutting out a piece of gill and floating it in water.

IDENTIFICATION (Figs. 12-1, 12-2, 12-3)

The following respiratory system structures can be identified.

Gill arches. (**Branchial arches**) This series of incomplete cartilagenous rings extends dorsoventrally just medial to the branchial blood vessels. The skeletal elements provide support for the individual gill units.

Gill rakers. These are a set of short spike-like projections, situated medially from the gill arches. They filter the respiratory water, directing food caudally toward the esophagus.

Gill rays. These are a series of long cartilaginous spikes fanning laterally from the gill arches. They act as a support for each interbranchial septum.

Interbranchial septum. This term refers to the combination of nervous, muscular, connective, and blood vascular tissue supported by each gill arch and its associated gill rays. The tissue serves to anchor the primary lamellae.

Primary lamellae. (**Gill filaments**) These are small plate-like sheets of tissue anchored in rows along the lateral ridges of each face of the interbranchial septum.

Secondary lamellae. These are tiny plates on the primary lamellae; they can be seen only with a hand lens. They are arranged in stacks or rows perpendicular to the primary lamellae and are the site of gas exchange.

Demibranch. (**or Hemibranch**) This term refers to the unit composed of the lamellae associated with one face of the interbranchial septum.

Holobranch. This term refers to the entire gill unit comprised of interbranchial septum and two sets of lamellar tissue, one associated with each face of the septum.

Pharyngeal cavity. This cavity is located caudal to the transverse mucosal folds of the buccal cavity. Characterized by its longitudinal mucosal folds, the pharynx extends caudally to the level of the fifth internal gill slit on each side.

Gill chamber. (**Gill pouch or branchial chamber**) This cavity encompasses the gill structures. It includes the area between the external and internal gill slits.

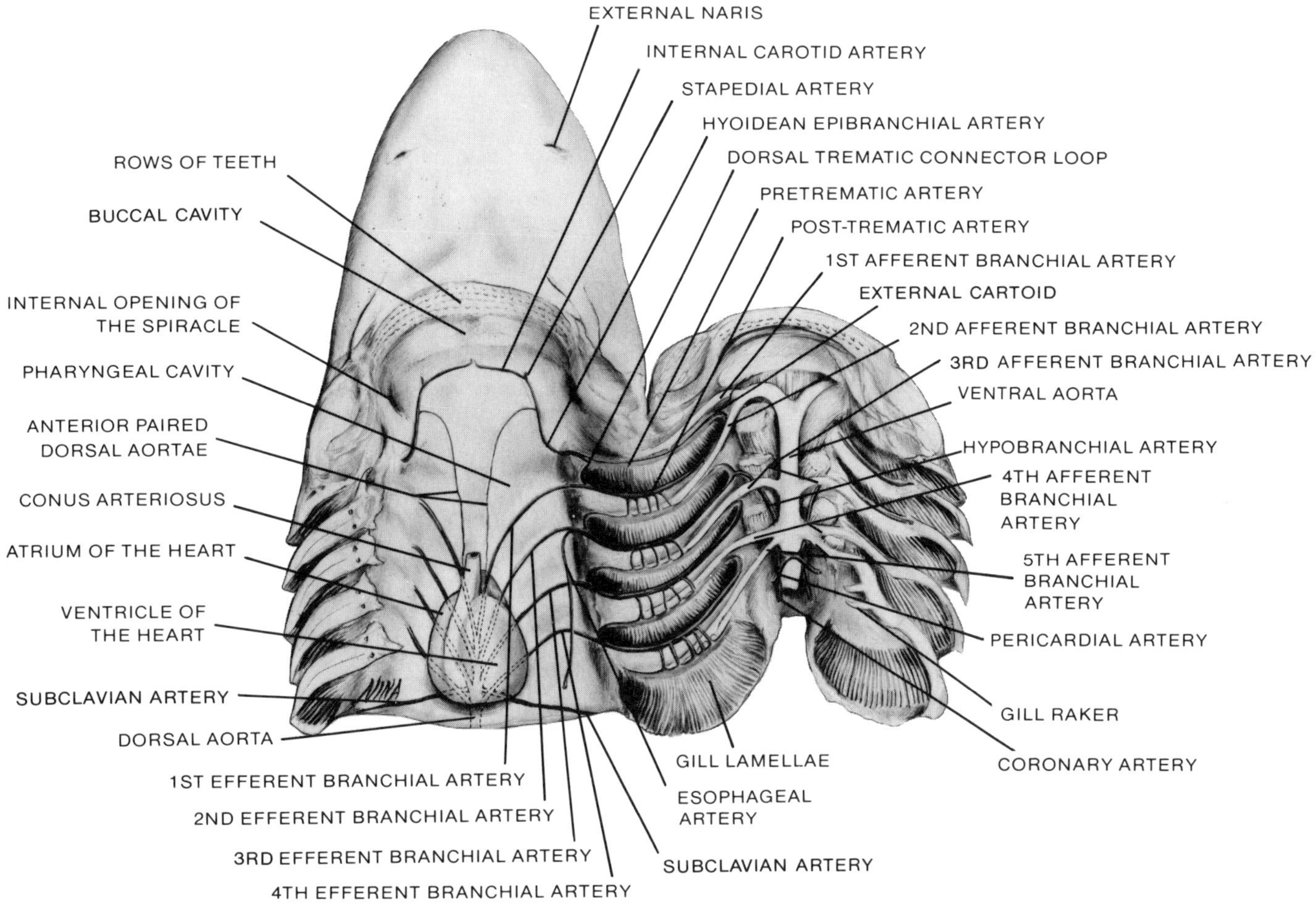

Figure 12-1. Ventral view of the pharynx and gill circulation of a dogfish, after lateral incision through the gills and reflection of lower jaw structures.

Parabranchial chamber. (Not shown in Fig. 12-1) This space, located lateral to the gill filaments, is the site of the parabranchial pump of the respiratory system. During each cycle, water flows laterally from the gills into these chambers.

Spiracle. This is a round aperture located dorsocranial to the five internal gill slits. This opening is believed to be a remnant of a gill opening found in more primitive chordates.

Pseudobranch. This is an abbreviated hemibranch found on the interior cranial surface of the spiracle; it can serve as an accessory respiratory exchange site.

FUNCTION

The exchange of respiratory gases between an organism and its environment is essential for the organism's survival. In elasmobranchs the gills and, in some cases, the spiracle are the points of respiratory exchange. Oxygen is loaded and carbon dioxide unloaded via passive diffusion across the blood-water barrier.

In addition, fast-swimming sharks use the Bernoulli effect to induce continuous flow into the open mouth and out through the lateral gill slits, which are held open. (Fast flow past the gill slits on the body's exterior surface lowers water pressure there, so inside water flows out.)

A dual pump circulates water over the gills. Water enters the pharynx mainly through the mouth. Then the mouth closes and the buccal or orobranchial pressure pump forces water from the pharynx between the lamellae of the gills. Next, the parabranchial chamber expands, drawing oxygen-depleted water from the lamellar area into the parabranchial chamber. The water then flows outward

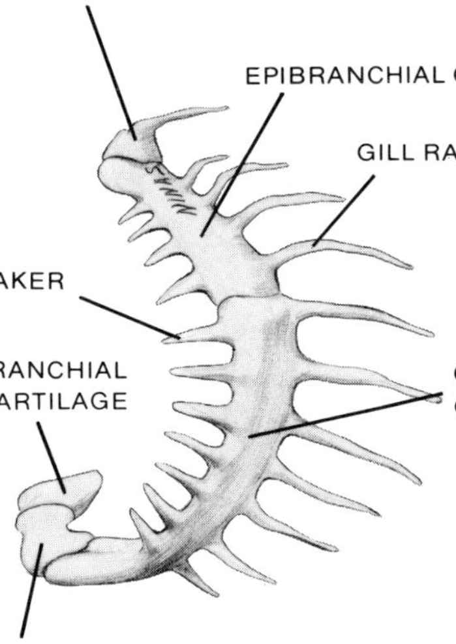

Figure 12-2. Craniolateral view of an intact half of a gill arch, showing component cartilagenous structures.

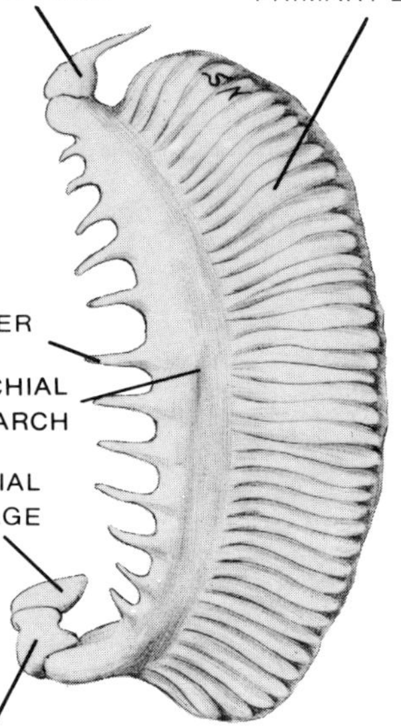

Figure 12-3. Craniolateral view of an intact half of a gill arch with a full complement of lamellar gas exchange tissue.

through the gill slits. The duality of the pumps assures a more constant water flow.

The blood capillaries in the secondary gill lamellae are arranged to carry blood inward, in a direction opposite to that of the sea water flowing *outward* from the pharynx. This counter-current flow maximizes gas exchange between the two fluids. Cardiorespiratory coordination tends to optimize carbon dioxide and oxygen exchange. Thus, as the pharynx inflates with water, nerve receptors initiate vagus nerve inhibition of the heartbeat. Blood flow slows in the gills so there is more time for exchange to occur. Water expulsion releases the vagal inhibition, thereby allowing return to normal rates of blood flow.

The spiracle can function as an accessory respiratory organ. Its prominence varies with the behavior of various species of sharks. Bottom-dwelling sharks, which often lie stationary or grub in the mud, have large spiracles because pumping water through the mouth would bring sand into the mouth. It is also postulated that spiracles may assist in respiration while sharks feed.

The Nervous System

NEIL J. BAKER

13

The Nervous System: General Information

Students unfamiliar with the structure of nerve cells should consult the appropriate sections in a general vertebrate text.

The serial contraction of myomeres (discussed in Chapter 8), and the synchrony of heartbeat with respiration (Chapter 12) are two striking examples of activities which require a central control mechanism for coordination and timing. In vertebrates, two such control systems are responsible for the integration of body activity so that the various organs function harmoniously. The ***nervous system*** controls activity by relay of commands directly to a particular organ by means of nerve axons. Speed and point-to-point connections characterize the nervous system.

In contrast, the ***endocrine system*** functions by release of chemicals into the blood stream for distribution everywhere in the body. This takes much more time than transmission of nerve impulses. Furthermore, since the chemicals are transported throughout the body, the nature of the response depends on the sensitivity of each "target" organ.

In the shark, the function of the endocrine glands is not as well understood as in mammals. Therefore, the endocrines are more suitably discussed in the cat, and only the nervous system will be covered at this time.

The specific regions of the nervous system which control responses are called ***integration centers***. Used in this respect, "integration" refers to the processing of sensory information collected from the internal and external environments in order to determine and coordinate appropriate responses. Integration networks are located in the brain and spinal cord, which together comprise the ***central nervous system***. These two structures develop from the hollow dorsal nerve tube found in the early stages of embryos and in the most primitive adult chordates. In all adult chordates, the central nervous system retains a hollow core throughout its length.

The tissue of the brain and spinal cord is usually segregated into two types—gray matter and white matter. The gray matter, composed chiefly of nerve cell bodies and their dendrites, is situated internally around the hollow core of the spinal cord. In the spinal cord, it forms a continuous column running the entire length; in the brain, it tends to segregate into discrete units. Myelinated axons, extending from the nerve cells in the gray matter, comprise the white matter. White matter is located peripherally in the spinal cord and internally in most divisions of the brain.

The brain and spinal cord are connected to the periphery of the body by ***cranial*** and ***spinal*** nerves, respectively. Sensory information from the peripheral receptors is received through ***sensory (afferent)*** fibers of these nerves; motor directives reach the necessary organs through ***motor (efferent)*** fibers. Collections of nerve cell bodies along the paths of these nerves are called ***ganglia***. All these structures located outside the central nervous system are part of the ***peripheral nervous system***.

EVOLUTION

The circuitry of the nervous system is based on three fundamental components: ***afferent neurons*** which transmit sensory impulses to the central nervous system; ***integration networks*** or centers in the gray matter which process the impulses; and ***efferent neurons*** which transmit the resulting motor directives to organs which effect the response. The integration centers have been particularly crucial during vertebrate evolution, because more complex modes of sensory discrimination or motor response are impossible without an accompanying increase in the complexity of neural networks in the central nervous system.

In primitive vertebrates, the brain is little more than an anterior extension of the spinal cord. As a result, sensory impulses are processed largely in the relatively simple networks of the spine, and movements tend to be limited to highly stereotyped, involuntary acts called ***spinal reflexes***. This lack of dependence of the spinal cord system on the brain for control of activity may be seen in lower vertebrates such as sharks. If the spinal cord is transected just behind the brain, the basic swimming motion involving serial contraction of body musculature is essentially unimpaired. However, sharks transected in this manner do not change direction to approach food they see or smell, since those senses are analyzed by the brain. Motor directives resulting from such sensory information cannot reach the spinal cord in these animals.

Complex reflexes are made possible by the presence of ***associational neurons***, or interneurons, within the gray matter of the spinal cord. These nerve cells are located between the sensory cells bringing information into the cord and the motor cells that convey a reponse to the sensory information. The associational neuron system may connect on the same side or the opposite side of the spinal cord; it may extend anteriorly or posteriorly in the cord. This greatly increases the number of different responses a single type of stimulus may elicit. Motor neurons receive input from many associational neurons. Based on the cumulative impulses the motor neuron receives, it may or may not stimulate contraction of a particular muscle. For example, in swimming, impulses from receptors in contracted muscles on one side of the body may inhibit the contraction of a muscle on the opposite side.

Associational neurons also transmit information to centers in the brain which combine it with other information and transmit appropriate directives back to the spinal cord. However, since the cord is relatively autonomous in lower vertebrates, fibers connecting it with the brain are not as abundant as in higher vertebrates.

In the course of vertebrate evolution, more complex integration networks have developed in the brain; these take control over motor activity except for the simplest reflexes. In terrestrial vertebrates, spinal reflexes alone do not coordinate the complex movements of many-jointed limbs. On the other hand, in a cat transected behind the brain so that vital functions are not impaired, so-called "rhythm generators" in the spinal cord can control normal sequential muscle movements of the legs.

Indicative of the loss of spinal independence is the prominence of ***ascending*** (afferent) tracts, bundles of fibers relaying sensory information from the spine to the brain and ***descending*** (efferent) tracts carrying directives back from higher centers to motor neurons leaving the cord. The spinal cord, in effect, becomes a pathway between the brain and peripheral nerves, in addition to being a "switchboard" in its own right.

In the brain itself, the gray matter is arranged into aggregates of nerve cells with similar functions called ***nuclei***. These range greatly in size. Bundles of fibers in the white matter that run between these nuclei are called ***tracts***.

Specific nuclei hooked into various afferent pathways in the brain process sensory information and relay the revised data along tracts to the next nucleus. At the appropriate level, a motor response is evoked and transmitted to peripheral nerves through efferent tracts. Other nuclei may hook into this pathway to revise the motor directive before it reaches the peripheral neurons.

In humans, for example, stimuli associated with spinal reflexes need not ascend higher than the spinal cord to evoke responses. At the next higher level, regions in the lower brain stem (medulla, pons, midbrain) integrate information to control the muscles supporting the body against gravity. Finally, sensory data from many sources can lead the cerebral cortex to elicit skilled "voluntary" movements. However, the motor directives of the cortex are further revised through a feedback pathway with the cerebellum to coordinate them with data received from the cerebellum concerning the orientation of the body in space. Thus, crude, stereotyped movements can be handled by lower-level integrating pathways, whereas skilled movements are initiated by the cerebral cortex, on the basis of sensory data.

The hierarchical organization of the brain is the product of evolution. As new centers evolved, older ones may have been incorporated into pathways to perform subordinate roles. The older centers may retain control over primitive, stereotyped movements, or they may perform new activities.

In the dissections and discussions that follow, the brains of the dogfish shark and the sheep will be compared to illustrate these general trends in evolution with respect to specific structures. Structural and functional differences between cranial and spinal nerves will also be examined.

NEIL J. BAKER and MARY BRYAN McNABB

14

The Sensory System in the Dogfish Shark

Before the central nervous system itself is examined, the inner ear and the structures of the orbit will be exposed and identified. This will greatly facilitate the dissection of the brain and cranial nerves and will provide landmarks for identifications.

DISSECTION OF THE INNER EAR (Figs. 14-1, 14-2)

Much of the inner ear has the same fundamental structure in all vertebrates. It is most readily studied in the shark, however, because the ear parts are embedded in cartilage which can more easily be sliced away than the bone of a cat's skull.

Unfortunately, it is very easy to cut away the inner ear itself, so *do the following dissection slowly and carefully*, using a *fine* scalpel and *fine* forceps for most of the work. The dissection is difficult; students may have to compare their animals to see all of the canals of the ear intact.

Find the endolymphatic pores, situated 0.6 cm apart in the skin between the spiracles along the middorsal line. These openings lead almost immediately to the endolymphatic sacs beneath the skin. With a scalpel, carefully remove the layer of skin in this area. The sacs are embedded within a mass of brown mucus which should be cleared away with forceps. The mucus fills the oval depression in the chondrocranium in which the endolymphatic foramina are located.

Use a scalpel and forceps to slice and remove the skin from the rest of the chondrocranium. Adhering to the skull is a tough, white connective tissue which should also be removed to reveal the smooth surface of the cartilage.

Above the orbit, you will cut through many small branches of the superficial ophthalmic nerve which emerge above the chondrocranium through a series of foramina and innervate the skin in this area.

Identification (Figs. 14-1, 14-2)

Rostrum. The rostrum, discussed in Chapter 8, is an oval cartilage at the anterior end of the chondrocranium which supports the snout. Its concavity is filled with a gelatinous material.

Ampullae of Lorenzini. (**Fig. 8-1**) These form a flat, porous V-shaped sheet of tissue covering the cartilage on each side of the anterior end of the chondrocranium.

Superficial ophthalmic nerve. *(On one side of the rostrum, slip forceps under the narrow posterior portion of the V of the ampullae of Lorenzini and pull this sheet of tissue off of the cartilage.)* Branches of the portion of this nerve which emerge from the large foramen anterior to the orbit transmit sensory information from the ampullae of Lorenzini and the lateral line of the dorsal snout to the brain.

Superficial ophthalmic foramina. These are openings in the skull above the orbit. Branches of the superficial ophthalmic nerve reach the skin above the orbit through these foramina. (See Figure 8-6.)

Epiphyseal foramen. This is an opening within what appears to be a circular depression in the cartilage just posterior to the rostrum. A small amount of cartilage may have to be sliced away to see it. The ***epiphysis***, or pineal body, extends posteroventrally from this foramen. (See Fig. 8-5.)

THE INNER EAR

(See Fig. 8-6 for the shape and location of the otic capsule.)

Lateral to the middorsal line, cut away the epibranchial musculature which covers the posterior half of the otic capsule. The cartilage slopes down sharply below the muscle at this point, so the posterior margin of it is about 2.5 cm below the surface.

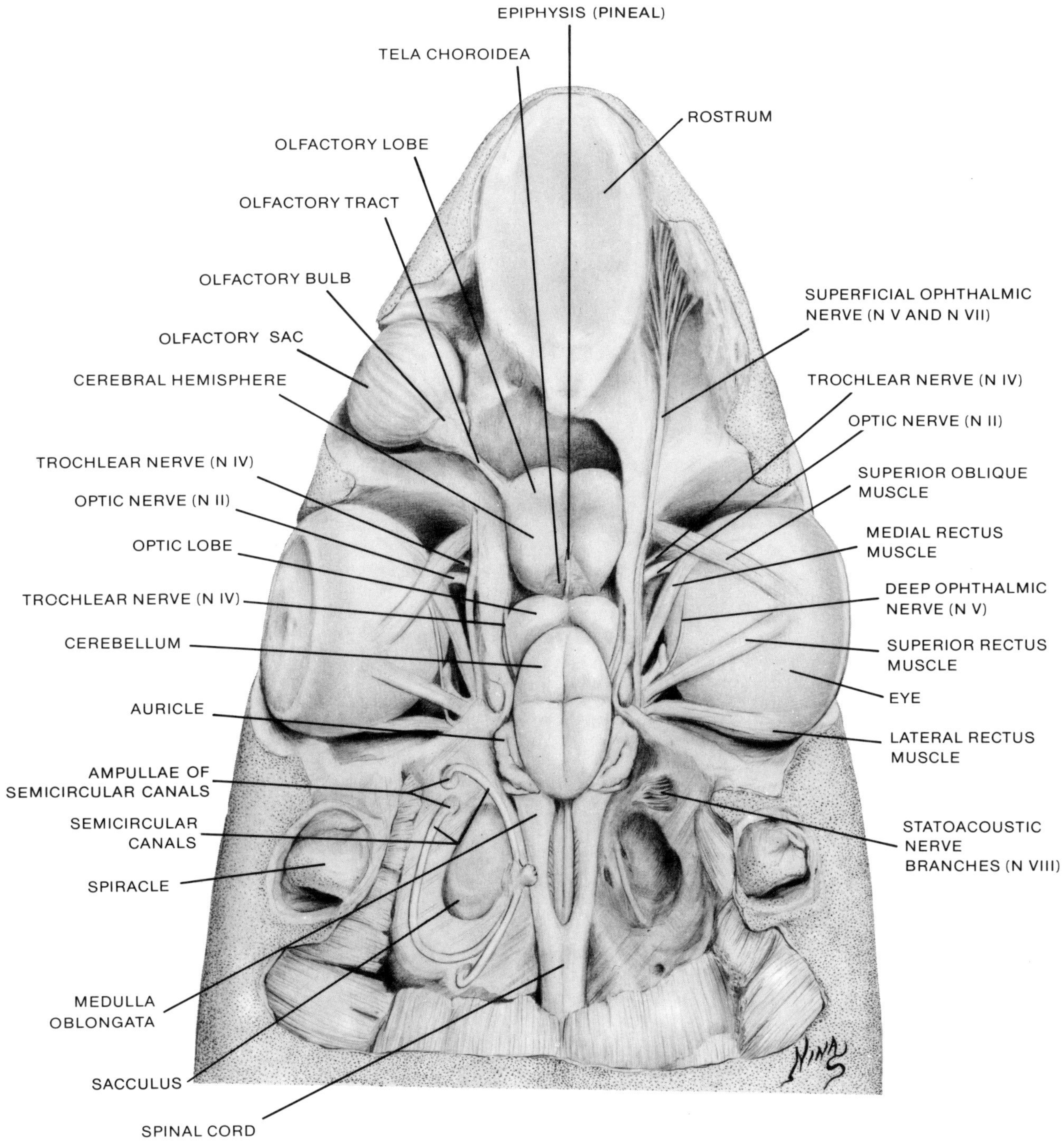

Figure 14-1. Dorsal view of a dogfish head with specific exposure of brain, eye, and ear structures.

Identification (Fig. 14-2)

Endolymphatic sac and duct. The duct leading from the endolymphatic sac descends into the cartilage and enters the ***sacculus***, a membranous sac of the inner ear.

Anterior semicircular canal. The proximal portion of this canal is located a short distance beneath the surface of the cartilage close to the endolymphatic duct. It extends anterolaterally from this point and curves around as it approaches the orbit. Remove cartilage just anterior to the endolymphatic duct in thin horizontal slices. The narrow chamber in which the anterior semicircular canal is located will be seen below the surface. The canal itself is a delicate, transparent tube resting in this chamber. Using a fine scalpel and forceps, follow the canal by chipping cartilage around it. (Take precautions to keep flying pieces out of your eyes.)

Posterior semicircular canal. This extends posterolaterally from the endolymphatic duct and curves ventrally just beneath the cartilage of the posterior wall of the otic capsule. The proximal portion of this canal near the endolymphatic duct is also situated close to the surface.

Sacculus. Carefully slice away the cartilage lateral to the endolymphatic duct until the membranous, triangular sacculus is exposed. Often the membrane is collapsed, but it may still be seen by pulling it carefully out of the cavity with forceps. If you remove the sacculus, you will see a mass of sand granules, the ***otoliths***, inside. (These are discussed below.)

Horizontal semicircular canal. This canal extends in a horizontal and lateral direction around the sacculus deep in the otic capsule. Its lateral side is located in the wall of the capsule just medial and ventral to the spiracle. (It may help to remove some of the muscle between the spiracle and the cartilage.)

Ampulla. This is a bulbous expansion present in each semicircular canal, in which the white ***crista*** may be seen. (These are discussed below.) The ***ampulla*** of the ***horizontal canal*** is located at the anterior end where it meets the anterior canal. Adjacent and slightly anteroventral to this structure is the ***ampulla*** of the ***anterior canal***. Finally, the ***ampulla*** of the ***posterior canal*** is located fairly deep in the cartilage just after the canal curves back toward the sacculus.

Utriculus. The division of the utriculus into two separate sacs is a feature unique to sharks.

Figure 14-2. Lateral view of the left inner ear of a dogfish.

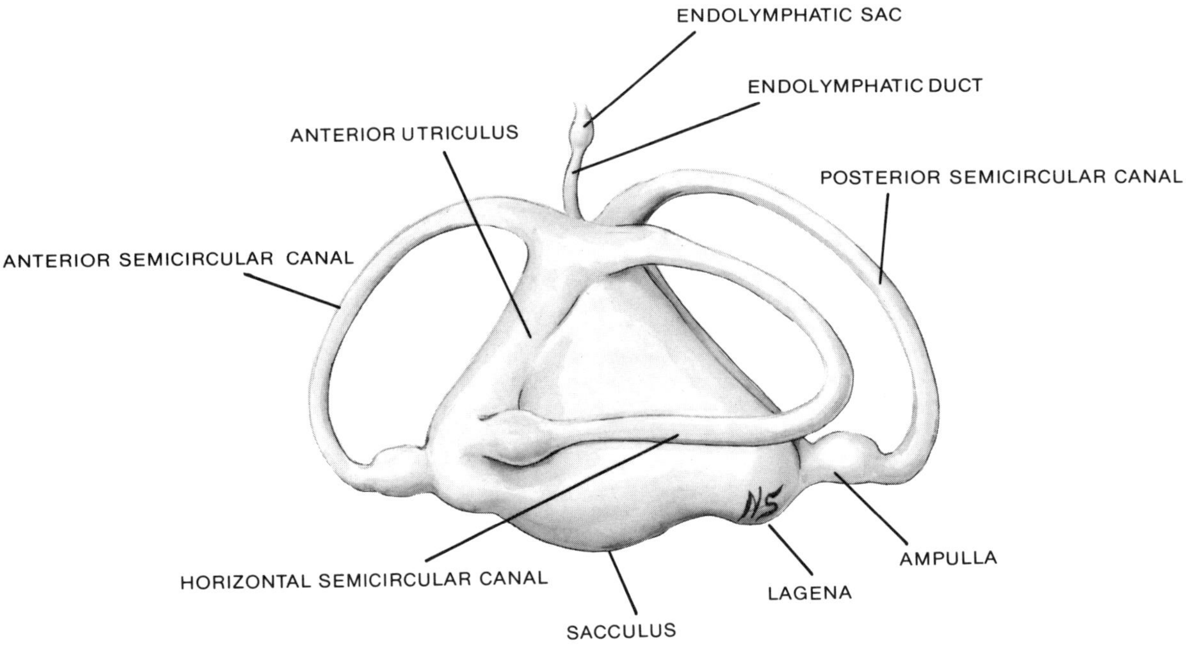

Anterior utriculus. This is a small chamber located at the union of the anterior canal and the horizontal canal. Remove the cartilage extending from the ampullae of these two canals to the anterior margin of the sacculus. Gently lift the horizontal canal near its ampulla to distinguish the utriculus from the sacculus.

Posterior utriculus. This is a membranous sac which receives both ends of the posterior canal. It is difficult to expose unless the entire inner ear is excised. If you do excise the ear, leave cartilage attached to its lower medial surface.

Lagena. This is a small membranous sac forming a bulge on the posterior side of the sacculus. It is difficult to expose. Remove cartilage medial and ventral to the horizontal canal to locate it.

DISSECTION OF ORBITS (Figs. 14-1, 14-3)

Although only the orbital structures will be dissected at this time, the diagrams include the exposed brain and several cranial nerves so these features will not be inadvertently destroyed. Also, the ventral view of the eye (Fig. 14-3) includes nerves which will not be discussed until a later dissection.

Remove the skin around the orbit which conceals the outer margins of the eyeball. Also slice away cartilage overlying the eye's dorsal surface until the superficial ophthalmic nerve is exposed as it passes dorsally along the medial wall of the orbit.

Wash out the mass of dried blood which may fill portions of the orbit. A large part of the orbit behind the eye is filled by a venous sinus which receives veins from the anterior portions of the head and brain and leads to the anterior cardinal sinus above the gill pouches.

With forceps, clear away the gelatinous connective tissue from around the eye. Be careful not to cut the trochlear nerve, which extends from the medial wall of the orbit, passing ventral to the superficial ophthalmic nerve, to the posterior side of the superior oblique muscle. An additional tough connective tissue which envelopes the eye may have to be removed from the muscles in order to see them.

In this dissection, it may help to cut away cartilage in the anterior and posterior walls of the orbit. Ultimately, the eye should be freed ***without cutting muscles or nerves*** *so that you can raise and shift it to see all of its surfaces. Nerves and muscles are frequently confused, so proceed with special care. The oculomotor is a very small nerve running along the ventral surface of the eyeball, and the mandibular branch of the trigeminal is immediately posterior to the lateral rectus muscle. The infraorbital is the large trunk seen on the floor of the orbit.*

Identification (Fig. 14-3)

Six muscles control movements of the eyeballs. The muscles are named according to whether they pull diagonally or in a straight line from their insertion points on the eyeballs.

Oblique muscles. These originate on the anterior portion of the medial wall of the orbit and insert on the eye.

Superior oblique. The superior oblique extends to the middorsal surface of the eyeball.

Inferior oblique. The inferior oblique inserts on the ventral surface of the eyeball.

Rectus muscles. (Recti) The recti are muscles which originate on and fan out from the posteromedial corner of the orbit.

Superior rectus. The superior rectus inserts just posterior to the superficial oblique on the dorsal surface of the eyeball.

Medial rectus. The medial rectus is a long muscle which crosses beneath the superior oblique to the medial side of the eyeball.

Inferior rectus. The inferior rectus inserts ventrally just posterior to the inferior oblique.

Lateral rectus. The lateral rectus inserts on the posterior side of the eyeball.

Orbital process. (See Chapter 8.) This is a conical cartilaginous projection of the upper jaw or palatoquadrate, medial to the eye. It supports the upper jaw and allows the palatoquadrate to be protruded when the animal bites.

Optic pedicel. This cartilaginous stalk extends to the eye from the posterolateral wall of the orbit, ventral and parallel to the superior rectus. It serves as a support on which the eye rotates.

FUNCTION

As previously noted, the ears of the shark are located within the chondrocranium. Each inner ear consists of three semicircular canals occupying three different planes, and three membranous sacs—the sacullus, the utriculus, and the lagena. The three

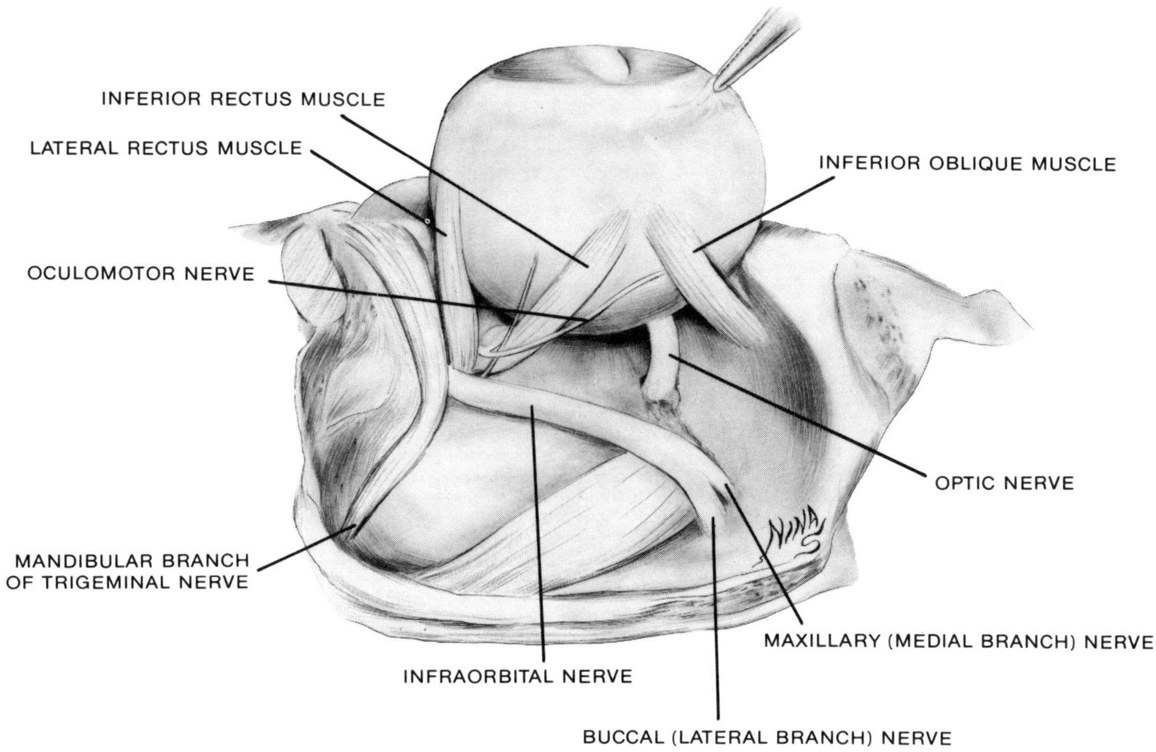

Figure 14-3. Ventral view of the right eye and orbit of a dogfish, with emphasis on structures underneath the eye.

canals and three sacs are filled with a fluid called endolymph and surrounded by another fluid called perilymph.

The ear of the shark has a variety of functions, some of which are not fully understood. The semicircular canals, which are in planes at right angles to one another, are used to detect acceleration and deceleration of the head in each of the three planes of space. Along each canal is a large bulb-like expansion, the ampulla, housing a patch of sensory epithelial cells bearing hair-like cilia; each patch is called a crista. The ends of the hairs of the crista are embedded in a cupula, which swings like a gate in response to the flow of endolymph in the canal. If the head accelerates in the plane of the canal, the endolymph flows past the cupula because of inertia. The cupula bends the sensory hairs, which in turn set in motion the cellular events that culminate in altered patterns of nerve impulses going to the brain. By comparing the patterns of information from the six semicircular canals, the shark can sense the motion of its head and the direction of any turns.

Two endolymphatic ducts, which open into pores on the surface of the head, provide a fluid-filled channel to the inner ear. In a newborn shark, these ducts allow grains of sand to enter the three sacs of the inner ear. These otoliths or "ear stones" aid in the detection of gravity. In the three planes in each ear are ciliated sensory patches called maculae, which respond to the pressure of the sand grains. Depending on the degree of response from each of the six patches, the shark can assess straight-line acceleration and deceleration as well as its position with respect to gravity. In addition, these internal ear sacs, particularly the lagena, may be used for hearing. In some sharks, there is evidence of directional hearing with sensitivity from 10 to 640 Hertz. Certain sharks appear to be especially attracted to low frequency sounds.

The Lateral Line, Ampullae of Lorenzini, and Pit Organs

(Identification of the lateral line and ampullae of Lorenzini is included in Chapter 8.)

The lateral line system is the best known of the group of sensory organs associated with the surface of the head. The cephalic canals of the system contact the environment through lines of pores. These pores lead to clusters of sensory neuromast cells, located within the canals. The mucus-filled pores are used to detect differences in the velocity of the water currents around the animal. By determining water movements, the shark can detect passing animals in the dark and even motionless objects in the water if the shark itself is moving.

The ampullae of Lorenzini are distinct in structure and function from the lateral line system. The ampullar pits sunk in the skin are filled with a gelatin-like material. The pits contain both structural and sensory cells. The ampullae of Lorenzini probably have multiple functions. Perhaps their most important role is detecting differences in electrical potential in the environment around the shark. In addition, the ampullae can apparently sense hydrostatic pressure changes, temperature changes, and salinity gradients.

The head of the shark has been called "one giant taste bud" due to the presence of the pit organs over much of its surface. These chemoreceptors, tiny sensorial crypts between the skin denticles, are used for "tasting" the water, recognizing food, and avoiding noxious materials.

Olfaction

The shark's olfactory powers have been shown at ranges of 200 to 500 feet with very dilute solutions of human blood in sea water. The mechanism for olfaction in the shark is basically similar to that found in other vertebrates. Each olfactory receptor is an actual nerve cell which is stimulated by some property of molecules it contacts. It is unclear whether the receptor cell is stimulated by molecular shape, binding frequencies, permeability to the odor molecule, or other characteristics. In any case, when an olfactory receptor cell is stimulated, impulses travel down its axon to a mitral cell located in the olfactory bulb. The mitral cell, which receives axons from thousands of receptor cells, seems to process the olfactory information and send nerve impulses to the olfactory lobe of the brain, and thence to other higher centers.

The actual olfactory receptors are found in the corrugated inner walls of the two olfactory sacs. These sacs are connected to the external environment by a pair of nostrils equipped with a unique feature that allows for continual sampling of the water. Each nostril of the shark has "Schneiderian folds," two pinches of skin which overlap the center of the nostril and create two openings for water movement. Since the dogfish's olfactory system lacks the internal connection with the pharynx seen in amphibians, there is a chance for temporary stagnation of the water being sampled for smell. The folds create a channel for the water; it is thus kept in continual motion, allowing for more sensitive olfaction.

Another feature of the sense of smell in the shark is its directional nature. With the two nostrils of the shark on the extreme sides of the head, differential sampling of the passing water can be performed. By comparing the intensity of smell received by the two nostrils, the dogfish can "home in" on food. This directionality also explains the appearance of hammerhead sharks.

Sight

In the marine environment, much sunlight is absorbed within a few feet of the surface. The eye of the shark is adapted for vision in dim light. The retina of the shark, and of most vertebrate animals, contains two types of light-sensitive cells: rods, which are sensitive in dim light, and cones, adapted to high-intensity light. Both are connected via bipolar cells to the ganglion cells, which transmit nerve impulses to the brain. In part, the acuity of a vertebrate eye depends on the ratio of light-sensitive cells to the ganglion cells that are connected to the brain. The more light-sensitive cells sharing a single ganglion cell, the fuzzier the image the eye produces. This is because a wider area of the retina activates a single ganglion cell. The shark eye is extremely poor in bipolar and ganglion cells—thus, it has a very low acuity. But the eye is extremely rich in rod cells, indicating great sensitivity in dim light. The shark eye also has a reflective layer of guanine crystals behind the retina, so that any light that happens to bypass the light-sensitive cells will be reflected. Thus light has a second chance to activate a rod or cone. All these measures adapt the shark's eye to performing efficiently in dim light.

The range of the shark's vision has been estimated at about fifty feet under the best conditions. When the shark's sperical lens is at rest, it is focused for this distance. To focus on closer objects, the shark must move its lens outward, nearer to the cornea.

For the final phase of homing in on food in clear water, sight seems to have precedence over smell and the lateral line system. When confronted with two four-inch cubes—one of fish, the other of wood painted to look like fish—a normal shark in muddy water never makes an error in selecting the food. In clear water, however, the shark makes about four errors in ten tries. Thus visual data, when available, may predominate over olfaction and the lateral line system.

NEIL J. BAKER

15

The Brain and Cranial Nerves of the Dogfish Shark—A Representative Lower Vertebrate

The brain of the dogfish shark is comparatively easy to expose for it is housed in a fairly large cavity in the chondrocranium. The cranial nerves pass through cartilage. They are therefore easily overlooked or damaged *unless the dissection is done slowly and carefully*. Use a *fine* scalpel and *fine* forceps wherever possible. It is likely that, because of time limitations or occasional mistakes, students will have to compare their dissections in order to see all the nerves.

Historically, the extraction of shark brains has actually had some practical importance. In the Middle Ages, according to Budker (1971), powdered shark brains were esteemed for their laxative and diuretic properties. In ancient times, Pliny's advice in dental care was "to rub the teeth once a year with the brains of a dogfish which had been cooked and kept ***alive*** in oil."

DISSECTION OF THE BRAIN (Fig. 15-1)

Remove any muscle remaining over the rear of the chondrocranium without cutting the spinal cord, which begins middorsally about 5 mm posterior to the endolymphatic ducts. It is located close to the surface at this point.

The epiphysis is easily destroyed unless it is exposed early in the dissection. Carefully slice away cartilage just posterior to the epiphyseal foramen (see Fig. 8-5) without cutting across the foramen itself, until you break through to the cranial cavity.

The epiphysis is an oval body attached to a slender white strand, the epiphyseal stalk, which originates in the roof of the diencephalon.

Slice away cartilage along the medial portion of the chondrocranium to expose the brain. A delicate transparent connective tissue called the meninx primitiva covers the brain.

The medulla is located between the endolymphatic ducts. Of all the brain centers, the cerebellum is situated closest to the roof of the cranial cavity. Use forceps when necessary to break away the cartilage above this structure.

In order to save the left and right trochlear nerves, which run along the dorsal surface of each optic lobe and enter the orbits, leave a wall of cartilage between each eye and the brain as shown in Fig. 15-1. Shave away the top of the wall down to the level of the superficial ophthalmic nerve.

To expose the olfactory sac on one side, find the external naris on the ventral surface of the snout; remove the skin around and above it. Then cut away the cartilage of the nasal capsule. The olfactory tract should be exposed by cutting the superficial ophthalmic nerve on one side and removing cartilage between the sacs and the cerebral hemispheres.

IDENTIFICATION

Structures listed below with an asterisk (*) are not included in the diagram and will not be dissected now. They are listed here for convenience in learning the components of each brain division; they will be discussed in later sections.

Telencephalon (Forebrain)

Olfactory sacs. These are located internally in the nasal organs; they are not really part of the brain. Part of the inner surface of the sacs is lined with olfactory epithelium in which the neurosensory cells (olfactory nerves) are located.

Olfactory bulbs. These are the most anterior part of the telencephalon, in direct contact with the olfactory sacs. Nerve cells of the bulbs receive olfactory impulses from their sacs; their fibers form the olfactory tracts.

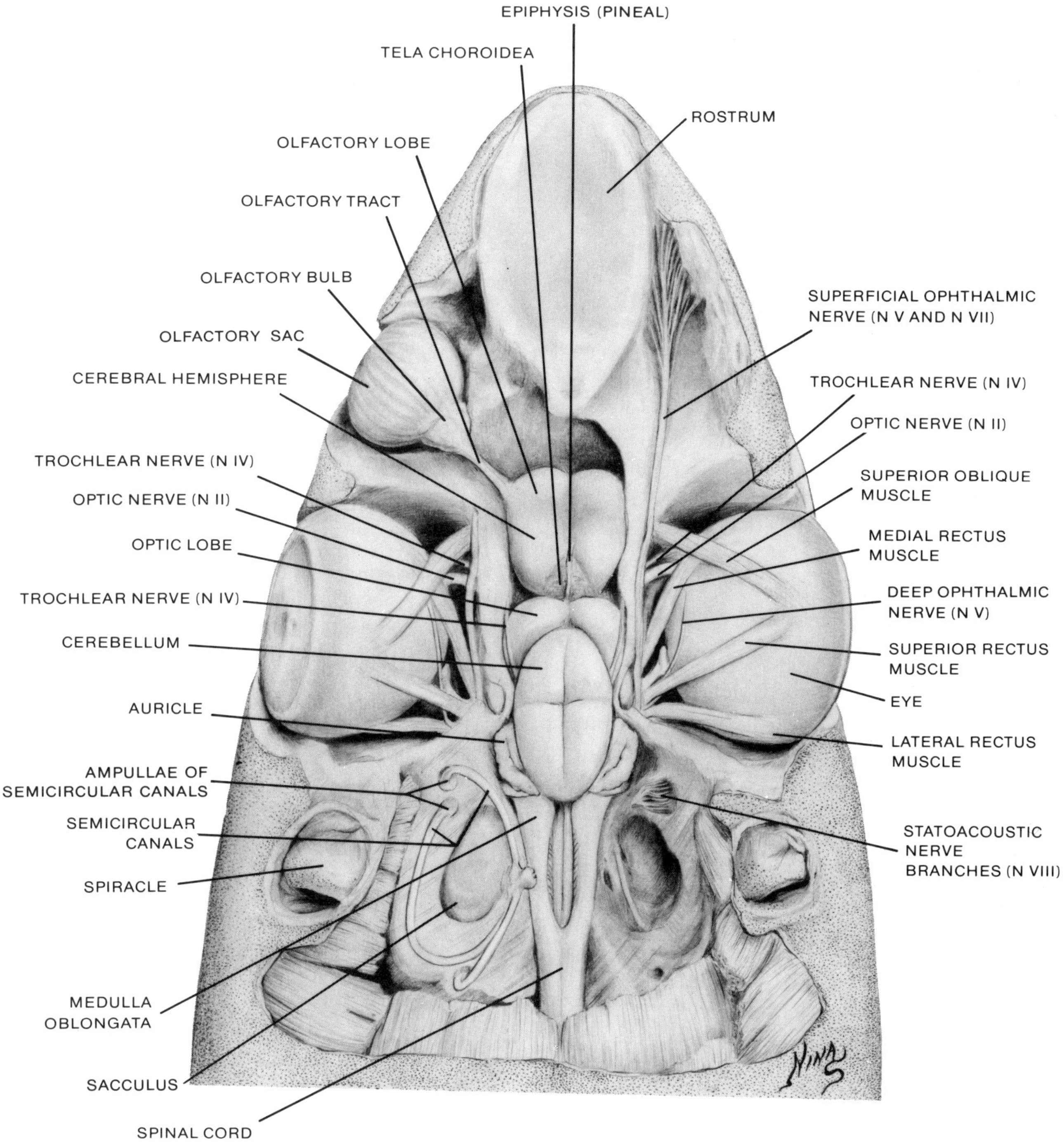

Figure 15-1. Dorsal view of a dogfish head with specific exposure of brain, eye, and ear structures.

Olfactory tract. This is a bundle of nerve fibers leading from the bulb to the olfactory lobes.

Cerebral hemispheres. In addition to integration of complex behavioral patterns, the cerebral hemispheres are involved in assembling and integrating olfactory information.

Lateral ventricles. The cerebral hemispheres are hollow. Their central cavities are the ***first*** and ***second ventricles*** and are continuous posteriorly with the central spinal canal.

Diencephalon (Forebrain)

Epithalamus. The epithalamus is the roof of the diencephalon.

Tela choroidea. This is a thin membrane of non-nervous tissue forming most of the epithalamus. From it, the highly vascularized folds of the ***anterior choroid plexus*** project into the cavity of the diencephalon.

Epiphysis. (Pineal body) The epiphysis may have both photoreceptive and endocrine functions. It projects from the rear of diencephalon, just posterior to the habenula.

Habenula.* The habenula is a ridge of tissue at the rear of diencephalon, concealed by the tela choroidea. The habenular nuclei are thought to integrate olfactory impulses received from the cerebral hemispheres.

Thalamus.* The thalamus represents the lateral walls of the diencephalon. It is small in sharks, serving as a relay station for impulses going to and from the cerebrum.

Hypothalamus.* This center, on the floor of the diencephalon, controls visceral functions.

Third ventricle.* A cavity of the diencephalon. Much of its roof is formed by the tela choroidea and the anterior choroid plexus, which are not nervous tissues.

Mesencephalon (Midbrain)

Optic lobes. These form the expanded roof of the midbrain, receiving the optic tracts.

Cerebral aqueduct.* This is a narrow canal connecting the third and fourth ventricles.

Metencephalon (Hindbrain)

Cerebellum. This is a large oval structure with four paired lobes. It receives sensory tracts, from the inner ear, lateral line, and other receptor areas. It is an important center for control of equilibrium and of muscular coordination.

Auricles. The auricles are two leaf-like lobes on each side of the cerebellum, partially covered by the brown posterior choroid plexus. They are evolutionarily the oldest centers for controlling equilibrium.

Myelencephalon (Hindbrain)

Posterior choroid plexus. This is brown vascular tissue which forms the roof of the fourth ventricle. The plexus has been removed in Fig. 15-1.

Fourth ventricle.* This is a cavity of the hindbrain which extends into the auricles.

Medulla oblongata. Basically, this is an anterior extension of the spinal column in which the gray matter is arranged into four columns (aggregations of cell bodies) similar to those shown in Fig. 15-2. Most of the cranial nerves arise in these columns. To expose the columns, remove the posterior choroid plexus; section the medulla (near the posterior part of the ventricle) and compare with Fig. 15-2.

Somatic sensory columns. These columns lie in the dorsal rims of the medulla. The expanded anterior portions, continuous with the auricles, receive cranial nerves from the inner ears, the lateral lines, and other sensory regions.

Visceral sensory columns. These consist of several bulges in the walls just ventral to the somatic sensory columns.

Visceral motor columns. These slender columns are lateral to the somatic motor columns and are not readily distinguished.

Somatic motor columns. These are two prominent ridges lying on the floor of the medulla.

Note: Discussion of structures on the ventral surface of the brain will follow discussion of the cranial nerves in order to prevent undue disruption of the various structures involved.

PRELIMINARY DISSECTION FOR CRANIAL NERVES (Fig. 15-3)

Dissection of these nerves in the dogfish shark affords an opportunity to see the ten cranial nerves found in all vertebrates. Before starting this dissection, read the discussion sections on both spinal and cranial nerves found in Chapter 17. In this section, the functional components with their periph-

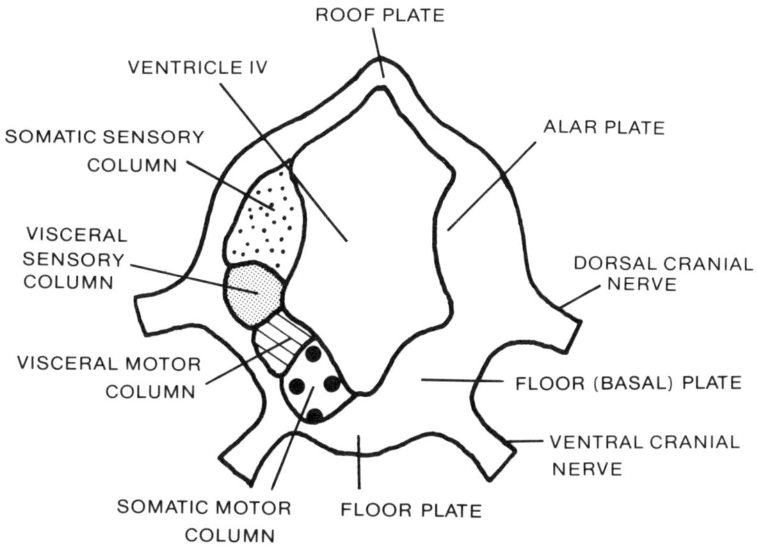

Figure 15-2. Idealized cross section of a mammalian medulla. (After Romer.)

eral distributions are listed first for each cranial nerve. A summary of components and nerve distributions is given in Table 15-1.

Although a cranial nerve may emerge superficially at a certain point on the surface of the brain, its fibers extend internally and may originate in gray matter in the brain some distance from this point. Information concerning such "superficial" or "fiber" origins is included in the dissection instructions. The term *originate* will be used, for convenience, with both sensory and motor nerve fibers, though sensory fibers, in reality, *terminate* in the gray matter; for the most part, their cell bodies lie inside dorsal ganglia, from which they carry information to the brain.

It may be helpful to refer to the illustrations of the chondrocranium in Chapter 8 to note the various foramina through which the nerves emerge.

Carefully lift the posterior side of one of the auricles. A large cranial nerve trunk, the trigeminofacial root, emerges from the dorsal rim of the medulla just behind the auricles. (This root is almost completely concealed by the auricle in the Fig. 15-3.) If the meninx primitiva has not been completely removed, you may have trouble seeing the root.

Cut away the muscle over the remaining intact otic capsule and slice through the cartilage until the cavity of the sacculus is exposed. The anterior and posterior semicircular ducts will, of course, be destroyed. Do not cut much deeper than the level of the horizontal duct. Also, do not cut cartilage directly along the side of the medulla, or cranial nerve connections will be destroyed.

Remove the saccular membrane, observe the sand grains, and then wash out the sand grains from the cavity. Identify the two branches of the statoacoustic nerve (NVIII) which supply the inner ear. One long branch runs parallel to the medulla in the medial wall of the cavity. At the anterior end, a shorter branch extends laterally. Both have many delicate smaller branches. Identification of these branches is not crucial at this point. Therefore, if you have trouble finding them, go on to the next dissection.

DISSECTION AND IDENTIFICATION: CRANIAL NERVES (Fig. 15-3)

Find and identify as many of the following structures as time (and inclination) will allow. The cranial nerves are designated NO, NI, NII and so forth.

NO: Terminal nerve. Functional components of this nerve are unknown. It is partly transparent and is difficult to find because its fibers have no myelin sheath. It extends from the surface of the olfactory sacs medial to the olfactory tracts and enters the fissure between the cerebral hemispheres.

NI: Olfactory nerve. This is a sensory nerve from the olfactory sac. It originates from olfactory neurons with cell bodies located in the olfactory epithelium of the sacs. The short fibers extend to the olfactory bulb without forming a compact unit and are therefore not visible to the naked eye.

NII: Optic nerve. This is a special sensory nerve extending from the retina. It is the thick, white,

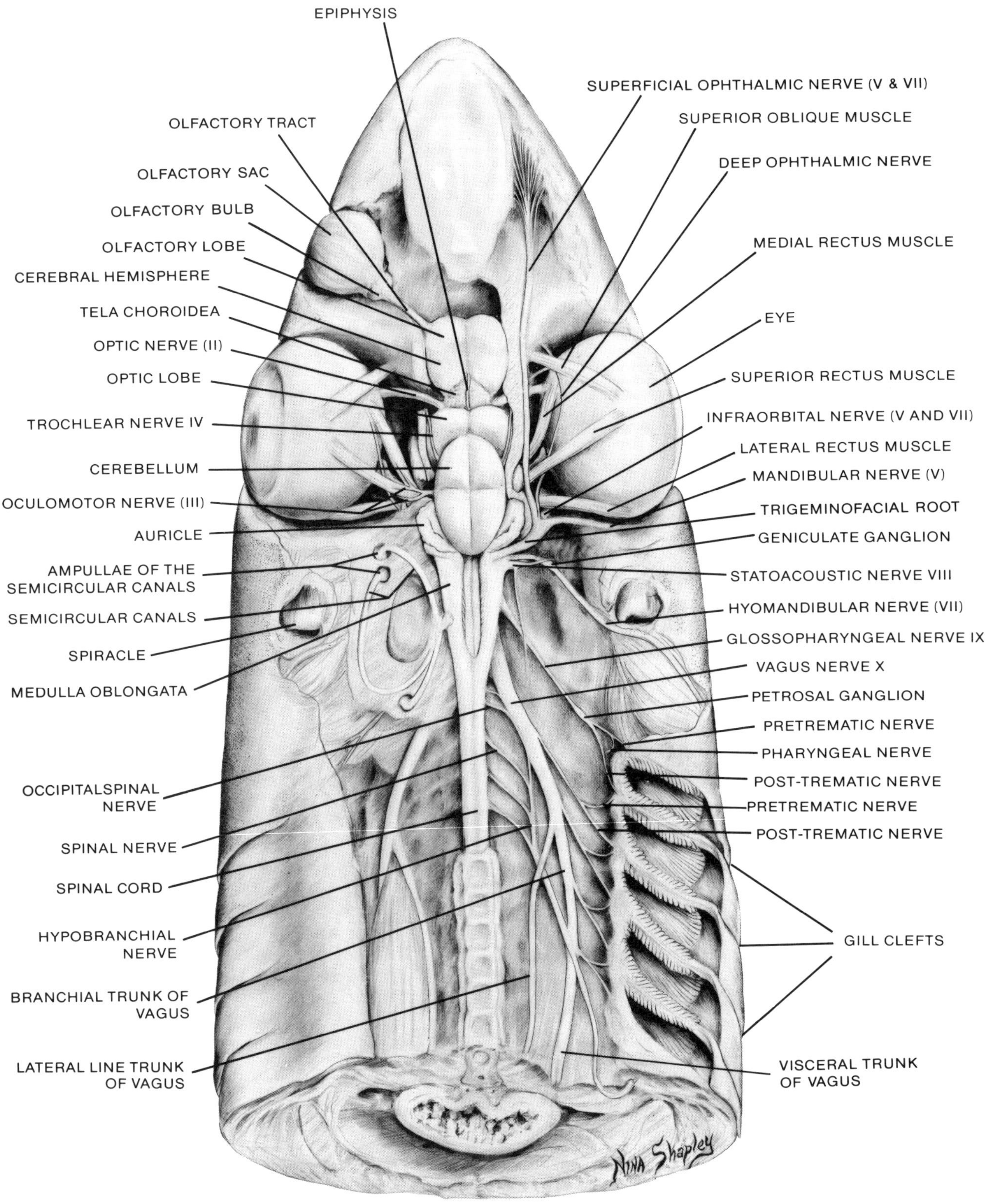

Figure 15-3. Dorsal view of a dogfish head with detailed exposure of brain, eyes, ears, and cranial nerves.

myelin-ensheathed nerve found in the anterior orbit. Push the brain to one side or remove cartilage in the medial wall of the orbit to expose the extension of NII to the ventral diencephalon. It is, in a sense, really part of the brain inasmuch as the neural retina, from which NII's fibers originate, arises as an outpocketing of the wall of the embryonic brain.

NIII: Oculomotor nerve. This is a somatic motor nerve to inferior oblique and superior, medial, and inferior rectus muscles. The fibers of the oculomotor nerve originate in the midbrain extension of the somatic motor column.

Expose only in left orbit; the orbital process on this side is cut to reveal the nerve. Press the brain to one side at the middle of the cerebellum to see the nerve deep in the cranial cavity passing to the orbit from the ventral midbrain. Follow the nerve into the orbit by removing the superficial ophthalmic nerve and cutting cartilage in the medial wall just posterior to the orbital process.

As soon as it enters the orbit, the oculomotor nerve divides into three branches, two of which innervate the superior and medial recti, respectively. The third branch turns ventrally between the superior and lateral recti and forks around a blood vessel which enters the eye adjacent to the inferior rectus (Fig. 15-4). The short branch of the fork, concealed by connective tissue, innervates the inferior rectus; the long branch continues embedded in connective tissue along the ventral surface of the eye to the inferior oblique muscle.

NIV: Trochlear nerve. This is a somatic motor nerve to superior oblique muscles. It arises from the floor of the midbrain, in the anterior extension of the somatic motor column, and emerges superficially through the groove between the cerebellum and the optic lobes.

NVI: Abducens nerve. This is a somatic motor nerve to the lateral rectus muscles. It is very difficult to find, since it originates from the ventral surface of the medulla and innervates the ventral side of the lateral rectus immediately on entering the orbit. (See Fig. 15-4.)

The fibers of all subsequent nerves originate in medullary columns; therefore, the fiber origins may be inferred from the corresponding functional component. For example, somatic motor fibers originate in the somatic motor column.

Various components of NV and NVII originate superficially from the brain in a combined trunk called the ***trigeminofacial root***. The portion of this root attached to the medulla has already been identified.

Using a fine forceps and scalpel, carefully remove cartilage to expose the root as it curves anteriorly toward the orbit and divides into the superficial and deep ophthalmic nerves and infraorbital and mandibular nerves.

NV: Trigeminal nerve. This is a somatic sensory nerve innervating the skin of the head and jaws. In addition, it has a visceral motor component to mandibular arch muscles and to certain glands. Cell bodies of sensory nerves are located in the ***gasserian*** or ***semilunar ganglion***, which is combined with the mass of tissue forming the trigeminofacial root. Fibers from the ganglion enter the somatic sensory column. NV has four divisions, as follows:

Superficial ophthalmic nerve. This is a somatic sensory nerve from the skin on the dorsal surface of the head, above the eye. This nerve has both NV and NVII components. Most of the NV fibers branch off the main trunk to emerge from the chondrocranium through the foramina above the orbit.

Deep ophthalmic nerve. This is a somatic sensory nerve from the skin on the dorsal snout. It is a thin white nerve which enters the orbit with the superficial ophthalmic nerve and passes along the medial surface of the eye, embedded in connective tissue beneath the dorsal eye muscles. Anteriorly, it turns medially to leave the orbit near the superficial ophthalmic nerve.

Maxillary nerve. This is a somatic sensory nerve from the skin on the ventral snout. (See Fig. 14-3.) It is combined with the buccal nerve (part of NVII) in the ***infraorbital trunk*** on the floor of the orbit. Although it separates medially from the buccal nerve in the anterior end of the orbit, the fibers of the two nerves become intertwined for most of their peripheral distributions. Do not confuse the infraorbital nerve with the suborbital (or preoribital) muscle which lies ventral to it and aids in anchoring the upper jaw to the brain case.

Mandibular nerve. This is a somatic sensory nerve from the skin of the lower jaw and a visceral motor nerve to the mandibular arch muscles.

Remove cartilage and tissue concealing the union of the mandibular and infraorbital nerves as they join the trigeminofacial root; follow the mandibular nerve out from this union. Carefully separate the nerve from muscle and tissue in the posteroventral wall of the orbit. (Unfortunately, the nerve is the same color as the muscle.)

NVII: Facial nerve. This nerve supplies the lateral line organs of the head, the spiracle, and the hyoid arch. Note that the anterior lateral line nerve, a distinct unnumbered cranial nerve, runs with NVII to some extent. NVII has three branches.

Superficial ophthalmic nerve. (See above.) This is a special sensory nerve from the lateral line and the ampullae of Lorenzini. NVII fibers form the bulk of this nerve.

Buccal nerve. This is a special sensory nerve from the lateral line and the ampullae of Lorenzini. (See Fig. 14-3.)

Hyomandibular nerve. This is a special sensory nerve from the lateral line and the ampullae of Lorenzini. It is also a visceral sensory nerve from the mouth lining and taste buds, and a visceral motor nerve to the hyoid muscles.

While exposing the hyomandibular nerve, be careful not to cut the statoacoustic nerve; its short branch extends over the hyomandibular.

Find the distal portion of the hyomandibular nerve, situated beneath the skin and connective tissue on the posterior margin of the spiracle and lateral side of the shark. It extends to the spiracle from the ventrolateral medulla, crossing obliquely beneath the otic capsule just anterior to the sacculus. Sensory cell bodies are located in the ***geniculate ganglion***, an expansion of the nerve near the brain.

Although the hyomandibular nerve may appear to branch off of the trigeminofacial root, it actually emerges from the medulla just posterior to the root. This origin may be difficult to expose. The trigeminofacial root is actually wider than is shown in the diagram; it conceals the origins of the hyomandibular and the statoacoustic (NVIII) nerves.

NVIII: Statoacoustic nerve. This is a special sensory nerve from the inner ear. It originates in the medulla, directly adjacent to, and slightly above, the hyomandibular, in a fairly wide trunk. The long ***saccular*** branch, extending parallel to the medulla, crosses over the glossopharyngeal nerve; the short ***vestibular*** branch crosses over the hyomandibular nerve. The vestibular branch goes to the anterior part of the inner ear, while the saccular branch goes to the posterior inner ear.

NIX: Glossopharyngeal nerve. This nerve supplies the first gill arch. It originates from the brain just anterior to the vagus nerve and crosses obliquely along the floor of the otic capsule, posterior to the cavity of the sacculus, to the first branchial arch. (To determine the angle at which the nerve leaves the brain, insert a probe into the first gill slit.) Sensory cell bodies are located in the elongate ***petrosal ganglion*** situated near the first gill arch.

At the ganglion, the nerve divides into three branches, which are characteristic of all branchial nerves.

Pretrematic nerve. This is a visceral sensory nerve from the anterior wall of the first gill pouch. Carefully separate this thin branch from the thicker posttrematic nerve to which it adheres within connective tissue for a short distance. To expose and follow these nerves, first remove the skin from the gill pouch and slit through its dorsal roof as shown in Fig. 15-3.

Posttrematic nerve. This is a visceral sensory and motor nerve, both from and to the posterior wall of the first gill pouch. It enters the interbranchial septum between the first and second gill arches along with the pretrematic branch of the next branchial nerve. It has several accessory branches.

Pharyngeal nerve. This is a visceral sensory nerve from the pharyngeal lining. It is difficult to dissect. It splits off from the pretrematic nerve and curves ventrally to disappear under the cartilage. It passes anteriorly from this point to innervate the pharyngeal lining.

NX: Vagus nerve. This nerve originates from a broad, fan-like series of rootlets in the medulla. It divides into a lateral line branch supplying the main lateral line (considered by some to be a separate "posterior lateral line nerve"); branchial nerves (similar to NIX) supplying the last four gill arches; and a visceral branch supplying the heart, digestive tract, and other internal organs.

To follow the vagus nerve caudally, a large amount of the dorsal musculature must be removed. The branches are approximately the same color as the surrounding muscle, so follow the nerve carefully by means of blunt dissection.

When the boundary of the epaxial muscle and gill pouches is reached, remove the shredded muscle and cut through the dorsal roof of the gill pouches as shown on the right side of Fig. 15-3.

The Branches of the Vagus Nerve

Lateral line branch. This is a special sensory nerve from the main lateral line canal. It branches off medially just before the vagus disappears under a narrow strip of muscle at the level of the first gill pouch. (Fig. 15-3, right side.) This branch continues caudally, parallel to the spinal cord, following the main lateral line canal.

Branchiovisceral branch. This is embedded in the connective tissue on the *ventral* surface of the strip of muscle mentioned above.

Make an incision along the connective tissue between the gill pouches and the epaxial muscle. Lift the muscle, carefully separate the vagus nerve from it, and then remove the muscle completely. (The connective tissue forms the walls of the anterior cardinal sinus.)

Branchial nerves. These four nerves are concealed in the connective tissue on the floor of the anterior cardinal sinus medial to the gill pouches. Each nerve branches from the vagus nerve somewhat anterior to its corresponding gill slit and divides into the same three divisions as those of NIX. Many of the pretrematic and posttrematic branches are difficult to find; they may have several accessory branches. (Note: pharyngeal branches are not shown in Fig. 15-3.)

Visceral branch. This is a visceral sensory and motor nerve to and from the heart, digestive system, and other internal organs. Be careful not to confuse this nerve with the hypobranchial nerve which crosses over it at the last gill pouch. The visceral branch may be followed onto the esophagus and stomach, where it divides into a number of smaller branches.

The Occipitospinal and Hypobranchial Nerves (Fig. 15-3)

Since the ***occipitospinal*** and ***hypobranchial*** nerves emerge posterior to the edge of the chondrocranium in sharks, they are not included in the series of cranial nerves. In higher vertebrates, derivatives of these nerves emerge through cranial foramina and comprise cranial nerves XI and XII.

Occipitospinal nerves. (Occipitals) These are somatic motor nerves which contribute to the hypobranchial nerves. They are homologous with the ventral roots of anterior spinal nerves in lower forms. Usually two or three in number, the occipitospinals may be seen under the cartilage in the angle between the vagus nerve and the medulla. The first occipitospinal nerve is actually *ventral* to the roots of the vagus nerve. Although these nerves appear to join NX, they pass along its ventral surface and eventually join together to form the proximal portion of the hypobranchial nerve.

Hypobranchial nerve. This is a somatic motor nerve to the musculature in the ventral gill arch region. It is formed by the union of the occipitospinal nerves and the first three spinal nerves, which it receives as it passes posteriorly. Because these contributing branches are so small, it is difficult to follow them out as they unite to form the nerve. Therefore, find the large combined trunk as it crosses the vagus nerve at the level of the last gill pouch. After crossing the vagus, the hypobranchial nerve curves downward and runs anteriorly beneath the gills. Follow it inwards toward the brain in order to find the contributing nerves.

Spinal nerves. These are discussed in detail in Chapter 17.

Dorsal roots. (Not shown in Fig. 15-3.) The dorsal roots and their unions with the ventral roots are extremely difficult to expose. Slice away the cartilage from the dorsal surface of the vertebrae and attempt to find the proximal portion of the roots, which emerge from the dorsolateral surface of the spinal cord.

Ventral roots. Push the spinal cord to one side and note the fan-like group of "rootlets" which arise from the cord to form the ventral root. The spinal nerves, formed by the union of dorsal and ventral roots, are most easily found by following the hypobranchial nerve, as explained above.

VENTRAL BRAIN (Fig. 15-4)

If the ventral brain is to be dissected, do not attempt to expose it through the ventral surface of the upper jaw and snout as shown in the diagram. That is an extremely difficult dissection. Instead, cut through the cranial nerves and olfactory tracts and pull the brain up out of the cranial cavity. In the region of the diencephalon, be careful not the damage the hypophysis, which projects downward into a depression in the floor of the cranial cavity. Also, the abducens nerve, which originates from the ventral medulla, may adhere to the floor of the cavity as you remove the brain.

Identification

Structures which have been identified in previous dissections are labelled in Fig. 15-4 but are not included in the text.

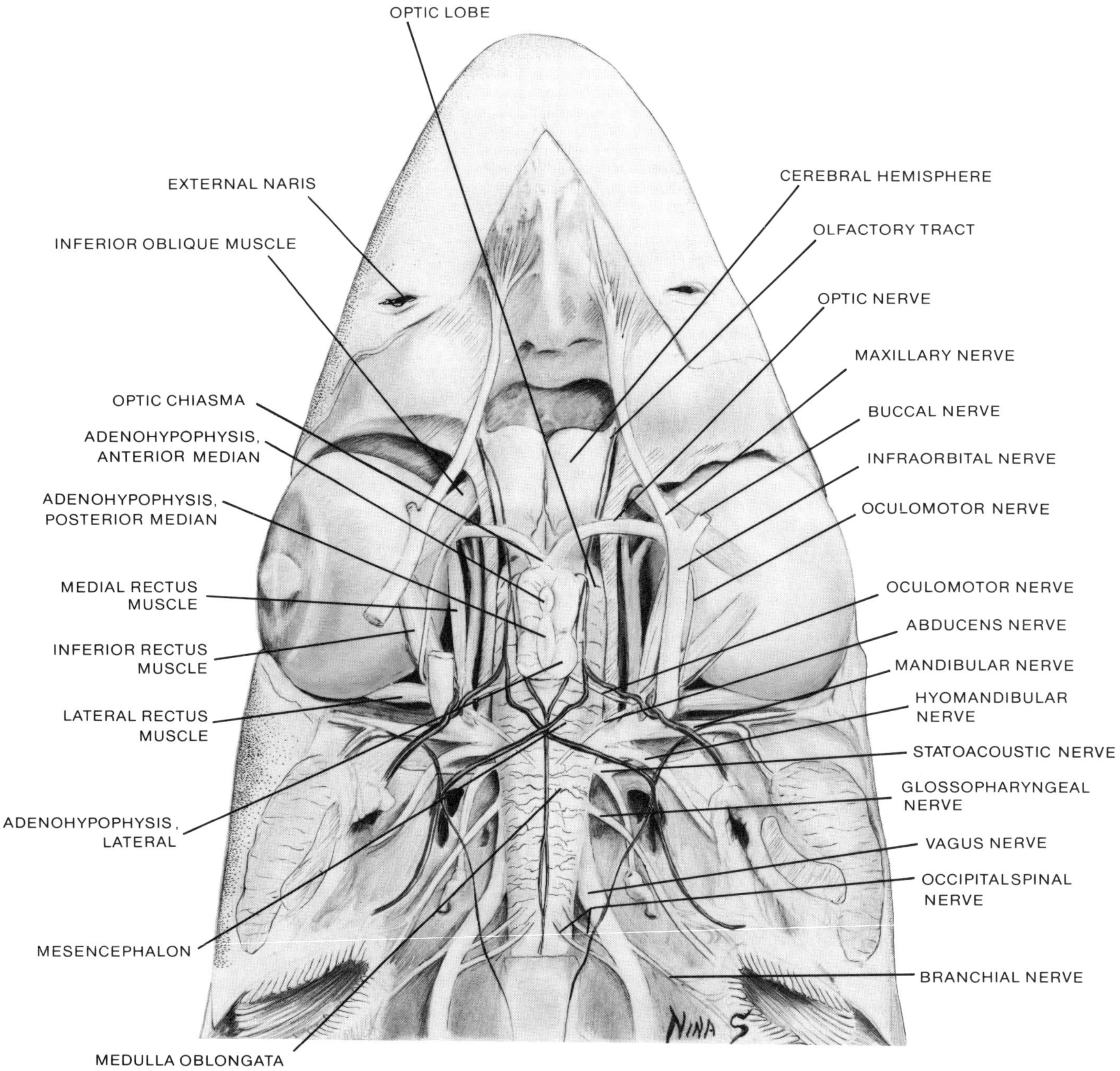

Figure 15-4. Ventral view of a dogfish head with specific exposure of brain, eyes, cranial nerves, and head arteries. (See also color version at end of book.)

Optic chiasma. This is located on the ventral surface of the diencephalon. In sharks, the optic tracts cross over (decussate) completely at the chiasma.

Hypothalamus. This is located on the floor of the diencephalon. The anterior portions, not visible externally, form the highest center for control of the autonomic nervous system. Some parts of the hypothalamus are listed below.

Supraoptic nucleus. (Not shown in Fig. 15-4.) This is located above the optic chiasma. It contains neurosecretory cells with long axons that extend into the neurohypophysis, where their hormones are secreted.

Infundibulum. In mammals, this is a stalk from the hypothalamus and it comprises part of the pituitary gland of the adult. In sharks, the infundibulum constitutes the major portion of the hypothalamus.

Pituitary gland. (Hypophysis) This is an endocrine gland consisting of two portions having different embryonic origins. The function of the shark pituitary is not well understood.

The terminology which has been applied to the parts of the hypophysis has been extremely variable and confusing. We have decided that the following is the most consistent and meaningful usage.

Neurohypophysis. This is derived embryonically from the posterior part of the infundibulum. It is a small region located beneath the adenohypophysis. Because of this placement and the small size, it is difficult to identify.

Adenohypophysis. This is derived embryonically from an out-pocketing of the mouth called Rathke's pouch. It consists of several lobes of tissue in anterior, medial, and lateral positions. An abundance of hormones are secreted from the adenohypophysis in response to stimuli from the hypothalamus.

Saccus vasculosus. (Not shown in Fig. 15-4.) This is a vascular sac between the posterior part of the adenohypophysis and the ventral surface of the midbrain. Common in deep sea fishes, it is thought to monitor fluid pressure.

Abducens nerve (NVI). This is a somatic motor nerve to the lateral rectus. It arises from the ventral medulla and, on entering the orbit, innervates the ventral surface of the lateral rectus muscle (see above).

Table 15-1. Summary of shark cranial nerves.

(The terminal nerve is not included in this table, since so little is known about it. It is present in most vertebrate classes and is probably the remnant of a nerve which innervated the mouth area of the most primitive vertebrates.)

Nerve and Ganglion		*Functional Components*	*Peripheral Distribution*
N I	*Olfactory Nerve*	special sensory	olfactory sac
N II	*Optic Nerve*	special sensory	retina
N III	*Oculomotor Nerve*	somatic motor	inferior oblique, and superior, medial, and inferior rectus muscles
N IV	*Trochlear Nerve*	somatic motor	superior oblique muscle
N V	*Trigeminal Nerve (gasserian ganglion)*		
	superficial ophthalmic nerve	somatic sensory	skin on dorsal surface of head
	deep ophthalmic nerve	somatic sensory	skin on dorsal surface of snout
	maxillary nerve (part of infraorbital)	somatic sensory	skin on ventral surface of snout
	mandibular nerve	somatic sensory visceral motor	skin of lower jaw mandibular arch muscles
N VI	*Abducens Nerve*	somatic motor	lateral rectus muscle

Table 15-1. Summary of shark cranial nerves. (continued).

	Nerve and Ganglion	*Functional Components*	*Peripheral Distribution*
N VII	*Facial Nerve*		
	superficial ophthalmic nerve	special sensory	lateral line system, ampullae of Lorenzini
	buccal nerve (branch of infraorbital)	special sensory	lateral line system, ampullae of Lorenzini
	hyomandibular nerve (geniculate ganglion)	special sensory	lateral line system, ampullae of Lorenzini
		visceral motor	hyoid arch muscles
		visceral sensory	mouth lining, taste buds
N VIII	*Statoacoustic Nerve*	special sensory	inner ear.
N IX	*Glossopharyngeal Nerve*		
	pretrematic nerve	visceral sensory	anterior wall of first branchial arch
	postrematic nerve	visceral sensory visceral motor	posterior wall of first branchial arch
	pharyngeal nerve	special sensory	pharyngeal lining, taste buds
(N IX has a small additional branch to the main lateral line canal)			
N X	*Vagus Nerve*		
	lateral line branch	special sensory	main lateral line
	branchial nerves	four branches to last four branchial arches. Each has three rami like those of N IX, with the same functional components.	
	visceral ramus	visceral sensory visceral motor	heart, digestive tract, other abdominal organs.
	Occipitospinal and Hypobranchial Nerves		
	occipitospinal nerves	somatic motor	contribute to hypobranchial nerve
	hypobranchial nerve	somatic sensory somatic motor	hypobranchial musculature

NEIL J. BAKER

16

The Brain and Cranial Nerves of the Sheep—A Representative Mammal

Because the dissection of the brain of the cat is difficult, a discussion of the sheep's brain is inserted here for comparison with the dogfish brain.

DORSAL VIEWS OF THE SHEEP BRAIN (Figs. 16-1 and 16-2)

Identify the following structures in Figs. 16-1 and 16-2. As you proceed, compare the general structural organization of brain divisions in the shark and in the sheep.

Telencephalon

Cerebral hemispheres. Dorsally, only the neocortex is visible. Due to the expansion of this portion of the hemispheres, the diencephalon and midbrain are concealed. The surface area is greatly increased by folding. The folds are called ***gyri***; the grooves between them are ***sulci***.

Corpus callosum. **(Fig. 16-2)** This is a commissure—a transverse tract of fibers—connecting the two halves of the neocortex, allowing each side to receive information stored in the other. Separate the dorsal portions of the hemispheres slightly until you see this long, white band of tissue within the fissure between the two halves.

Diencephalon and Midbrain

To expose the dorsal surface of the diencephalon and the midbrain, cut slowly through the **corpus callosum** *until you can spread the cerebral hemispheres apart as shown in Fig. 16-2. Do not spread the hemispheres completely apart.*

Identify the following structures.

Pineal body. **(Epiphysis)** This is a knob-like projection from the rear of the epithalamus. Secretions of this endocrine organ are thought to suppress the onset of puberty in humans until the body is sufficiently developed in other respects. The pineal in many other vertebrates plays a crucial role in cyclical activity patterns; for example, it is the site of the "biological clock" in birds.

Tela choroidea. This vascularized tissue and its anterior choroid plexus cover most of the epithalamus. Remove it to reveal the narrow ***third ventricle*** below.

Habenula. These are short white bands of tissue forming the posterolateral sides of the V-shaped rim of the third ventricle.

Midbrain. The roof, or tectum, of the midbrain has four swellings called the ***corpora quadrigemina***, consisting of two sets of lobes, the superior colliculi and the inferior colliculi.

Superior colliculi. These are two large rounded masses posterior to the pineal body. They are located on the optic pathway which controls pupillary and accomodation reflexes of the eye. They may have other complex functions which are not yet understood.

Inferior colliculi. These are two small white lobes just posterior to each superior colliculus. The cerebellum must be pulled back slightly to see them. They integrate and relay auditory impulses and control auditory reflexes.

Metencephalon: cerebellum. Because of the expansion of the cerebellum, almost all of the medulla is concealed.

Myelencephalon: medulla oblongata. Only the most posterior portion of the medulla is visible. As in sharks, the anterior portion is covered by the posterior choroid plexus. The medulla contains centers for control of the heartbeat, respiration, swallowing, and salivation. Also, most of the cranial nerves originate in medullary columns.

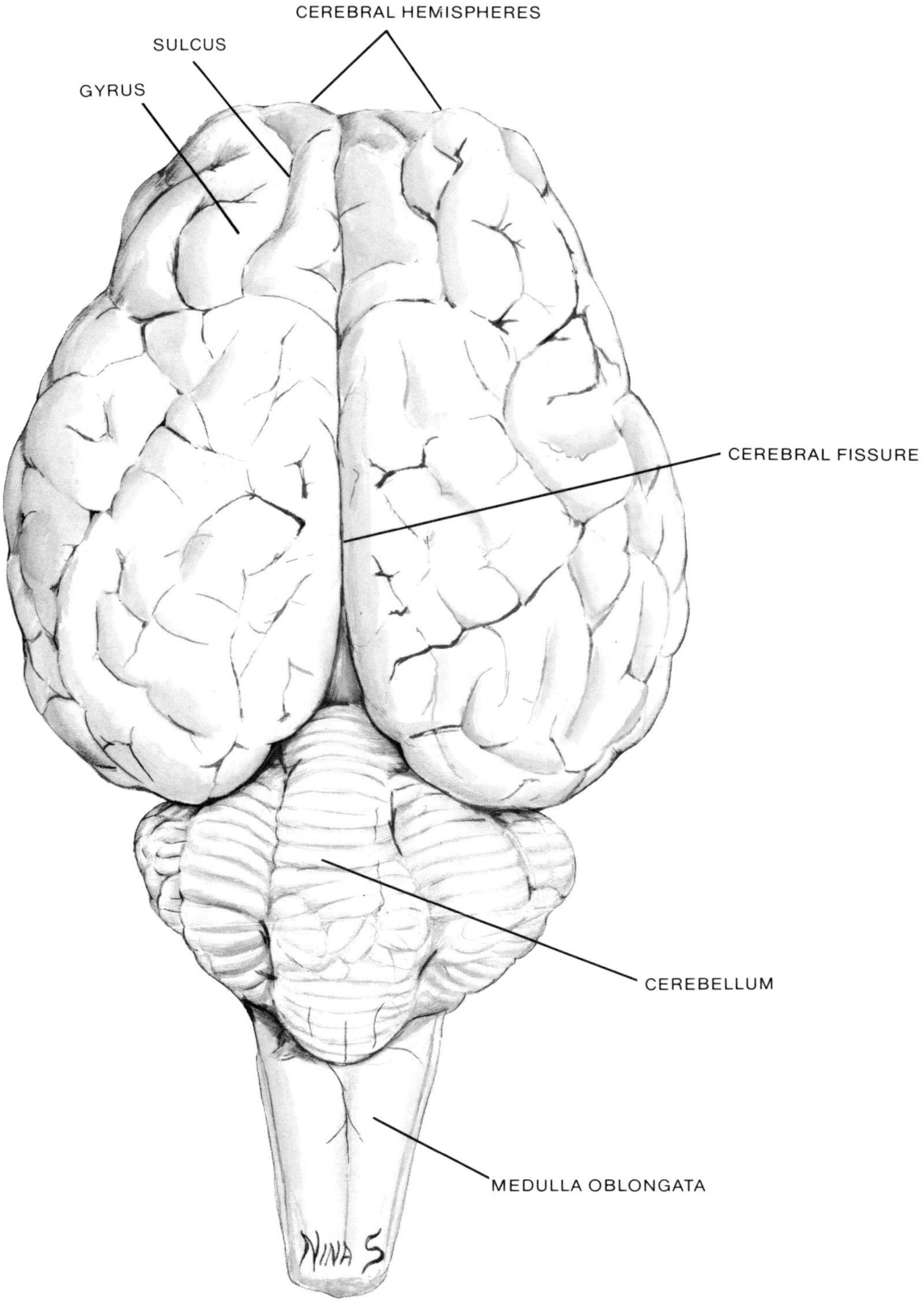

Figure 16-1. Dorsal view of an intact sheep brain.

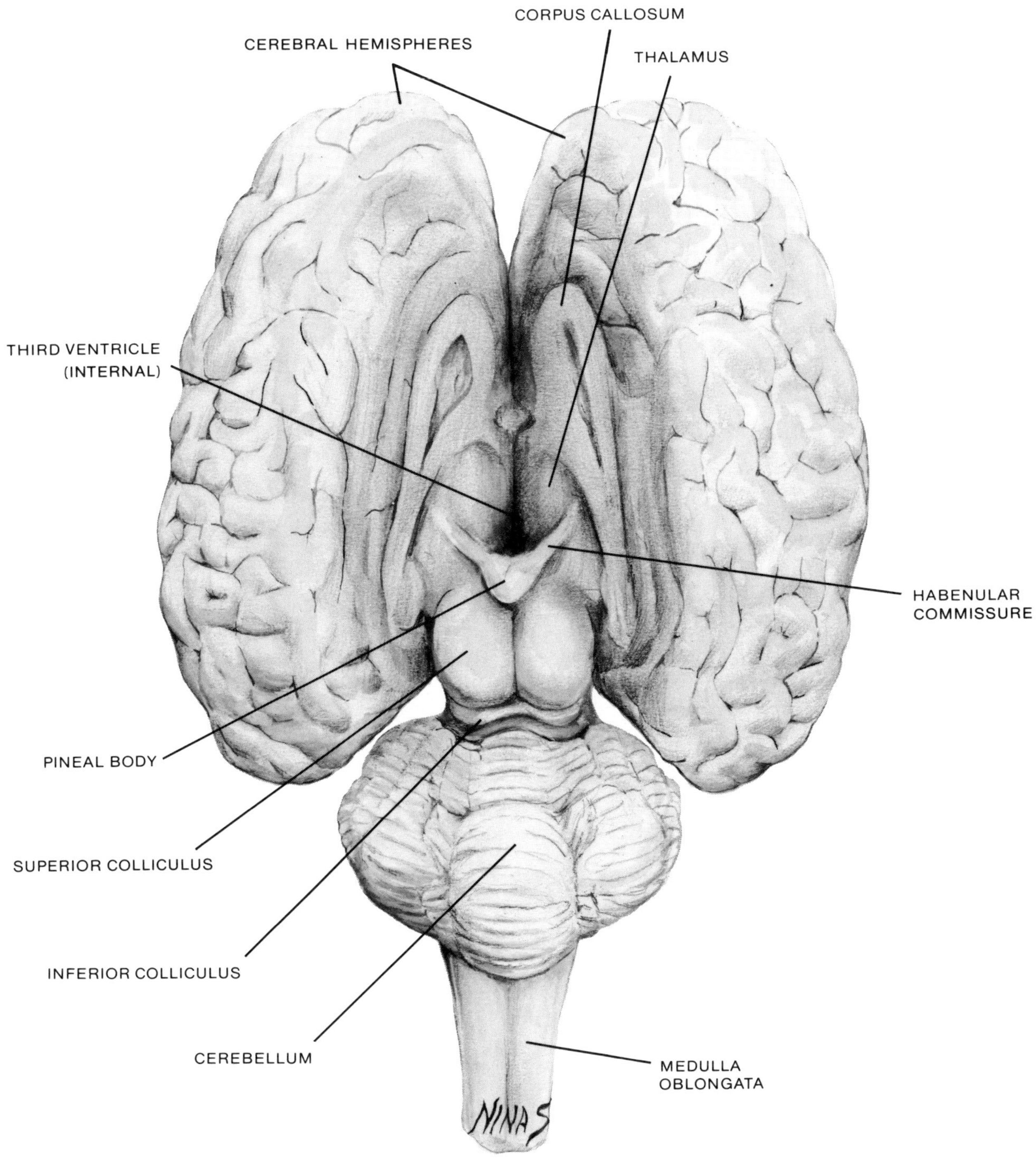

Figure 16-2. Dorsal view of a sheep brain after separation of the cerebral hemispheres and cutting of the corpus callosum.

VENTRAL VIEW OF THE BRAIN (Fig. 16-3)

Telencephalon

Olfactory bulbs. **(Fig. 16-1)** These are pear-shaped structures extending from the anterior end of the cerebral hemispheres. They receive the olfactory nerves.

Lateral olfactory tracts. These are whitish bands of fibers extending posterolaterally from the olfactory bulbs to the pyriform area. They may be difficult to see.

Pyriform area. **(Paleocortex)** This area is homologous to the olfactory lobes of the shark. It is separated from the lateral portions of the ventral cerebrum by a long, deep furrow called the *rhinal fissure*.

The ***archicortex***, another portion of the hemispheres originally associated with olfaction in lower vertebrates, cannot be seen superficially.

Diencephalon

Hypothalamus. This is the floor of the diencephalon.

Optic chiasma. The tracts appear to cross, or decussate, at the anterior margin of the hypothalamus and continue along the optic tracts to the thalamus. In reality, the decussation is partial in cats.

Pituitary. **(Hypophysis)** (Not shown in Fig. 16-3) This is the "master" endocrine gland of the body. Since this rectangular mass of tissue is suspended below the hypothalamus by a fragile stalk (the infundibulum), it may not be present in many specimens. As the pituitary is removed, the wide oculomotor nerves will be seen below its posterior end.

Mammillary bodies. These form the posterior end of the hypothalamus. The mammillary bodies integrate information received via the fornis, another fiber tract.

Midbrain and Hindbrain

Cerebral peduncles. These are two curved columns of fibers. Some of the fibers are part of the pyramidal system, the major motor pathway which leaves the cerebral hemispheres.

Metencephalon: pons. This is a region of transverse fibers at the anterior end of the medulla which connect the cerebral cortex and the cerebellum.

Myelencephalon: medulla. This is located posterior to the pons. To each side of the median fissure is a band of fibers which form the ***pyramidal tracts***. Most of the white matter of the medulla consists of tracts connecting the brain and the spinal cord.

Cranial Nerves (Fig. 16-3)

Peripheral distributions of these nerves and their functional components are discussed in Chapter 15. The olfactory nerves (NI) cannot be seen grossly. All cranial nerves are paired.

NII: Optic nerve. Only the stumps leading to the optic chiasma can be seen.

NIII: Oculomotor nerve. This is a large nerve originating from the medial portion of the cerebral peduncles.

NIV: Trochlear nerve. This is a long, thin nerve visible dorsolateral to the pons.

NV: Trigeminal nerve. The stumps of this nerve may be seen lateral to the pons.

NVI: Abducens nerve. This nerve originates at the posterior end of the pons.

NVII: Facial nerve. This nerve originates near the abducens and turns laterally to leave the surface of the brain with NVIII.

NVIII: Vestibulocochlear nerve. (Not shown in Fig. 16-3) This nerve, called the ***statoacoustic nerve*** in the shark, arises ventrolaterally from the medulla, posterior to the level of the abducens.

NIX: Glossopharyngeal nerve; NX: Vagus nerve. These nerves originate from rootlets along the side of the medulla posterior to NVIII. The more anterior rootlets belong to NIX. In some preparations, the roots of NIX and NX may not appear as separate trunks.

NXI: Accessory nerve. This is a longitudinal nerve originating from a series of rootlets along the posterolateral medulla and spinal cord.

NXII: Hypoglossal nerve. This nerve originates in a series of rootlets on the ventral surface of the medulla.

SAGITTAL SECTION (Fig. 16-4)

Carefully make a saggital section, staying as close to the medial line as possible. The less precise the section, the less it will match the diagram. After separating the two cerebral hemispheres, cut through

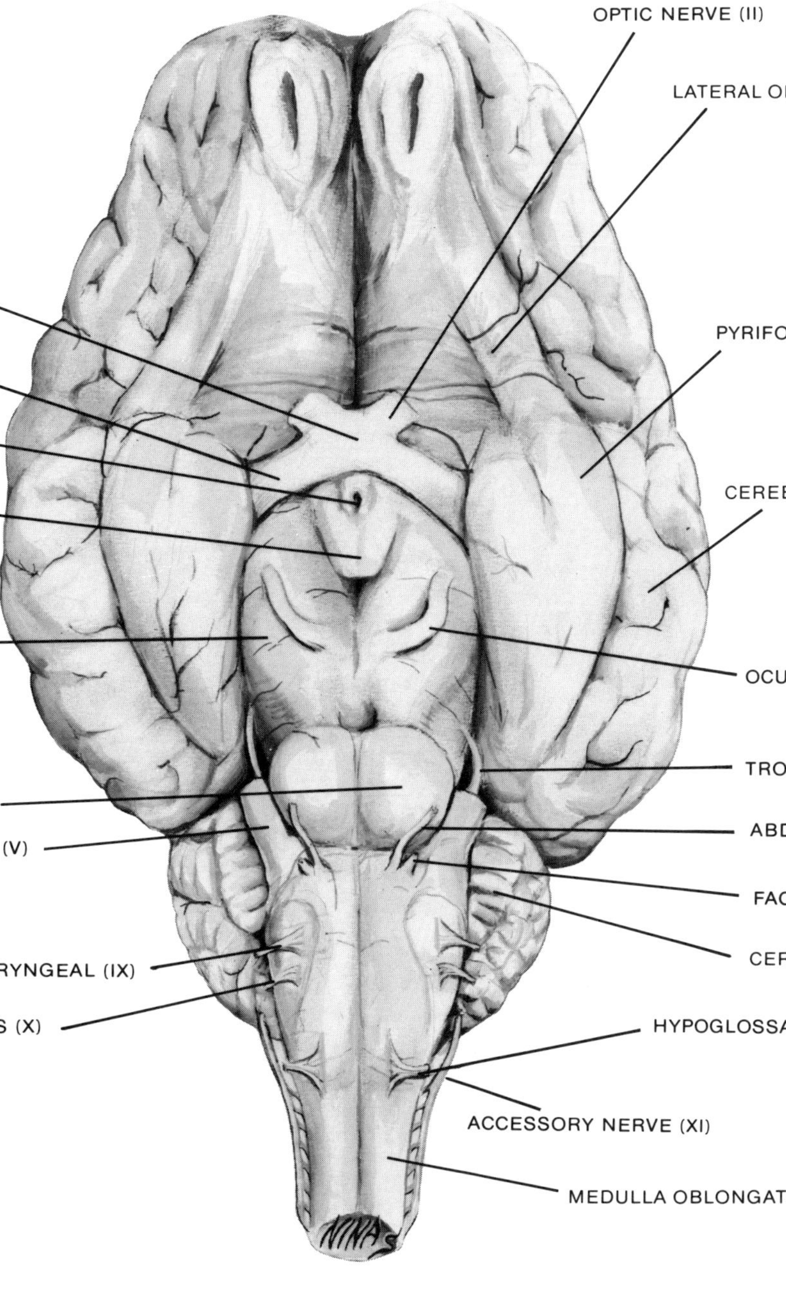

Figure 16-3. Ventral view of an intact sheep brain, including cranial nerves.

the pineal, superior colliculus, cerebellum, and medulla to provide a mid-sagittal section.

Identify the following structures.

Telencephalon

Corpus callosum. The expanded and curved anterior end of the corpus callosum is united with its expanded posterior portion by a relatively thin longitudinal region.

Fornix. This is a tract leading from the archicortex to the mammillary bodies in the hypothalamus. It curves downward from the posterior end of the corpus callosum and turns out of the plane of the section near the anterior commissure.

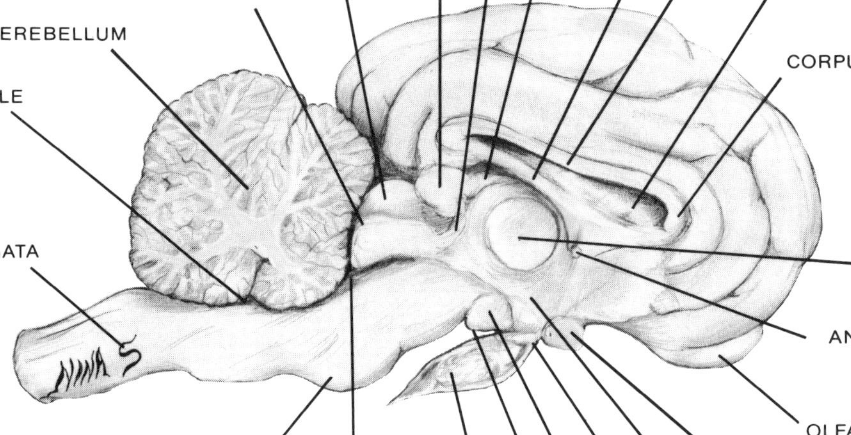

Figure 16-4. Sagittal section of a sheep brain.

Anterior commissure. This is a small, round bundle of fibers near the ventral end of the visible part of the fornix.

Septum pellucidum. This is a thin sheet of gray matter surrounded by the corpus callosum and the fornix. Lateral to the septum on each side are the ***lateral ventricles***, which connect to the third ventricle by openings in the wall anterior to the massa intermedia (see below).

Diencephalon

Epithalamus. Identify the habenula and the pineal body.

Thalamus. The thalamus—which forms the sides of the diencephalon—is so enlarged in mammals for the relay of sensory information to the cerebral cortex that the two sides meet to form the ***massa intermedia.*** This circular mass varies greatly in size in different specimens.

Hypothalamus. Located on the floor of the diencephalon, this is the posterior oval portion which includes the mammillary bodies.

Third ventricle. Because of the expansion of the thalamus, the third ventricle is limited to a narrow canal surrounding the massa intermedia. It extends from the cerebral hemispheres to the aqueduct of the midbrain. Since it is so narrow, it may be partially or entirely unexposed if the brain was not cut accurately. In that case, the area looks like a solid mass of tissue.

Mesencephalon

Midbrain. Identify the colliculi in the roof of the midbrain. These structures are composed of nuclei associated with the auditory and optic systems.

Cerebral aqueduct. This is a narrow canal connecting the third and fourth ventricles.

Metencephalon

Cerebellum. Note that the gray matter of the cerebellum is external to the white matter, as in the cerebral hemispheres. The white matter resembles the branches of a tree.

Pons. This is the anterior end of the medulla.

Myelencephalon. Identify the medulla.

NEIL J. BAKER

17

Function of the Central Nervous System

The earliest vertebrates moved by swimming. Probably because a swimming, elongate animal finds it most advantageous to scan where it is going, the *special* sense organs developed near the anterior end of the body. The special sense organs of this end include the olfactory sacs, the eyes, the inner ears, and the ampullae (see below).

In conjunction with these sense organs, a major concentration of nervous tissue developed in the anterior end of the central nervous system to assemble and correlate the complex data received from the sense organs. As noted in the discussion of the lamprey, the primitive brain evolved three major divisions: the ***forebrain*** or prosencephalon, associated with olfaction; the ***midbrain*** or mesencephalon, which receives visual stimuli; and the ***hindbrain*** or rhombencephalon, for equilibrium in space and, in higher vertebrates, hearing.

In subsequent evolution, each division except the midbrain subdivides into the structures shown below.

Forebrain. Telencephalon (cerebral hemispheres)
Diencephalon (epithalamus, thalamus, hypothalamus)

Midbrain. Mesencephalon (superior, inferior, colliculi)

Hindbrain. Metencephalon (cerebellum, pons)
Myelencephalon (medulla oblongata)

The above organization, characteristic of adult sharks, is seen in the embryos of higher vertebrates. The functions of each of the brain divisions will be compared in the shark and in mammals to illustrate major trends in evolution.

FOREBRAIN: CEREBRAL HEMISPHERES AND DIENCEPHALON

In sharks, the cerebral hemispheres were thought to be mainly olfactory centers. For instance, J. Z. Young (1962, p. 168) said, "If they are removed, the sense of smell is lost but there is no disturbance of posture, locomotion, or behavior." More recent experiments suggest that important behavioral deficits do occur if the hemispheres are injured or destroyed. It is likely, therefore, that among various fishes the forebrain began to assume a predominant role in controlling complex behavior.

In reptiles, a new center called the ***neocortex*** or neopallium appears in the cerebral hemispheres, forming a sheet of gray matter situated superficial to the white matter. In mammals, the growth of the neocortex is chiefly responsible for the tremendous expansion of the cerebral hemispheres. The posterior side of the hemispheres extends back to conceal the diencephalon and midbrain. The paleocortex and archicortex, portions of the cortex which are perhaps evolutionarily older, assume a more ventral or medial position. And in mammals, deep within the hemispheres, the corpus striatum of sharks has developed into discrete nuclei called the ***basal nuclei*** or basal ganglia, which direct relatively stereotyped movements.

The neocortex itself controls the bulk of motor activity, including all of the more complicated and skilled movements. This domination by the cerebral cortex is demonstrated by the presence of a large efferent pathway called the ***pyramidal tract***, which transmits motor directives directly to the spinal cord without intervening relay. The pons, another prominent tract at the anterior end of the medulla, connects the motor portions of the neocortex with the cerebellum. The latter region modifies motor directives to produce proper timing and coordination.

In addition, the neocortex is the final terminus for some sensory data, including that from visual, auditory, and olfactory sources. The complex networks in the neocortex result in very precise visual, olfactory, and auditory discrimination in mammals.

Certain areas in the neocortex have been identified as specific motor or sensory centers. There are, in addition, "unassigned areas" associated with such functions as memory, intelligence, and personality. In lower mammals, these unassigned areas are limited to small regions; in humans, however, they form the bulk of the neocortex.

The thalamus in the mammalian diencephalon is greatly enlarged to serve as a relay center for sensory information going to the neocortex. Three of its principal nuclei, or bodies, are: the ***lateral geniculate nucleus*** which relays optic stimuli; the ***medial geniculate nucleus*** which relays auditory stimuli; and the ***ventral nucleus*** which relays somatic sensory stimuli. The function of these nuclei is actually more complex than simply projecting sensory information; they modify and organize impulses before relaying them. In fact, some crude "awareness" of optic or auditory stimuli may occur at this level.

MIDBRAIN

The evolution of the midbrain clearly shows the development of the hierarchical organization of brain centers discussed above. In sharks, a wide variety of sensory fibers are projected in the midbrain, including optic, acoustico-lateral, general cutaneous, olfactory, and taste fibers. Motor pathways leading away from the midbrain are comparatively small. This is consistent with the idea, introduced above, of a significant degree of spinal independence in control of movement. Most motor directives are transmitted through the ***reticular formation***, a mass of nerve cells and fibers in the brain stem that are not organized into nuclei or tracts.

In mammals, the cerebral cortex is the major center of sensory integration and motor control. The midbrain is greatly reduced in relative size and importance. Its roof is divided into the superior and inferior colliculi, which are incorporated into visual and auditory pathways. In some mammals, these nuclei still retain certain primitive functions in visual discrimination. Dogs, for example, may still distinguish between light and dark without the optic centers of the neocortex. Humans, however, are completely blind without the neocortex.

The reticular formation serves as a primitive, low-level motor pathway. More important, the reticular formation stimulates or suppresses the general activity level in the cerebral cortex, to yield sleep, wakefulness, or awareness of sensory input at the level of the cerebral cortex.

HINDBRAIN: CEREBELLUM AND MEDULLA

The cerebellum evolved as a center integrated with the inner ear and the lateral lines for control of equilibrium. Input to this center is received from the inner ear, the proprioceptors, and tactile, auditory, and visual sources. The output initiates no muscular activity of its own but modifies all motor directives so they are in accord with the existing orientation of the body and its appendages.

In birds and mammals, the cerebellum is greatly enlarged due to the increased complexity of muscular activity. New regions of gray matter, such as the neocerebellum, provide feedback connections with the neocortex and other parts of the brain. This gray matter, like that in the cerebral hemispheres, is located ***external*** to the white matter.

The medulla is basically an anterior extension of the spinal cord in both sharks and mammals. It differs far less from the cord than do other centers.

ACCESSORY STRUCTURES OF THE CENTRAL NERVOUS SYSTEM

The system of ventricles in the brain and spinal cord is derived from the hollow core of the primitive central nervous system. In the shark brain, the ventricles are fairly large and the cerebral aqueduct has large expansions into the cerebellum and midbrain. In the human brain, however, these expansions are occluded by reduction of the midbrain and expansion of the cerebellum.

In the roof of the third and fourth ventricles are the highly vascularized tissues of the ***choroid plexus***. These tissues contribute to the ***cerebrospinal fluid*** which circulates in the ventricles. The fluid maintains a constant environment for the brain, cushions it from injury, and transports nutrients.

Meninges

In sharks, the brain is covered by a single vascular membrane called the ***meninx primitiva***. Cerebrospinal fluid carrying various waste products exits the brain through foramina in the medulla and circulates in the space between the meninx and braincase. Here it diffuses into venous sinuses within the meninx.

In mammals, three meninges cover the brain. Most exterior is a tough connective tissue sheath called the ***dura mater***, which is penetrated by large venous sinuses. The meninges extend into the fissure between the cerebral hemispheres and between the hemispheres and the cerebellum to form a vertical and a horizontal septum, respectively.

The internal meninges consist of two membranes separated by a space filled with cerebrospinal fluid. Beneath the dura mater is the ***arachnoid membrane***; innermost, adhering to the brain, is the ***pia mater***, which connects to the arachnoid by delicate threads of tissue.

Cerebrospinal fluid leaves the ventricles through foramina in the medulla to the subarachnoid space. Exchange with blood in the venous sinus occurs via villi on the arachnoid membrane.

CRANIAL NERVES

In the earliest vertebrates, segmentation of body musculature is believed to have extended into the head region. In *Branchiostoma*, the dorsal and ventral roots of spinal nerves do not fuse, but emerge separately from the cord. Each cranial nerve of higher vertebrates may be derived from one of the separate dorsal or ventral roots of primitive nerves which innervated these ancient anterior segments. However, the development of the head, particularly with the evolution of the eye, ear, and gills, is associated with the disruption of the original segmentation. The cranial nerves retain the primitive spinal form, with no fusion of dorsal and ventral roots. (This does not apply to cranial nerves I, II, and VIII, which are developed in relation to special sensory organs of the head.)

Cranial nerves III, IV, and VI may have been derived from primitive ventral roots, for they originate in the ventral brain stem and contain somatic motor fibers. Primitive dorsal roots are believed to be represented by nerves V, VII, IX, X, and XI, since they may contain visceral motor, visceral sensory, and somatic sensory fibers.

This evolutionary argument may not be entirely accurate. As Romer (1970) contends, the arrangement and composition of nerves V, VII, IX, and X may have been affected by the arrangement of gill slits, which develop in the embryo entirely independently of the myomeres. In any event, the comparison with spinal nerves may help in learning the components of cranial nerves. Romer organizes the cranial nerves into three categories: those which supply the special sensory organs (I, II, VIII, and the lateral line nerves); the somatic motor or ventral root nerves (III, IV, VI, and XII); and the branchial or dorsal root nerves (V, VII, IX, X, and XI). Autonomic fibers (parasympathetic nerves) accompany a few of the nerves.

In the course of evolution, the peripheral distributions of these nerves remain remarkably constant. The following discussion compares shark and mammalian cranial nerves.

Special Sensory Nerves (I, II, VIII, portions of VII, IX, X)

NI (Smell). The olfactory nerves are unique in that the neurons themselves are the sensory receptors. They are, therefore, located in the periphery (in the epithelium of the olfactory sacs in sharks) rather than in or near the brain or spinal cord.

NII (Vision). The optic nerve is in reality a brain tract, since the eye and its nerve develop in the embryo from an outpocketing of the diencephalon.

NVIII (Hearing and equilibrium). In mammals, one branch of this nerve is enlarged to supply the cochlea of the internal ear. In all vertebrates, the sensory patches in the semicircular canals and various sacs of the inner ear are served by this nerve.

Lateral line nerves (portions of VII, IX, X). The fibers which supply the lateral line organs and taste buds run with different cranial nerves in various fishes, amphibians, and terrestrial vertebrates.

Somatic Motor Nerves (III, IV, VI)

These "ventral root" nerves innervate the extrinsic eye muscles in all vertebrates. Although they are comparable to ventral roots, they include a small number of somatic sensory proprioceptive fibers, since there must be feedback from every muscle concerning its state of contraction. How these "ventral root" nerves acquired this sensory component is not well understood.

Branchial Nerves (V, VII, IX, X, XI)

The branchial or "dorsal root" nerves innervate the visceral arches and their evolutionary derivatives in both sharks and mammals. In sharks, each nerve has three branches corresponding to the pretrematic, posttrematic, and pharyngeal rami of NIX. Except for the autonomic fibers mentioned later, the visceral motor fibers of these nerves have such unique structures and peripheral distributions (to gill arches) that they are called ***special visceral motor***, rather than autonomic nerves.

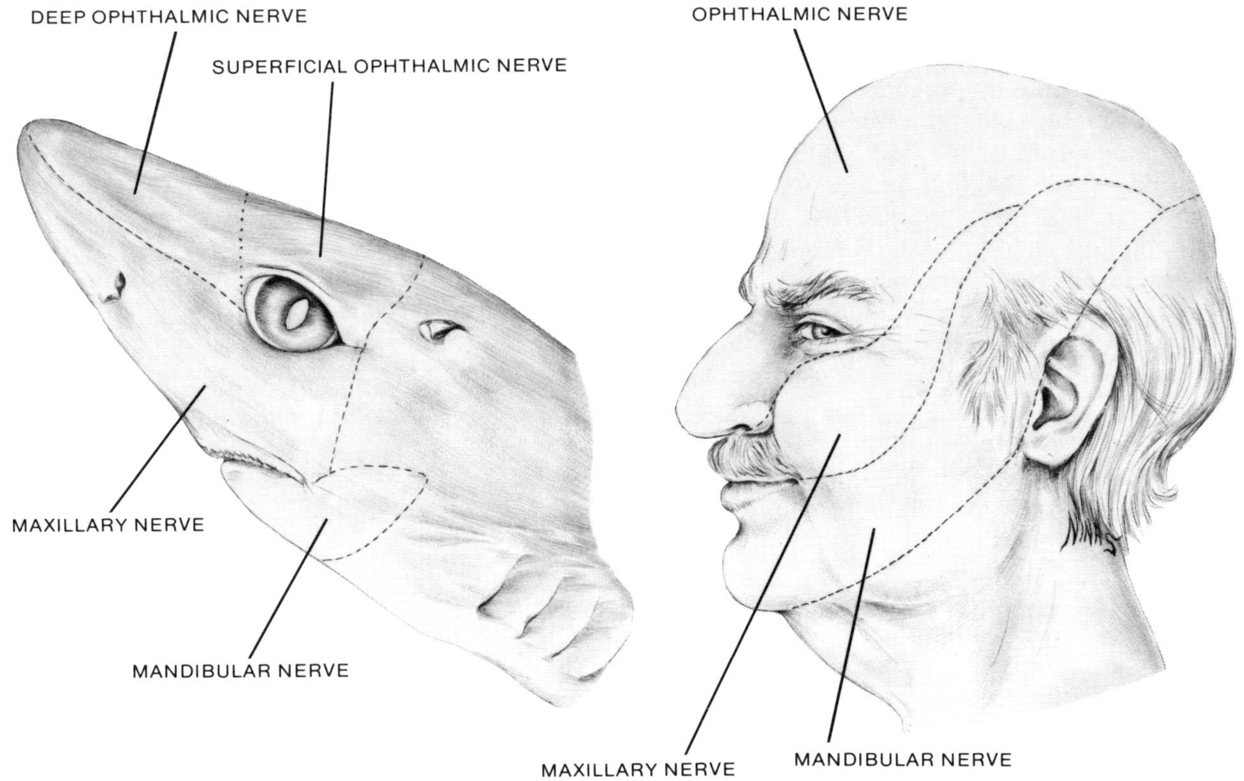

Figure 17-1. A comparison of facial nerve distribution of sharks and humans.

NV: Trigeminal nerve. (Somatic sensory, visceral motor, visceral sensory) The theory that the jaws evolved from part of an ancient anterior gill system is supported by the manner in which the trigeminal nerve branches around the mouth; the maxillary nerve passing dorsal to the mouth corresponds to the pretrematic branch, while the mandibular innervating the ventral jaw corresponds to the posttrematic ramus.

Because it is positioned anterior to the branchial region and the pharynx—which, as part of the digestive tract, is a "visceral region"—NV has lost its visceral sensory components. The distribution of these fibers to the jaws and head in sharks and humans is shown in Fig. 17-1. (In mammals, the deep and superficial ophthalmics combine to form one ophthalmic nerve.)

In both mammals and sharks, the visceral motor portion of the mandibular branch innervates the mandibular muscles which close the lower jaw; in mammals, these muscles are also used for chewing. As explained earlier, these muscles are derived from branchial arch constrictors.

NVII: Facial nerve (Visceral sensory, visceral motor, somatic sensory) This nerve innervates the derivatives of the spiracular or hyoid gill arch. In sharks, the large branch posterior to the spiracle corresponds to the posttrematic ramus, while other branches correspond to the pretrematic and pharyngeal rami.

The large posttrematic branch of sharks—the hyomandibular—innervates various muscles of the hyoid arch, including the dorsal constrictor muscle. In mammals, the nerve serves the muscles which control facial expression; it is for this function that NVII is named.

NIX: Glossopharyngeal nerve (Visceral sensory, visceral motor, somatic sensory) NIX innervates the first branchial arch of sharks. In mammals, it is relatively small and innervates the pharyngeal lining and pharyngeal muscles.

NX: Vagus nerve (Visceral sensory, visceral motor, somatic sensory) In sharks, the vagus has incorporated originally independent branchial nerves; it supplies the last four gill arches. Fibers from these nerves in mammals innervate muscles of the pharynx and larynx derived from the gill arch musculature. In both sharks and mammals, the vagus is crucial for the control of the internal organs (see Chapter 15).

As previously stated, the posterior margin of the skull in mammals is situated so that the derivatives of the occipitospinal nerves and the hypobranchial nerve are incorporated as cranial nerves XI and XII, respectively.

XI: Accessory nerve (Visceral motor) This nerve is derived from the occipitospinals and several visceral motor fibers of the vagus. It innervates the sternomastoids and cleidomastoids and the trapezius muscles of the back.

XII: Hypoglossal nerve (Somatic motor) This nerve innervates the muscles of the tongue that are derived from the hypobranchial musculature.

SPINAL NERVES

In primitive vertebrates, each myomere of the segmented musculature required for swimming was innervated by its own nerve. This provision resulted in a "segmental" arrangement of nerves along the spinal cord. Although in higher vertebrates living on land the muscles have gradually lost their segmentation, the spinal nerves still leave the cord in a segmental fashion to innervate musculature derived from their corresponding myomeres.

As the spinal nerves extend from the cord toward peripheral structures, the nerve fibers may gradually lose their segmental quality and run together. This occurs particularly in limb or fin regions, where adjacent spinal nerves leave the cord, interweave to form a network called a "plexus," and then, as a few large "nerves," extend to supply the limb. Such multifiber innervation of fins and limbs probably resulted from the incorporation of musculature from several myomeres in these appendages. In sharks, the ***brachial plexus*** supplies the pectoral fins while the ***lumbosacral plexus*** supplies the pelvic fins.

Each spinal nerve is connected to the cord by a ***dorsal root*** and a ***ventral root***. The traditional view was that each root contains only one type of nerve fiber: the dorsal root was thought to have sensory, or afferent, neurons; the ventral root was thought to have motor, or efferent, axons. Recent observations on vertebrates ranging from fishes through mammals suggest much greater complexity and variability.

Nevertheless, it is true that sensory nerve cell bodies are located outside the central nervous system; the cell bodies of many of them are gathered together in swellings called ganglia on the dorsal roots. Motor neuron cell bodies are located in the spinal cord, their axons extending out in most cases over ventral roots, but in some cases over dorsal roots.

The gray matter of the spinal cord is divided into dorsal and ventral regions. The dorsal gray matter contains associational neurons which receive sensory impulses from the dorsal root ganglion nerve cells and connect with motor neurons or to the brain via fibers in the white matter of the cord. The ventral portion of the gray matter contains the nerve cell bodies of motor neurons.

Functional Components of Spinal Nerves (Fig. 17-2)

Both sensory and motor neurons may be further classified into either the somatic or the visceral division of the nervous system. The ***somatic division*** is concerned with adjustments and movements required in relation to the external environment; the ***visceral division*** regulates the internal environment, including the digestive, excretory, reproductive, endocrine, and circulatory systems. Sensory and motor components of these divisions are described below with respect to mammalian physiology; however, the information applies generally to all vertebrates.

The dorsal root of a spinal nerve contains both somatic and visceral sensory fibers. ***Somatic sensory fibers*** transmit information from general cutaneous receptors (touch, pain, temperature, pressure) in the skin, and from proprioceptors, or stretch receptors, in tendons, joints, and striated muscle. By reporting on the degree of muscle contraction or the angulation of joints, the proprioceptors supply information concerning the current positions and states of tension of body parts. Appropriate motor responses could not be made without this information. ***Visceral sensory fibers*** monitor the internal organs and transmit information related to pain, stretching, H^+ concentration, and the like. Ventral roots also contain nerves of both divisions. ***Somatic motor fibers*** transmit motor impulses to striated ***voluntary muscles***. ***Visceral motor fibers*** comprise the ***autonomic nervous system***, which controls the activity of the heart, smooth muscle, glands of the skin, and internal organs. The autonomic nervous system is discussed later in this chapter.

As in the medulla (Fig. 15-2), the gray matter of the spinal cord is subdivided into columns. Each column is associated with a single spinal nerve component. They are arranged in the following order, from the

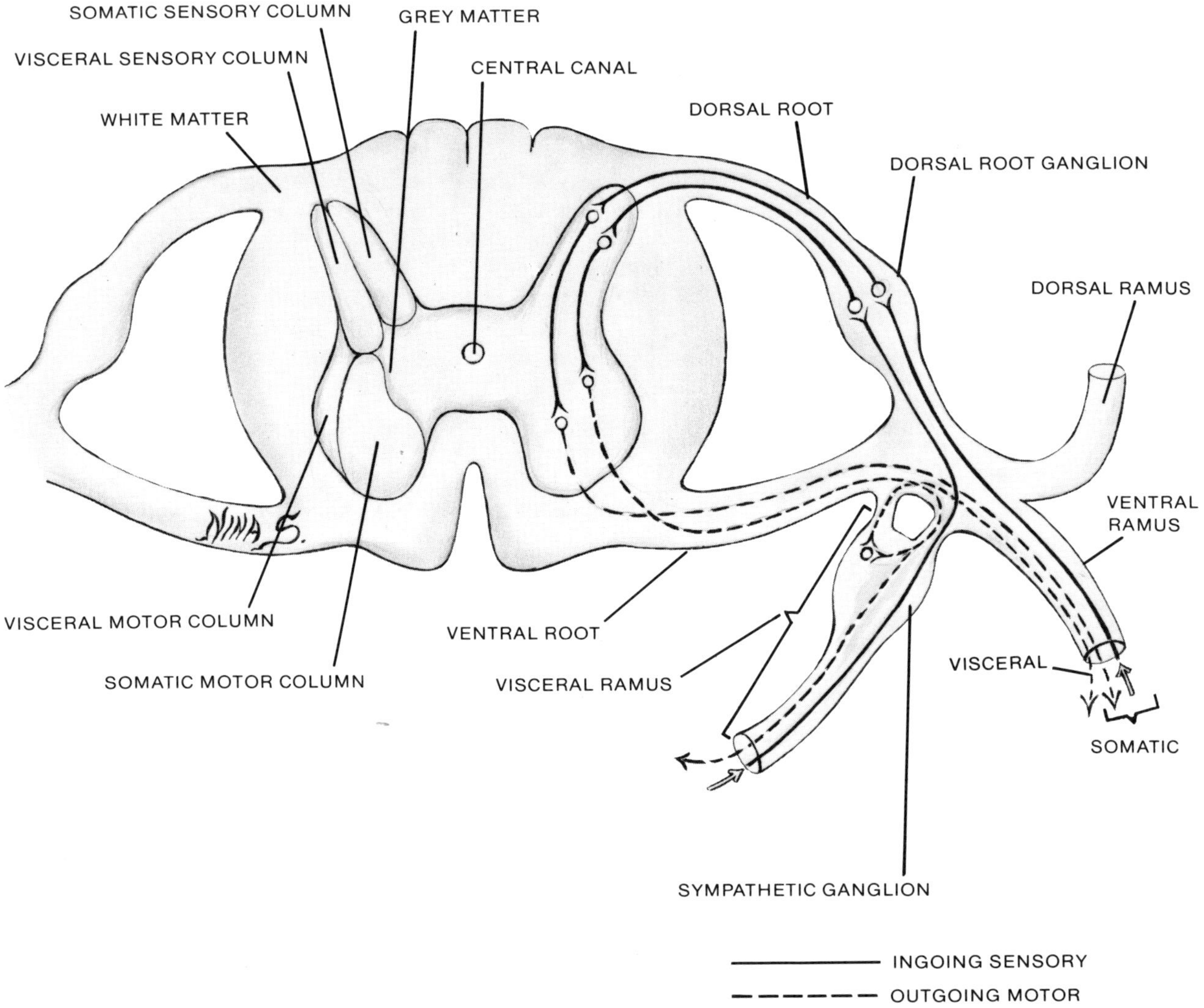

Figure 17-2. Diagram of spinal nerve components and cross section through spinal cord. Sensory and motor columns of the gray matter are indicated.

most dorsal position: somatic sensory; visceral sensory; visceral motor; and somatic motor (see Fig. 17-2).

THE AUTONOMIC NERVOUS SYSTEM (VISCERAL MOTOR NERVES)

The following discussion is based on mammalian physiology and anatomy.

Somatic motor nerves transmit motor impulses directly from the spinal cord to muscles without intervening relays. In contrast, visceral motor nerves employ a two-step relay to the periphery. In the first step ***preganglionic fibers*** carry impulses from the spinal cord to a ganglion. ***Postganglionic fibers*** then transmit to the end organs (Fig. 17-2).

These visceral motor fibers may be categorized into two divisions. The ***sympathetic division***, associated with energy expenditure and reactions to stress, includes such effects as acceleration of heart rate, increase in force of the heartbeat, vasodilation of vessels in voluntary muscles, and inhibition of digestive activity. The ***parasympathetic division***, associated with conservation and restoration of energy stores, includes effects such as deceleration of heart rate, decreased force of heartbeat, and stimulation of digestion. Although the divisions may act independently, they may also function in combination according to the degree of stress to which the organism is subjected. Central control of these divisions is believed to originate in the ***hypothalamus***.

The two divisions may also be distinguished according to various anatomical characteristics. Fibers of the sympathetic division leave the spinal cord only in the trunk, or thoracic and lumbar regions (see Waterman, 1971). Each preganglionic fiber enters the ventral root and passes through the visceral ramus to a ***sympathetic ganglion*** situated near the cord. The ganglia of each segment are connected by fibers to form the ***sympathetic chain*** on each side of the spinal cord.

At that point, several different paths may be taken, two of which are as follows. First, the preganglionic fibers may synapse with cell bodies in the ganglion. Postganglionic fibers then return to the spinal nerve and ultimately innervate glands in the skin and smooth muscle in peripheral blood vessels. In the second path, preganglionic fibers pass through the sympathetic ganglion to one of three ***subvertebral ganglia*** situated beneath the spinal cord. Postganglionic fibers innervate the internal organs. Additional secondary ganglia, called ***cervical ganglia***, supply postganglionic fibers to the eye and the salivary and lacrimal (tear) glands. Most postganglionic fibers of the sympathetic system release norepinephrine (noradrenalin) at the receptor organ and are therefore said to be ***adrenergic fibers***. A few, however, liberate acetylcholine.

In mammals, the adrenal medulla is also associated with the sympathetic system. Adrenal medullary cells are modified postganglionic units that secrete norepinephrine and epinephrine on stimulation by preganglionic fibers. Most of the secretion is epinephrine which stimulates the heart and metabolism. (Sharks have a series of small structures on the surface of the kidney which correspond to the adrenal medulla.)

Parasympathetic fibers originate either from the brain, in cranial nerves, or in the sacral portion of the spinal cord, in spinal nerves. Cranial nerve parasympathetic fibers are found in nerves X, III, IX, and VII. NX, the vagus nerve, carries 80 percent of all parasympathetic fibers. The oculomotor nerve (NIII) has a small branch which supplies the ciliary muscles of the eye and regulates pupil size. Parasympathetic fibers in NIX innervate the parotid gland, while those in NVII innervate the remaining salivary glands and the lacrimal glands. In sharks, only NX and NIII have parasympathetic fibers.

The spinal nerves of the parasympathetic division differ in two respects from those of the sympathetic division. First, the preganglionic fiber is very long. Second, the ***parasympathetic ganglion*** is not situated close to the spinal cord but is located in or near the receptor organ. A short postganglionic fiber then innervates the organ and releases acetylcholine. Such fibers are therefore termed ***cholinergic***.

Mammals and birds have highly-developed autonomic nervous systems. Both the sympathetic and parasympathetic systems innervate many internal organs, such as the heart. According to Noback (in Waterman, 1971) this system makes possible the fine adjustments and corrections necessary for the maintenance of a constant body temperature. Also, such fine control may be essential if the animal is to adjust to the stresses of the highly variable terrestrial environment.

Sharks, on the other hand, are poikilotherms—that is, their body temperature varies with that of environment. In addition, they inhabit a fairly stable environment. Therefore, fine control over the internal environment may not be as crucial a problem. As a result, the autonomic nervous system is less complex, with nonsegmental, irregular ganglia and little dual innervation of internal organs. The heart, for example, appears to have only parasympathetic innervation, via branches of the vagus nerve.

The Mudpuppy

CARL GANS and RONALD A. NUSSBAUM

18

The Mudpuppy: External Morphology

During the Devonian Period of earth history, a group of bony fishes gave rise to the first amphibians. These now extinct amphibians were relatively large, stout-bodied creatures. Nothing like them survives today. The key transformation from fishes to amphibians was the shift from fins to limbs; many other "amphibian" specializations, such as lungs and ribs, already existed in the ancestral fishes.

The earliest amphibians ultimately gave rise to reptiles and the reptiles, in turn, to birds and mammals. The earliest amphibians also gave rise to the modern amphibians, namely the salamanders, frogs, and caecilians. Many of the characteristics of the modern forms, seemingly intermediate between fishes and reptiles, are really secondary adaptations for the conditions under which contemporary amphibians live. Examples of such secondary developments are found in the skin, the reproductive pattern, the method of ventilation, feeding, and locomotion.

MODERN AMPHIBIANS

With a few exceptions, the surviving amphibians show three important characteristics that have great impact on their whole biology. First, much or all of their skin is moist. The combination of horny keratin and lipids that water-proofs the skin of higher vertebrates is absent. Second, the development of eggs and embryos occurs in water just as in bony fishes. Amphibian embryos lack the extra-embryonic membranes (chorion, amnion, and allantois) that characterize reptiles, birds, and mammals. Accordingly, amphibians are called *anamniotic*. Finally, amphibians typically have a *biphasic (two-stage) life cycle*. The first, or larval, stage undergoes metamorphosis into an adult or into a juvenile with an adult body form. Ordinarily the larval stage is aquatic and the adult terrestrial.

Contemporary amphibians are categorized in three major groups. Some 3,000 species of frogs occur on all continents from north of the Arctic Circle to the southern-most tips of the large continents and from sea level to high on tropical mountains. Their larval stage is referred to as a tadpole. Its tail is lost at metamorphosis, and the hindlimbs of the adult frog tend to be enlarged for jumping and swimming. A second group includes the caecilians, which are represented by some 160 species of tropical, completely limbless, subterranean and aquatic forms. Caecilians range through the Tropics, except for New Guinea and Australia. Finally, there are the salamanders, a third group of amphibians consisting of about 350 species and found in Europe, extreme northern Africa and Asia, as well as the Americas. Except for the Americas, salamanders are restricted to temperate regions.

In some ways, the salamanders are the most generalized of living amphibians, inasmuch as they retain a tail and most have four functional limbs. Perhaps their greatest diversity occurs in North America, where one sees different forms occupying pools, lakes, streams, and various parts of the terrestrial environment. Though they are often inconspicuous, they may be extremely common.

THE MUDPUPPY

The name ***mudpuppy***, or "waterdog," is obviously a lay term and was invented by early settlers in North America. The Latin name for the genus is ***Necturus***; five species occur, all restricted to the United States and Canada. *Necturus maculosus* is both the most common in nature and the species most often used for study in biology courses. It occurs throughout most of the eastern United States, except for the states of the southeastern coastal region.

Mudpuppies are atypical amphibians in that they do not undergo metamorphosis. Instead, the larval stage grows to adult size and the gonads mature while it is still aquatic, a process referred to as ***paedogenesis*** or ***neoteny***. Consequently, some of the metamorphic changes ordinarily seen in the skin and ventilatory and locomotor patterns of adult salamanders are not present in the mudpuppy. However, the mudpuppy is both larger and more common than most species of salamanders that metamorphose, hence its use in courses on comparative anatomy of the vertebrates.

Mudpuppies live in a variety of aquatic habitats, including ponds, lakes, streams, and rivers. They are mainly active at night, hiding during the day under submerged logs, rocks, and cutbanks.

Courtship and mating normally occur in the autumn. In mudpuppies, the sperm is stored by the female in a specialized cloacal structure called the ***spermatheca***, where it remains until ovulation and (internal) fertilization occur in the spring. The female then deposits up to 200 large (5–6 mm diameter), pigmentless eggs in a protected nest chamber. The chamber is often beneath a stone or other submerged object. The eggs are attached individually to the undersurface of the roof of the nest chamber. The female usually stays with the embryos during the incubation period (about 40 days), but it is uncertain whether her attendance has any function.

Hatchlings average 25 mm total length but grow to about 36 mm before the yolk is used up and they start to feed. Mudpuppies become sexually mature in their sixth year, having achieved lengths of near 200 mm. They continue to grow throughout life; the average length of *Necturus maculosus* is close to 330 mm, though specimens as long as 435 mm have been observed.

EXTERNAL ANATOMY (Fig. 18-1)

Body divisions. The body may be roughly divided into the ***head***, which carries the ***external gills***, and the ***trunk*** from which arise two pairs of ***limbs*** and the ***tail***.

External nares. These are forward-facing openings on the anterolateral tips of the blunt snout. They differ from those of the shark in forming the external openings for continuous tubes that end in the ***buccal cavity***. The sensory chambers of the olfactory system open into these passages.

Labial folds. Open the mouth and see how the lips interlock on the sides of the mouth. This permits the mudpuppy to open to the front of the mouth and suck in water without leakage along the sides. This is critical in ***suction feeding***. In salamanders that metamorphose, the mouth becomes much wider and edged by lips. Such animals catch prey with the aid of a tongue (Fig. 20-1).

Eyes. Very small lidless eyes lie on the anterior half of the head, just dorsal to the edge of the mouth. Generally, in metamorphosed salamanders and frogs the eyes develop lids.

External gills. There are three external gills on each side. Each gill is largely a mass of connective tissue, with many blood vessels and covered with a thin epithelium. The proximal and dorsal portions of

Figure 18-1. External morphology of the mudpuppy (*Necturus*).

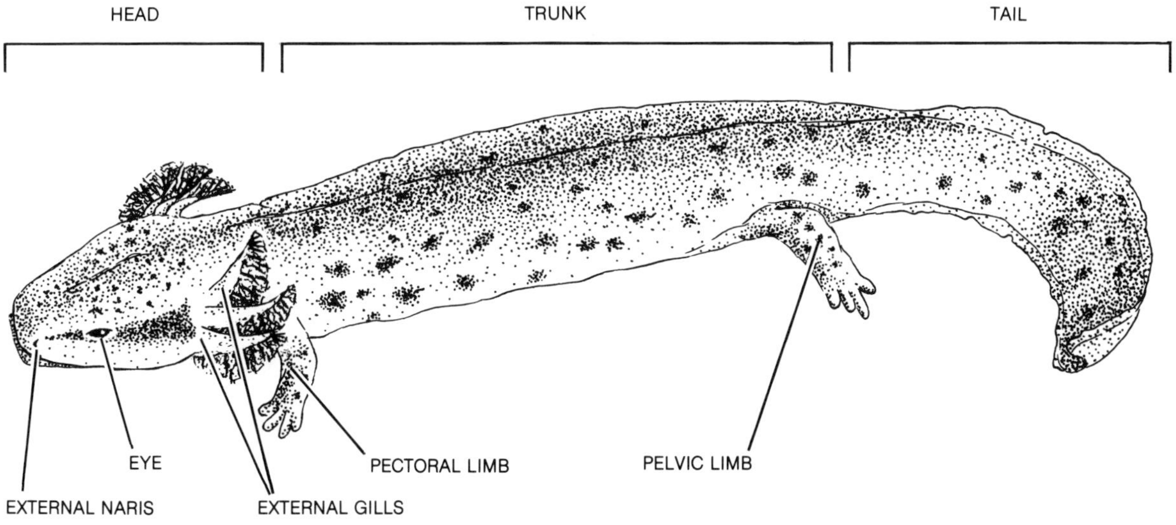

each gill are solid, but the distal and ventral edges are divided into filamentous masses that contain capillaries of the blood vascular system. The gills are movable by internal muscles that attach to their bases.

These gills are larval specializations derived from the ectoderm and adjacent mesoderm and are *not* homologous with the internal gills of fishes. The size of the filamentous portion of the gills varies with the amount of dissolved oxygen existing in the waters in which the mudpuppies have been raised (in other words, low oxygen leads to gills with a large surface area, and high oxygen concentrations to gills with a small surface area).

External gill slits. Two narrow openings between the external gills provide an exit for water from the sides of the buccal cavity. To see them, bend the gills forward and look into the gular fold. External gills are a larval specialization found otherwise only in some other amphibians and in lungfishes.

Gular fold. This ventral flap marks the separation of head from body. It extends ventrally between the external gill slits. In salamanders that metamorphose, the fold fuses; hence it is yet another example of a larval characteristic in the neotenous mudpuppy.

Lateral line organs. The mudpuppy has several lines of sense organs along the sides of the head and a single (disjunct) lateral one along each side of the trunk and tail. These lateral line organs are never as large and regularly spaced as they are in the dogfish; rather, they consist of short groupings of pits. The pigmentation and mucous coating of the skin obscures them, but after the skin has dried they can be seen as short and irregularly-directed lines on the snout, around the eye, and along the sides.

Trunk. The mudpuppy has a stout, flattened trunk and lacks a true neck. A dorsal depression marks the attachment of skin and muscles to the vertebral column. Along the sides of the trunk one may see a series of ***costal folds*** separated by depressed ***costal grooves***, each of which is internally continuous with a myoseptum so that the number of folds corresponds to the number of vertebrae. Recall that a ***myoseptum*** is a sheet of connective tissue between adjacent muscle masses. Preserved specimens often show other external folds that are artifacts.

Limbs. Both ***fore (pectoral)*** and ***hind (pelvic) limbs*** are small, scarcely as long as the diameter of the body. Each has four toes, of which the third is the longest. The mudpuppy uses these flattened limbs to walk slowly along the bottom of a pond or stream. It folds them along the side of the body during brief rapid swimming, which is performed by lateral undulation of the trunk and tail.

Cloacal aperture. This longitudinal slit, its lips breaking into radiating folds, provides the single exit for alimentary and urogenital products. The lips contain internal glands, and those of males in breeding condition tend to be swollen.

Tail. The muscular tail is bordered by a thin, peripherally-rounded ***caudal fin*** that lacks internal skeletal supports.

Skin. The skin of the mudpuppy (as of all vertebrates) consists of an ***epidermis*** and an underlying ***dermis***. The epidermis has numerous large glands that grow down into the dermis. These glands can be seen as round, transparent spots in the skin as the skin is removed (try this on the skin of the gular region). There are two kinds of glands, the ***mucous glands***, which are generally smaller, and the ***granular glands***. The former provide a slimy protective coating over the outside of the animal, while the granular poison glands protect the mudpuppy against the attacks of some predators. The occurrence of numerous glands and the thin, relatively undifferentiated nature of the epidermis represent further larval specialization; in many salamanders such specializations normally disappear at metamorphosis.

As you skin your animal (Chapter 19), examine the underside of the skin and the surface of the trunk to note the pigmentation of the skin and the regional distribution of glandular spots. After fixation of dead specimens only the ***melanophores*** (chromatophores containing melanin) remain obvious; however, in living specimens other pigment cells provide a very rich coloration of browns and olives that matches the color of the animal's back to that of local stream or lake bottoms. As in the dogfish, the dorsal color is darker than the ventral. Pay particular attention to the skin along the caudal fins. The bases of the fins are filled with a strip of fatty tissue, and the sacs of many enlarged poison glands may be seen on the inside of the fold. No salamander retains any bony scales in the skin; however, bony plates occur in the dermis of some caecilians and perhaps certain frogs.

If cross sections of the skin are available, it is easy to demonstrate microscopically (1) the two types of glands, both consisting of inpocketings of the epidermis into the dermis, (2) the dense network of collagen fibers within the dermis, and (3) the dermal network of capillaries important in gas exchange across the skin (see Chapter 21). Also, note the ***pigment cells (chromatophores)*** with their branching processes projecting between the other cells of the integument. The pigment granules within the skin may be concentrated toward the center of the cells (causing paling of the skin) or dispersed into fine processes (darkening the skin). This process is under endocrine control, though it may also involve a direct response of the individual pigment cells to light.

CARL GANS and RONALD A. NUSSBAUM

19

The Mudpuppy: Musculoskeletal System

The musculoskeletal system of the mudpuppy is divided into ***somatic*** and ***visceral*** components. In adult amphibians, the skeleton has become almost entirely ossified, so that cartilage is seen only at some articulating points, in the sensory capsules, in the pectoral and pelvic girdles, and in the sternal region. Finally, much of the branchial (gill) skeleton remains cartilaginous. The transition from embryonic cartilage to a more bony condition proceeds continuously throughout the larval period and is not immediately tied to metamorphosis in amphibians. However, there are certain skeletal elements and bones that only develop after metamorphosis in other salamanders. One of these is the maxilla; this bone is absent in the mudpuppy and most other paedomorphic salamanders.

The neat subdivision between somatic and visceral skeleton seen in the dogfish is obscured in the mudpuppy. The axial skeleton again consists of a skull and vertebral column; however, the skull of the mudpuppy incorporates both somatic and visceral components. In the skull we also see derivatives of the primary dermal bones. The appendicular skeleton and muscles are entirely somatic, though the shoulder girdle includes some dermal bones as well, and some of the originally visceral muscles now are associated with the pectoral girdle.

SKELETON: THE SKULL (Fig. 19-1)

The skull of the mudpuppy, like that of all amphibians, has two major components. The first, the ***chondrocranium***, appears early in embryonic development, forming the protective covering for the brain and the olfactory and otic capsules. At first, the chondrocranium consists entirely of cartilage; later in development certain parts ossify, thereby forming ***replacement*** or ***cartilage bones***. The second part of the skull, the ***dermal component***, consists of investing bones that ossify directly from the mesenchyme that composes the dermis. Dermal bones appear relatively late in development. They overlie the chondrocranium, adding strength and protection to the skull.

The chondrocranium is best studied during its early stages of development; specially prepared specimens should be used (Fig. 19-1). Much of the chondrocranium of the adult mudpuppy is obscured by the dermal bones; on the other hand, the following bones can easily be observed in skulls of adult specimens and are labeled in Figure 19-1.

Dorsal Aspect of the Skull

Premaxillae. These are the anteriormost elements of the skull; they are paired, V-shaped, dermal bones that bear teeth.

Frontals. These large, flat, paired, dermal bones lie just posterior to the premaxillae.

Parietals. These large, flat, paired, dermal bones cover the posterior portion of the cranium.

Ventral Aspect of the Skull

Vomers. These paired, anterior, tooth-bearing dermal bones, lie along the roof of the mouth, posterior to the premaxillae. The vomers (sometimes referred to as prevomers) articulate with the parasphenoid posteriorly, and with the palatopterygoids posterolaterally.

Ethmoid. This small, unpaired cartilage bone lies dorsal to the vomers and can be seen between their anterior ends.

Palatopterygoids. These paired dermal bones bear teeth anteriorly. They articulate with the vomers anteriorly and with the quadrates posteriorly.

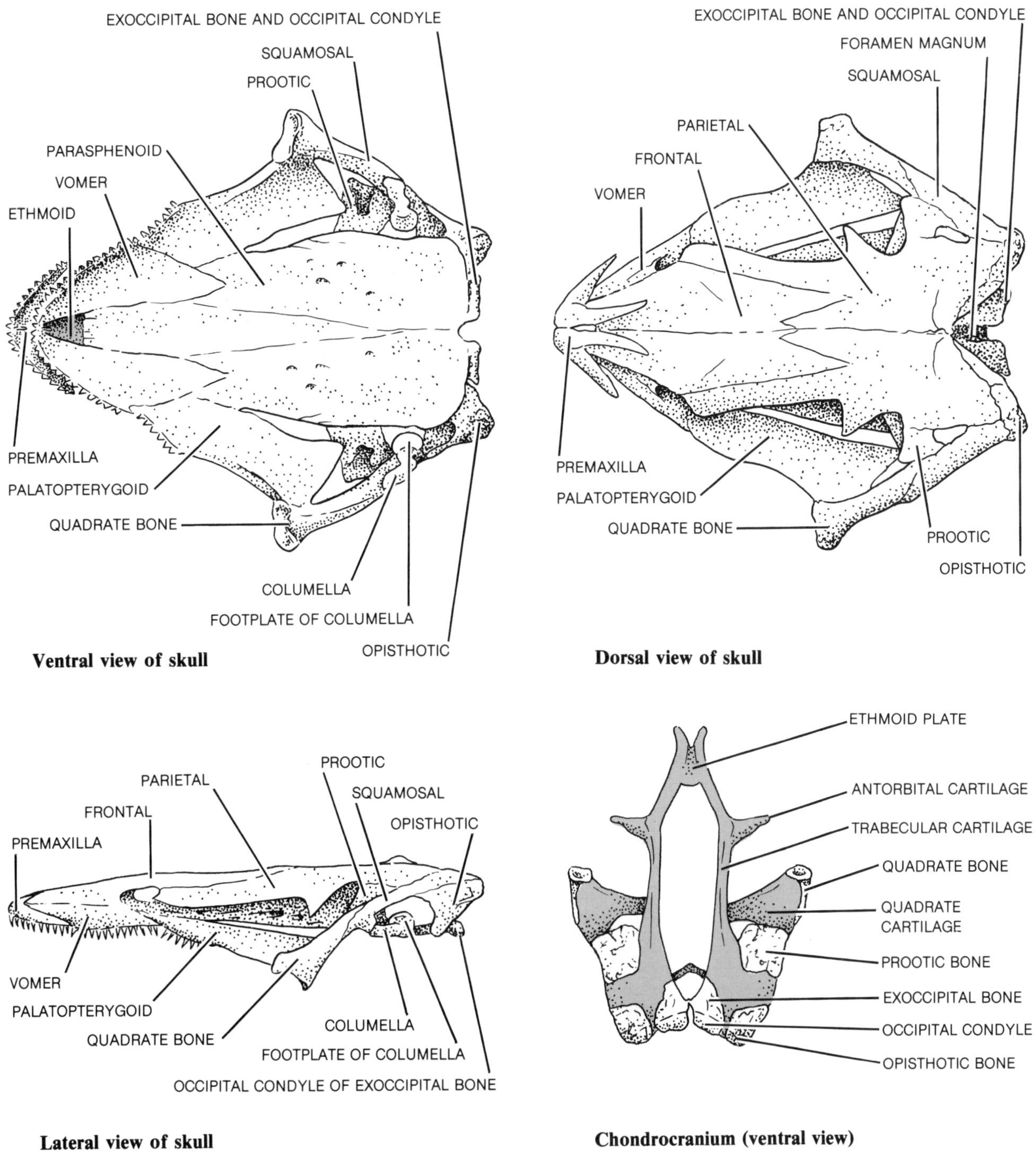

Figure 19-1. Ventral, dorsal, and lateral views of a mudpuppy skull; ventral view of the chondrocranium.

Quadrates. These paired cartilage bones articulate with the palatopterygoids anteriorly and with the squamosals posteriorly. The quadrates articulate with the lower jaw.

Squamosals. These elongate dermal bones run in a dorso-medio-posterior direction from their articulation with the quadrates. Dorsally, they articulate with the opisthotic bones.

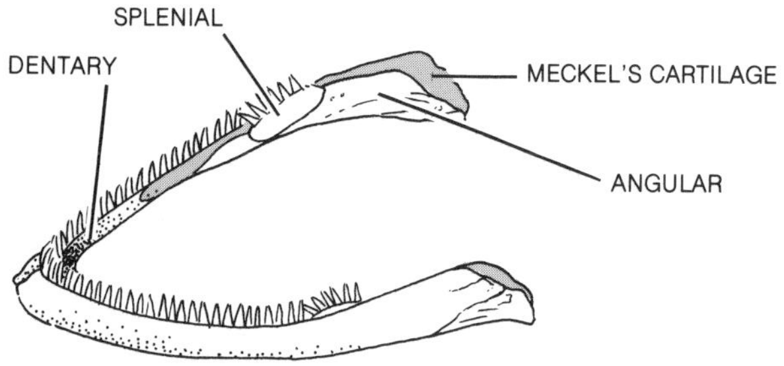

Lower jaw (¾ view)

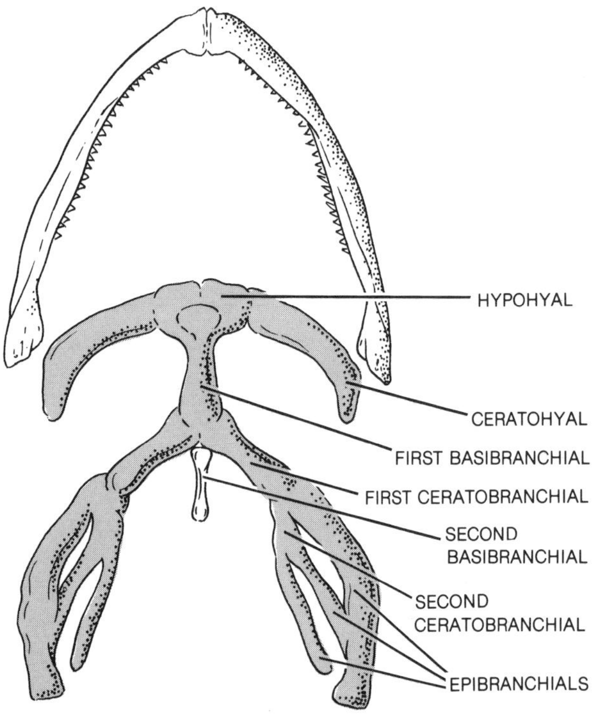

Lower jaw and hyoid apparatus (ventral view)

Figure 19-2. Lower jaw of a mudpuppy (three-quarter view); ventral view of the lower jaw and the hyoglossal skeleton.

Prootic Bones. These cartilage bones form in the wall of the otic capsules, anterior to the fenestra ovalis and columella.

Opisthotic Bones. These cartilage bones form in the wall of the otic capsules, posterior to the fenestra ovalis and columella.

Parasphenoid. This large, plate-like, unpaired dermal bone forms the posterior, ventral floor of the skull.

Exoccipitals. These cartilage bones form in the chondrocranium posterior to the parasphenoid. They bear the ***occipital condyles***, which articulate with the anteriormost vertebra, the atlas. They also form the lateral borders of the large opening known as the ***foramen magnum***, through which the brain stem continues into the spinal cord.

Otic Capsules. These thin-walled cartilaginous structures on the posterolateral regions of the skull house the inner ears. As development proceeds, their anterior portions ossify to become the prootic bones and their posterior regions ossify to become the opisthotic bones. Each otic capsule has a ventrolateral opening called the ***fenestra ovalis***. The ***footplate*** of the ***columella*** normally fills the fenestra ovalis; often the columella is partially dislodged when the specimen is prepared, and the opening to the inner ear is then exposed.

The Lower Jaw (Fig. 19-2)

Each half of the lower jaw, i.e., each of the two ***mandibular rami***, consists of a cartilaginous strand, ***Meckel's cartilage***, which is more-or-less enclosed by three dermal bones: (1) the ***dentary***, which bears teeth (the ramus is dentigerous) and covers much of the lateral surface of each mandibular ramus; (2) the ***splenial***, a small tooth-bearing bone that lies about midway along each side on the medial surface; and (3) the ***angular***, which is a toothless bone that forms the posteromedial surface of each ramus. On each side Meckel's cartilage forms the articulating surface with the quadrates of the upper jaw.

SKELETON: THE HYOGLOSSAL SKELETON (Fig. 19-2)

The hyoglossal structures are best observed in special preparations of the branchial apparatus preserved in fluid. The ***hyoid arch*** and ***visceral arches***, numbers 3–5 (gill arches 1–3), remain largely cartilaginous in the mudpuppy, although the ***second basibranchial*** is usually calcified in larger individuals. The second, or hyoid arch, consists of a median ***hypohyal cartilage*** and two lateral ***ceratohyals***. The hyoid arch is connected to the gill arches (3–5) by the ***first basibranchial***. The gill arches consist of ***first*** and

second ceratobranchials and three pairs of ***epibranchials***. The epibranchials support the gills. A pair of small ***laryngeal cartilages***, derived from the seventh gill arch, occurs in the larynx on either side of the glottis, through which air passes to the lungs.

SKELETON: THE VERTEBRAL COLUMN (Fig. 19-3)

There are basically four kinds of vertebrae in the mudpuppy, as in all salamanders. A single ***cervical vertebra (atlas)*** articulates with the skull. About 17 ***trunk vertebrae***, lead to a single ***sacral vertebra*** that articulates with the ilium via its ribs; as many as 30 ***caudal vertebrae*** support the tail. The development of the vertebrae is poorly understood in salamanders and their parts are not necessarily homologous to those of mammals.

The cervical vertebra is relatively short and ribless and has special anterior ***condylar facets*** that articulate with the occipital condyles of the skull. A short ***odontoid process*** of the atlas extends anteriorly between the condylar facets. A trunk vertebra consists of a dorsal ***neural arch*** and a ventral ***centrum***. The neural arch encloses the ***neural canal***, which houses the spinal cord. Neural arches of successive vertebrae articulate by means of anterior ***prezygapophyses*** and posterior ***postzygapophyses***. The posteromedial part

Figure 19-3. Lateral view of the skeleton in an outline of the mudpuppy's body; three-quarter views of the cervical vertebra and a trunk vertebra.

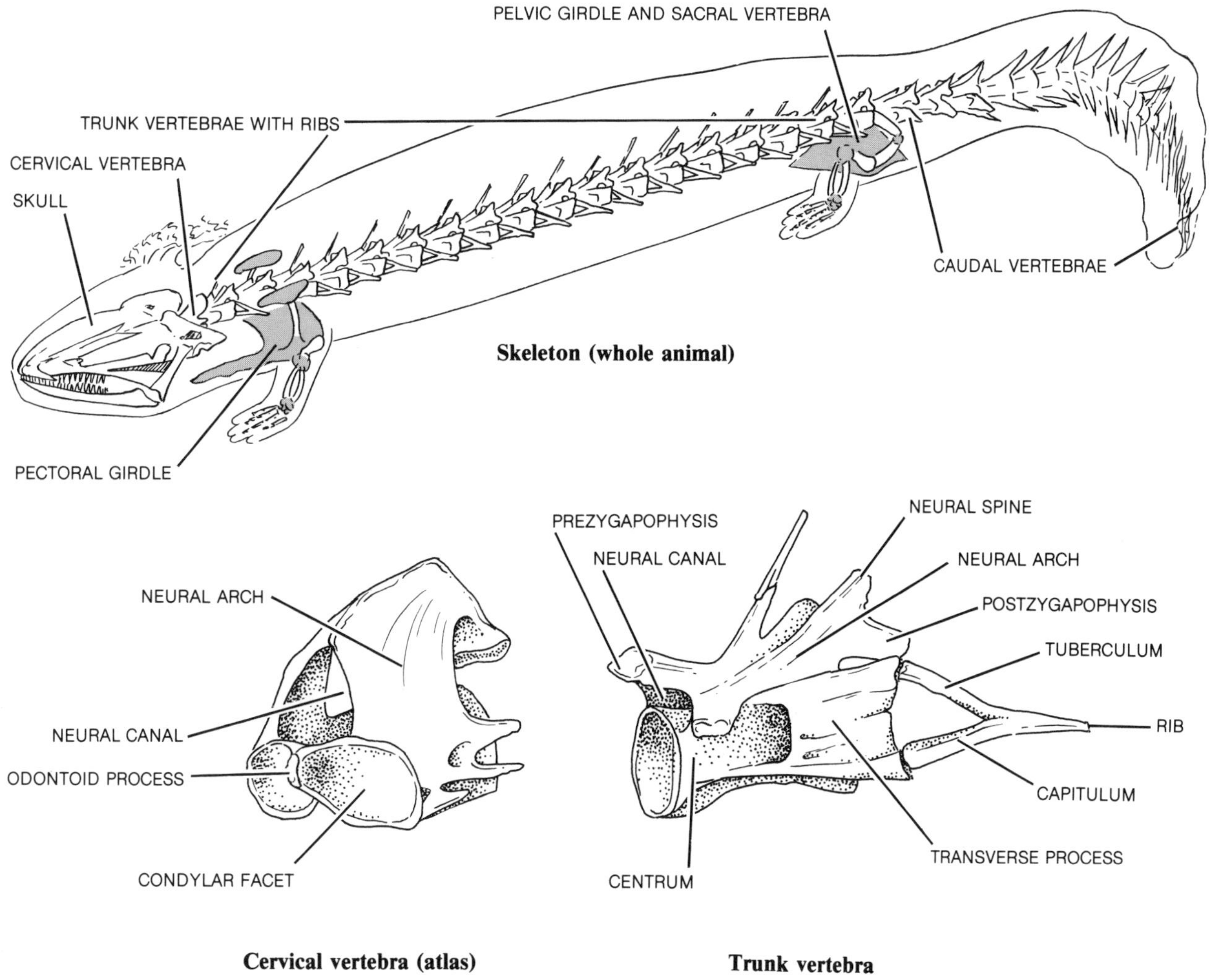

of each neural arch is drawn up into a low ***neural spine***. The centrum is biconcave (amphicoelous) and spool-shaped. Extending from the middle of each side of a trunk vertebra is a posterolaterally directed ***transverse process*** that consists of a dorsal ***diapophysis*** and a ventral ***parapophysis***. These processes articulate respectively with the dorsal ***tuberculum*** and ventral ***capitulum*** of the two-headed ***ribs***.

The single sacral vertebra differs from the trunk vertebrae in that its enlarged ***sacral rib*** articulates with the pelvic girdle. The ribless caudal vertebrae have long neural spines that increase in length posteriorly. The caudal vertebrae are compressed laterally, and each, except for the first two, has a ventral ***hemal foramen*** enclosed in a bony ***hemal arch***. An elongate ***hemal spine*** projects ventroposteriorly from each such arch.

SKELETON: THE GIRDLES (Fig. 19-4)

The Pectoral Girdle and Forelimb

The ***pectoral girdle*** is largely cartilaginous and is best studied from fluid-preserved specimens. It lacks a skeletal connection with the skull, gill apparatus, and vertebral column, and is held in place by muscles. Each half of the pectoral girdle is triradiate (radiates in three directions). A ***glenoid fossa*** lies at the juncture of the three arms and forms a socket in which the humerus of the forelimb articulates. The ventralmost element of the girdle is the ***coracoid cartilage***, a large, flat plate that extends medially to the midline, where it meets its mate from the other side. The anterior element of the girdle is the ***procoracoid cartilage***; this skeletal element is never replaced by bone, but remains cartilaginous, even in salamanders that metamorphose. The ***scapula***, the only ossified part of the pectoral girdle in the mudpuppy, extends dorsally and articulates with the ***suprascapular cartilage***.

The forelimb contains a proximal ***humerus***, an anteromedial ***radius***, an ***ulna*** lateral to the radius, six wrist bones or ***carpals***, four ***metacarpals***, and nine ***phalanges*** (two on each finger except for the third finger which has three).

The Pelvic Girdle and Hindlimb

The ***pelvic girdle*** is largely cartilaginous and, again, it should be studied from fluid-preserved specimens. The large, ventral ***puboischiac plate*** consists of anterior cartilaginous ***pubes*** and largely ossified, posterior ***ischia***. The bony ***ilium*** extends from the puboischiac plate to the sacral rib. All three pelvic girdle car-

Figure 19-4. Right half of pectoral girdle and right forelimb from a mudpuppy; anterior view of the pelvic girdle and one hindlimb.

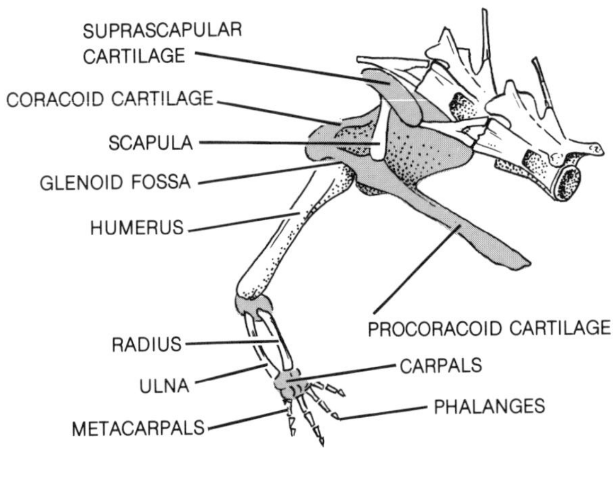

Pectoral girdle and forelimb

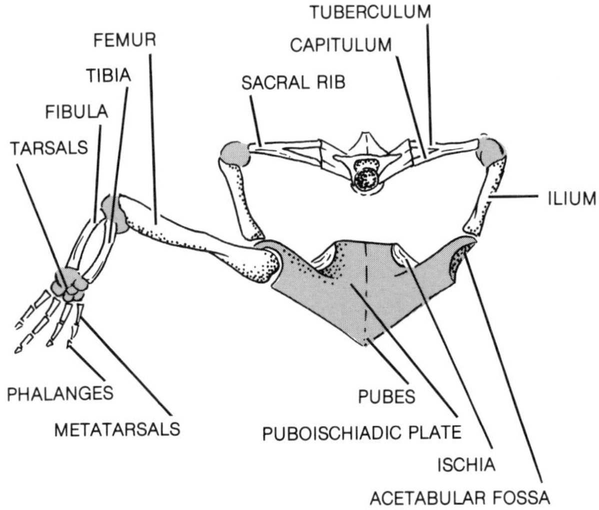

Pelvic girdle and hindlimb

tilages on each side contribute to the ***acetabular fossa***. The latter is the socket at which the hindlimb articulates with the pelvic girdle.

The hindlimb has a large, proximal, thigh bone, the ***femur***, which articulates distally with both the ***tibia*** and the ***fibula***. Six ankle or ***tarsal*** bones and four ***metatarsal*** bones occur. The hindfeet have nine ***phalanges***, arranged just as in the forefeet.

MUSCULAR SYSTEM

The muscles of the mudpuppy, and of amphibians in general, differ from those of the dogfish shark in being able to perform much more specific and restricted movements involving multiple jointed skeletal elements. The fibers of the body musculature tend to be longer than the equivalent ones in fishes. However, much of the total muscle mass is still axial, and the muscles of the limbs and their girdles compose a relatively small fraction of the total. This suggests that undulation of the back and tail, rather than limb movement, may be a primary means of locomotion, as it is in most fishes. However, in some other amphibians, such as metamorphosed frogs, the pattern is different and the limbs, rather than the tail, perform the primary propulsive role.

In the survey to follow, the body musculature is treated by regions under the headings of trunk, head and throat, and girdles. However, as in the shark, the skeletal muscles belong to two main categories. The first derives from somites and contributes most of the trunk and girdle musculature, as well as the epibranchial, hypobranchial, and eye muscles. The second includes the branchiomeric muscles that derive from the anterior extension of the embryonic hypomere (the ventral-most portion of the mesoderm). This difference is important as it is reflected in the innervation of the muscles (see your textbook).

The action of the muscles in the mudpuppy is here characterized based upon their performance when acting individually. However, muscles generally act in groups, and their effect and the sequence in which they are activated may then be different. As most motor patterns have never been directly recorded in the mudpuppy, we present only the simplest action patterns.

DISSECTION INSTRUCTIONS

General Considerations

Dissection of muscles has its own rules. The trick is to find out where the muscles lie, to separate one from the other, and to determine their fiber length and direction. The mudpuppy, being smaller than the dogfish shark or cat, will require an extra measure of care and scrutiny. Indeed, fiber color and direction are the critical aspects allowing separation of muscle bundles.

After skinning a region, pick away at the superficial connective tissue until you can see discontinuities in the underlying muscles. Then use a pointed needle or probe to scratch the connective tissue along this junction. You will generally find that the bundles separate quite easily. Be careful that you do not rip into the muscle fibers themselves; identification of the components becomes progressively more difficult thereafter.

Never cut a muscle away from its origin or insertion, as this makes it extremely difficult to reconstruct its original position (this may be necessary, for instance, when studying for an examination). However, it is often necessary to remove superficial muscle layers in order to understand the architecture of deeper ones. To do this, free the muscle from connective tissues, pass a blunt forceps deep into it, and cut with a scissors or scalpel at a right angle to the fibers in the belly of the muscle. Matching of the cut ends will allow you to reconstitute the previous arrangement. The appearance of the muscle in cross section will permit an estimate of the force that a muscle can produce. Remember that the force produced by a muscle may be about one Newton per square centimeter of cross section. You should thus be able to estimate the relative force contributed by the various muscle bundles.

The surface of muscles may be crossed by thin connective tissue that makes it difficult to determine the directions of the underlying muscle fibers. These muscle directions can be demonstrated much more easily by drying the surface of a muscle very slightly and then adding a few drops of household bleach (such as Clorox) with a pipet. Within minutes the loose connective tissues start to foam and dissolve away. Flush the surface with lukewarm water and dip the specimen again in preservative. The fiber direction and intersegmental raphes will now be much more obvious. Note that determination of fiber direction proceeds even more effectively when a surface stain, such as iodine or methylene blue solution, is applied. A hand lens or dissecting microscope will let you see these small muscle fibers more clearly.

Many of the muscles and underlying tissues are relatively thin; others are bulky, and a three-dimensional view will require more than a single perspective. Consequently, it is best to have two or three extra specimens of *Necturus* available for the entire class. The first of these should be deep frozen and then sectioned sagittally. Use a jig saw and cut just slightly to one side of the midline. Then, wash the specimen and make a shallow dissection of the surface to note the arrangement of different muscle elements. Divide a second frozen specimen into cross-sections: the first section is cut through the nasal cavity, just posterior to the external nares; the second through the eye; the third through the angle of the mouth; and further ones at the level of the shoulder (pectoral) girdle and throughout trunk and tail. Place each section in a covered, deep petri dish so that it stays moist and can be examined from both sides. Be sure to label each section, both for position and orientation.

All other dissecting instructions will be presented under the more specific structures.

Specific Dissections

To see the muscles, the animal must first be skinned. Place the mudpuppy dorsal side up and, using the tip of a sharp scalpel, make an initial, shallow (less than 1 mm deep) incision on the left *side of the middorsal line, from the level of the eye to the pelvic girdle. Then, attempt to peel away the skin toward the* right *side until you can grasp its edge with a pair of blunt forceps. Keeping the connection to the underlying musculature tight, use the tip of a pointed probe (or the back of a sharp scalpel, if you have a steady hand) to run parallel to the contact line between skin and muscle, always on the skin side. The stretched fibers will rip and a deep pocket will form. Continue the incision down the right side; however, try to keep the skin in one piece and pull it away from the surface as a complete blanket until the muscles of the back and right side are fully exposed.*

Now make three incisions at right angles to the initial cut, one toward each limb and another along the front edge of the gills. Also, stop the incision and trim the skin along the edges of the lips and the cloaca. When skinning the region over the eye and nostrils, take particular care with blood vessels and nerves, as there are several kinds of special attachments to deeper tissues. Remember, avoid cutting into the muscle; rather, trim along the edge of the skin.

When you reach the right gills cut them off flush with the body wall. Run a blunt probe beneath the skin, just anterior to the gular fold, and follow this with a scissors to cut a transverse opening. It is then easy to peel the skin forward and the same thing should be done just anterior to the girdle. Be careful as the interhyoid muscles in this region also run transversely, and it is easy to rip them. Note the difference between branchiomeric and somatic muscles.

If the specimen was initially injected through a ventral incision, strip the skin down to this and then extend the incision anteriorly, removing the skin from half the body. Now peel the skin toward the right pectoral and pelvic limbs and gradually skin these, being careful to cut the tendinous insertions at the level of each joint. You will have noticed that the skin separates much more easily from some regions than others. You are then cutting through cutaneous lymph sacs. Note that the skin is much thicker in the palmar and plantar regions than on the dorsal surfaces of the hands and feet.

As you move onto the tail, run the incision slightly to the left of the caudal fin. Cut the fin base to identify the fat body and many large glandular sacs distributed in the fin fold. Dissecting the skin away from the caudal musculature proves to be slightly more difficult than skinning the trunk. Each myoseptum here inserts on the skin; the tail engages in major undulations, a process that requires that the skin conform and not be able to slip away when the body bends.

EXAMINATION OF THE MUSCULAR SYSTEM

Trunk Muscles (Fig. 19-5)

The trunk musculature of the mudpuppy is subdivided basically in the same way as that of the dogfish shark. The ***epaxial*** group of muscles lie dorsal to and the ***hypaxial*** group ventral to the ***horizontal septum***. Identify the division line; it runs dorsal to the attachment of the limbs. The trunk and tail muscles are subdivided longitudinally by ***myosepta***, forming a series of ***myomeres***.

Figure 19-5. Lateral view of mudpuppy trunk muscles.

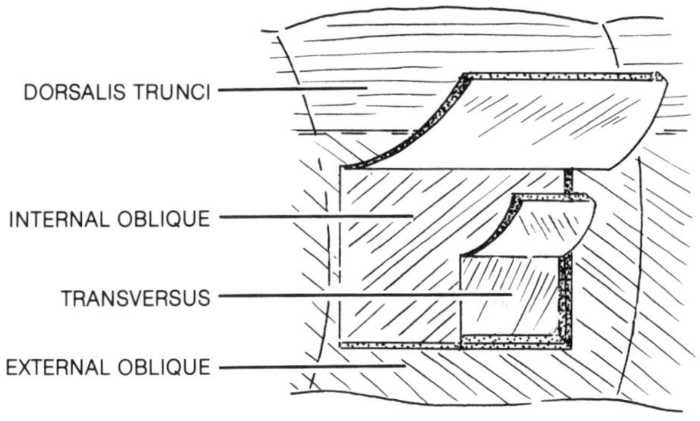

Expaxial musculature. The bulk of the epaxial musculature consists of the outer longitudinal bundles named the ***dorsalis trunci*** (Figs. 19-5 and 19-6). Fibers of the dorsalis trunci extend between successive myosepta and vertebrae. Deeper fibers of the epaxial series extend only between adjacent vertebrae. These fibers form more-or-less discrete units that are sometimes called ***interspinalis*** muscles. Action of the dorsalis trunci and the deeper fibers of the epaxial units flex the trunk and tail laterally and facilitate swimming and walking.

Hypaxial musculature. (Fig. 19-5) These muscles form a ventral rectus series and a lateral oblique series. The oblique series consists of three muscles: (1) the ***external oblique***, which is the outermost layer, with fibers running posteroventrally (see also Fig. 19-6); (2) the ***internal oblique***, which lies just beneath the external oblique and with fibers running nearly at right angles with those of the external oblique; and (3) the ***transversus***, which is the innermost layer, the fibers of which run transversely. Action of these lateral muscles of the body wall compresses the viscera and may aid in locomotion. You can visualize these muscles most easily by noting the external oblique and picking away its fibers along one myomere until a change in fiber direction indicates that the internal oblique has been reached. Then cut a window in the wall and examine the fiber direction of the transversus on the inside surface of the cut block.

The ***rectus abdominus*** (Figs. 19-6 and 19-8) is a parallel-fibered, longitudinal muscle that lies ventral to the oblique series on either side of a midventral aponeurosis named the ***linea alba***. The rectus abdominus originates on the pubis, and inserts on the rectus cervicis. Throughout its length, the rectus abdominus is interrupted by transverse myosepta. Action of the rectus abdominus will compress the viscera, bend the body ventrally, and, in concert with the rectus cervicis (mentioned below), pull the hyobranchium posteriorly.

Head and Throat Muscles (Fig. 19-6)

Levator mandibulae anterior. The superficial origins of these muscles can be seen on either side of the midline of the skull where they attach to the dorsal skull bones; their insertions on the lower jaw are hidden. They act to close the jaws.

Levator mandibulae externus. The larger of the two jaw-closing muscles, this muscle is observed lateral to the levator mandibulae anterior and is posterior to the eye. Its superficial fibers run posteroventrally toward the angle of the jaw. It, too, originates on the skull and inserts on the lower jaw.

Depressor mandibulae. This muscle, a jaw opener, is a narrow band of fibers running obliquely along the posteroventral margin of the levator mandibulae externus. It originates on the back of the skull and inserts on the angular bone of the lower jaw.

Branchiohyoideus. The branchiohyoideus is a large muscle on the side of the head posterior to the depressor muscle. It originates on the first branchial arch, just anterior to the gills, and inserts on the ceratohyal. It acts to pull the hyoid and branchial arches toward each other and probably assists in depressing the pharyngeal floor. Note that the muscle swings deep to the interhyoid and intermandibular muscles.

Levatores arcuum. These muscles form a thin sheet of radiating fibers dorsal to the gills. They originate on the fascia of the skin and dorsalis trunci and run ventrally to insert on the epibranchial cartilages of the gills. They move the gills.

Dilator laryngis. This small, flat muscle is found just posterior to the levator arcuum, and its fibers run parallel to those of the latter muscle. It originates on fascia of the overlying skin and underlying dorsalis trunci and inserts on the larynx and glottis. It dilates the larynx and opens the glottis.

Intermandibularis. This thin sheet of muscle is found just under the skin of the chin region. The fibers originate along the rami of the lower jaws and run transversely to insert on a medial raphe. Action of the intermandibularis elevates the floor of the mouth.

Interhyoideus. This muscle is continuous with and posterior to the intermandibularis. It originates on the fascia of the branchiohyoideus and the depressor mandibulae and inserts on the medial raphe. Its action is similar to that of the intermandibularis; it elevates the floor of the mouth and constricts the pharyngeal cavity.

Deep-Lying Muscles of the Throat (Fig. 19-6)

These muscles can be seen by deflecting the transverse intermandibularis and interhyoideus muscles. *This is best done by making a single longitudinal cut through both muscles, parallel to the medial raphe and a few millimeters to the right side of it.* Note that the geniohyoideus and most deeper muscles run at right angles to the superficial ones.

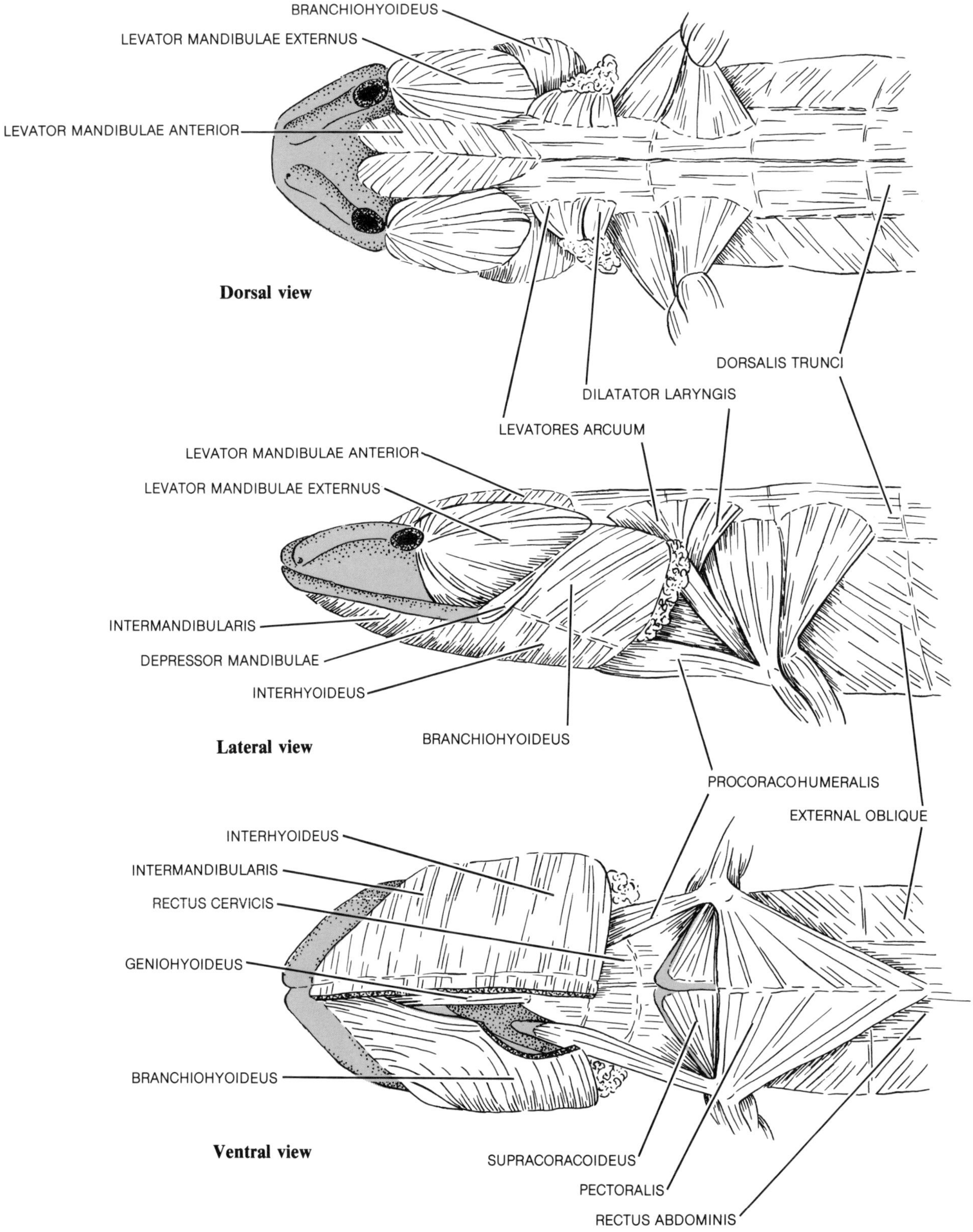

Figure 19-6. Dorsal, lateral, and ventral views of the head, throat, and outermost trunk muscles of the mudpuppy.

Geniohyoideus. The geniohyoideus is a paired, narrow, longitudinal muscle situated on either side of the medial line. It originates on the second basibranchial and inserts on the dentary of the lower jaw. It represents a continuation of the rectus cervicis. The geniohyoideus pulls on the hyobranchium and the jaw.

Rectus cervicis. This is a large, segmented muscle with longitudinal fibers. It originates on a myoseptum that separates it from the rectus abdominus. It inserts largely on the second basibranchial. Its action brings these structures together.

Subarcuals. These deep muscles (not illustrated here) are observed by cutting through the geniohyoideus to expose the branchial arches. Note that the anterior end of the tongue is free. The subarcuals are short, thin, longitudinal muscles running between the gill arches. They originate on the ceratohyal and on the first and second epibranchials. They insert on the first, second and third epibranchials. Their contraction draws the arches together.

Depressores arcuum. These small muscles (not illustrated here) run parallel to the gill arches, lateral to the subarcuals. They originate on the ceratobranchial and epibranchial cartilages and insert on gill fascia. Their contraction depresses or lowers the gills.

Eye muscles. The eye muscles of the mudpuppy are similar in arrangement to those of the dogfish shark. However, they are smaller and more difficult to dissect. To see them, cut carefully around the right eye and lift it away from its cavity.

Muscles of the Pectoral Girdle and Forearm (Fig. 19-7)

In lateral view, the following superficial muscles of the shoulder girdle are observed.

Pectoriscapularis. This is the anteriormost muscle running obliquely and posteroventrally from the gill region. It originates on the fascia of the dorsalis trunci and inserts on the anterior edge of the scapula. It pulls the scapula upwards and forwards.

Cucullaris. This narrow muscle lies almost parallel and posterior to the pectoriscapularis. It originates on fascia in the neck region and inserts on the scapula near the glenoid fossa. Like the pectoriscapularis, the cucullaris pulls the shoulder girdle forward and upward. Note that this is a branchiomeric muscle, homologous to the branchial levators of the dogfish.

Dorsalis scapulae. This is the next most posterior muscle. It is fan-shaped and runs vertically from its origin on the suprascapular cartilage to its anteroproximal insertion on the humerus. It acts to pull the humerus upward and forward.

Latissimus dorsi. This large muscle is the last in the radiating series of levators. It lies posterior to the dorsalis scapulae. It too is a fan-shaped muscle with fibers running anteroventrally from its origin on fascia on the dorsalis trunci and skin to its insertion on the proximoposterior aspect of the humerus. It pulls the humerus backward and upward.

The following muscles of the shoulder girdle, described in order of their exposure from front to rear, are best seen in ventral view.

Procoracohumeralis. This is an elongate strap muscle with fibers running largely posterolaterally from the origin along the procoracoid cartilage to the insertion on the humerus near the glenoid fossa. It pulls the humerus forward.

Supracoracoideus. This is the anteriormost of the two fan-shaped muscles on the ventral surface of the shoulder girdle and has the smaller exposure. Fibers originate on the coracoid cartilage and fascia and insert on the proximal end of the humerus. This muscle draws the humerus forward.

Pectoralis. This is the posterior of the two ventral fan-shaped muscles of the shoulder girdle; it broadly overlaps the supracoracoideus. It originates on the coracoid cartilage and fascia of the skin and the rectus abdominus and inserts on the proximal, medial region of the humerus. Its action draws the humerus medially and posteriorly.

Muscles of the Upper Arm (Fig. 19-7)

The process of skinning the distal portion of the extremity was described above. A small hand lens and a bit of iodine surface stain will make the muscles and connective tissues much easier to identify.

Humeroantibrachialis. This muscle lies along the anteroventral border of the humerus. Its fibers originate proximally on the anterior portion of the humerus and insert anteroproximally on the radius. It flexes the arm.

Coracobrachialis. This muscle is found on the inner surface of the upper arm immediately posterior to the latter. It originates on the coracoid cartilage and inserts on the distal end of the humerus. It pulls the humerus posteromedially.

Anconeus. The anconeus is the extensor of the forearm. It lies along the posterior border of the

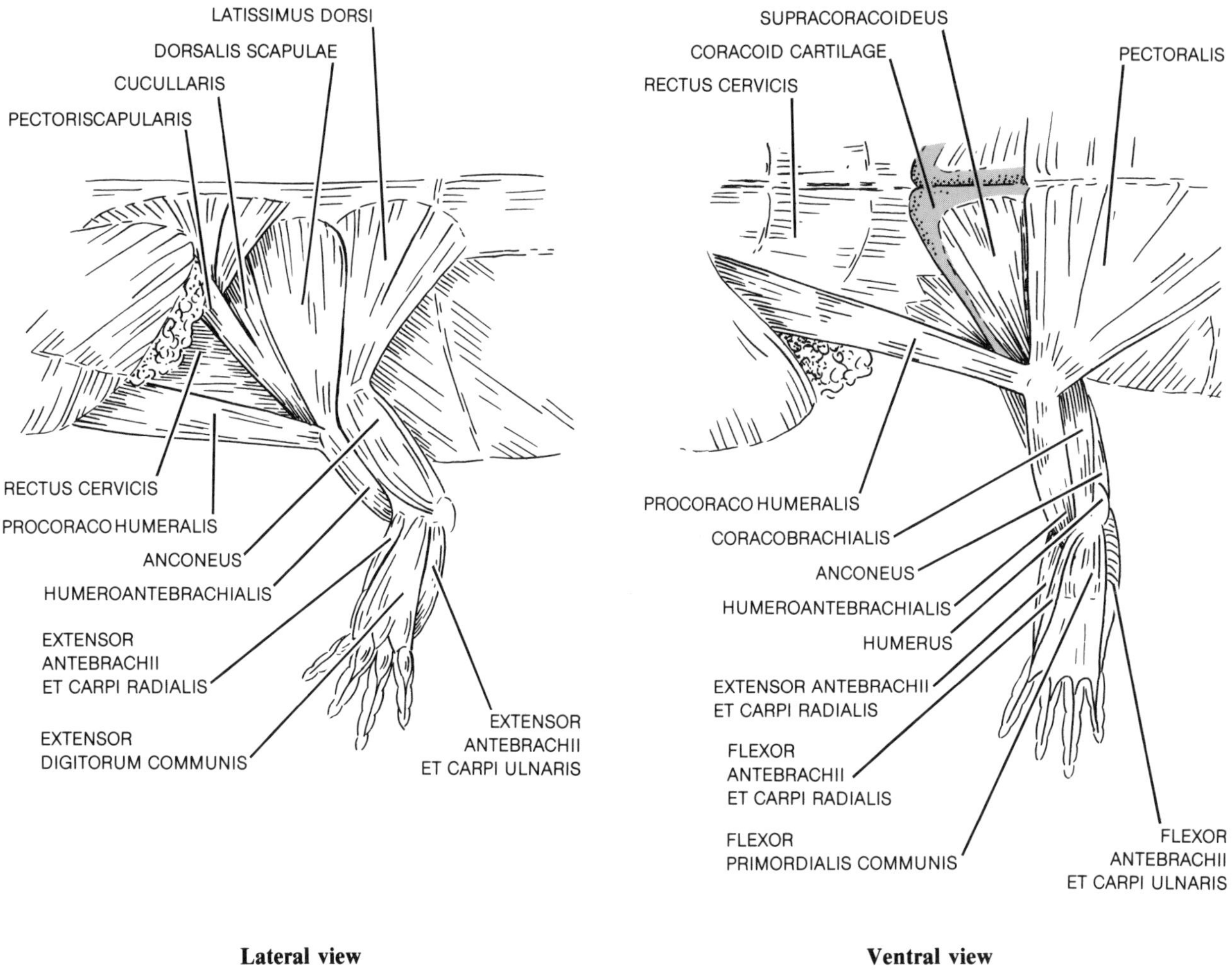

Figure 19-7. Lateral and ventral views of the muscles of the pectoral girdle and forelimb in the mudpuppy.

humerus. It originates by four heads on the shoulder girdle and humerus and inserts on the ulna via a stout fascial sheath that curves over the elbow.

Muscles of the Forearm and Forefoot (Fig. 19-7)

These muscles are relatively small. They should be examined with the aid of a hand lens.

Extensor antibrachii et carpi radialis. This small muscle lies along the anterior (preaxial) border of the forearm close to the radius. It originates distally on the humerus and inserts on the radius and carpus. It extends the ***manus*** (hand).

Extensor antibrachii et carpi ulnaris. This is a small, slender muscle observed along the postaxial border of the forearm in close association with the ulna. It originates on the distal end of the humerus and inserts distally on the ulna and on the carpus. It extends the manus and contributes to the extension of the forearm.

Extensor digitorum communis. This is a large muscle that lies on the dorsal or outer surface of the forearm between the two smaller muscles described immediately above. It originates on the distal end of the humerus and inserts proximally on the metacarpals. It extends the manus and digits.

Flexor antibrachii et carpi radialis. This slender muscle is seen on the anteroventral part of the forearm. It originates on the distal end of the humerus and inserts on the radius and carpus. It flexes the forelimb and manus.

Flexor antibrachii et carpi ulnaris. This small muscle is found on the posteroventral aspect of the forearm. Its origin is on the distal end of the humerus and its insertion is on the distal end of the ulna and the carpus. It flexes the forelimb and the manus.

Flexor primordialis communis. This is the large central muscle of the ventral surface of the forearm. The origin is on the distal end of the humerus, and it inserts by tendons on the four digits. It flexes the digits.

Muscles of the Pelvic Girdle and Hindlimb Ventral View (Fig. 19-8)

These muscles are described from front to rear.

Puboischiofemoralis externus. This muscle forms a large triangular mass consisting of several posterolaterally-directed bands that originate on the ventral surface of the puboischiadic plate and insert on the femur. It adducts the hindlimb, pulling it toward the axis of the body.

Puboischiotibialis. This muscle is posterior to the puboischiofemoralis externus. It originates on the ventral surface of the puboischiadic plate, and its fibers insert on the tibia. A band of connective tissue holds the muscle near the proximal end of the femur. The muscle adducts the thigh and flexes the hindlimb.

Pubotibialis. This is the next posterior muscle in ventral view. It lies just anterior to the puboischiotibialis along the front of the thigh. It originates on the pubis and inserts on the tibia, and its action pulls the thigh forward and extends the lower part of the hindlimb.

Publischiofemoralis internus. This muscle extends from its origin on the dorsal surface of the puboischiadic plate, passes anterior to the pubotibialis along the preaxial border of the hindlimb, and inserts on the femur. It pulls the thigh anteriorly.

Ischioflexorius. This slender muscle, which lies along the posterior border of the thigh, originates on the ischium and inserts on fascia on the plantar side (the sole) of the lower hindlimb. It mainly flexes the shank.

Extensor tibialis. This more distal muscle passes along the anterior border of the lower hindlimb. It originates on the distal part of the femur and inserts on the tibia and tarsus to extend the lower limb and ***pes*** (foot).

Figure 19-8. Ventral view of the muscles of the pelvic girdle and hindlimb in the mudpuppy.

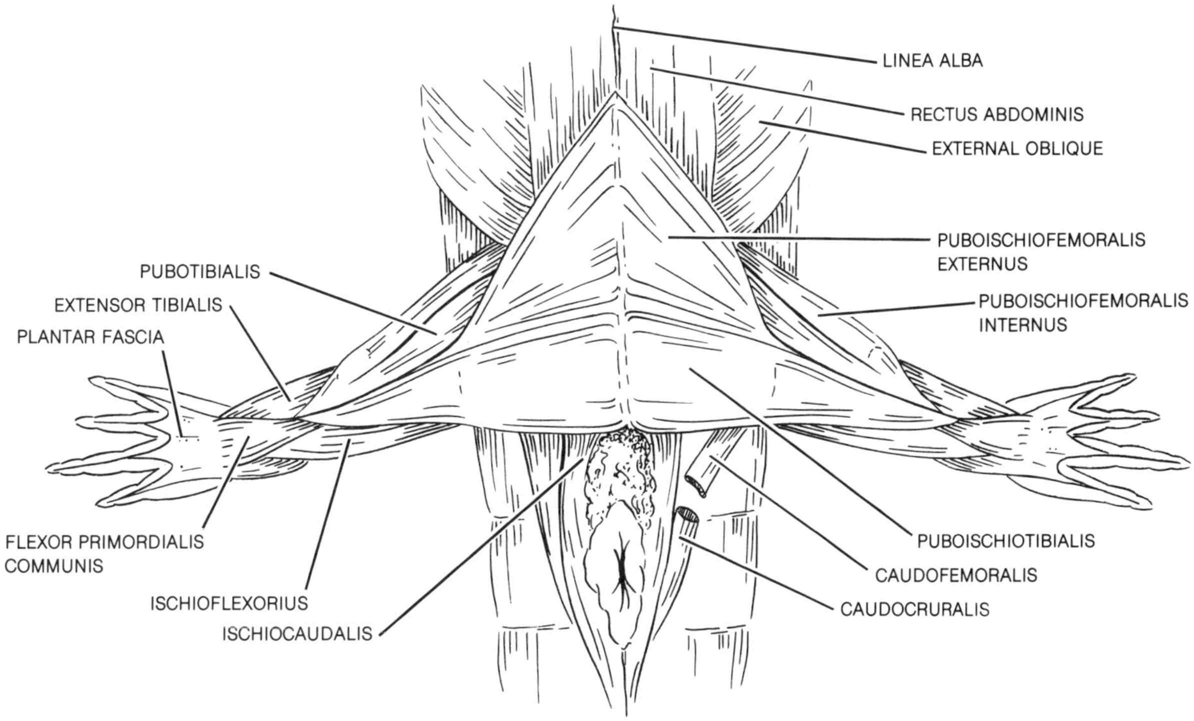

Flexor primordialis communis. This muscle occurs on the plantar surface of the lower hindlimb. It originates from the femur, passes between the puboischiotibialis and ischioflexorius, and inserts on the phalanges via tendons radiating from the ***plantar fascia***. It flexes the pes.

Ischiocaudalis. This is the next medial strap. It is an elongate muscle with fibers that run longitudinally from the insertion on the ischium, past the cloaca, to the origin on hemal spines of caudal vertebrae. The ischiocaudalis helps to flex the anterior part of the tail.

Caudocruralis. This strap-like muscle lies lateral to the ischiocaudalis; its parallel fibers originate on the hemal spines of the 4th and 5th caudal vertebrae and insert on fascia on the posterior surface of the puboischiotibialis. The caudocruralis pulls the thigh posteriorly and/or flexes the tail laterally.

Caudofemoralis. This muscle lies dorsal to the caudocruralis and is seen most easily by separating the caudocruralis from the hypaxial musculature and probing between the two. It originates on hemal arches and spines of a few anterior caudal vertebrae and inserts proximally on the femur. It pulls the thigh posteriorly and may flex the tail.

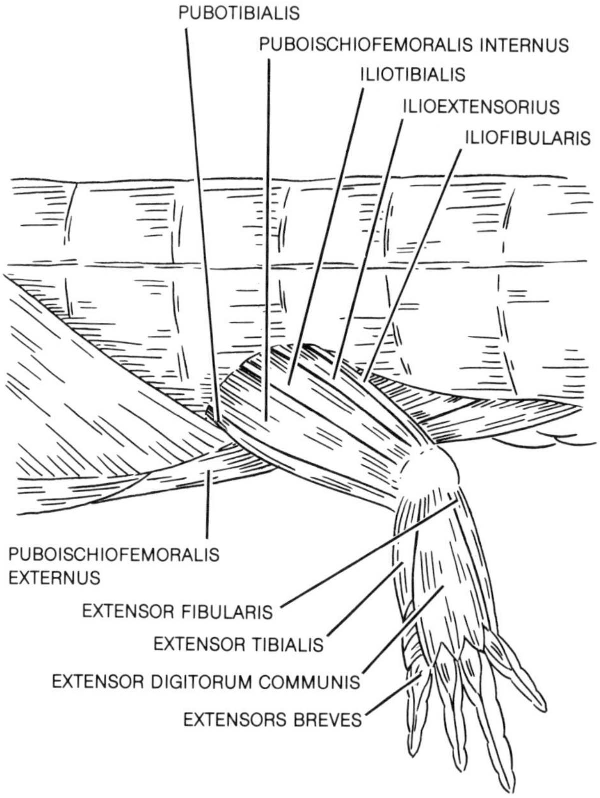

Figure 19-9. Lateral view of the muscles of the pelvic girdle and hindlimb in the mudpuppy.

Muscles of the Pelvic Girdle and Hindlimb; Dorsolateral Aspect (Fig. 19-9)

Posterior to the puboischiotibialis lies a shield-shaped glandular mass; dissect it away carefully to note the three strap muscles that fix the pelvic girdle.

Turn the animal on its side and identify the puboischiofemoralis internus.

Iliotibialis. This muscle lies just posterior to the puboischiofemoralis internus. It arises from the ilium and inserts proximally on the tibia. It extends the shank out of a flexed postion.

Ilioextensorius. This muscle lies parallel and posterior to the iliotibialis. It passes from its origin on the ilium to its insertion on the fascia of the knee joint. It mainly extends the shank.

Iliofibularis. Located posterior to the ilioextensorius, this slender muscle originates on the ilium and inserts on the proximal end of the fibula. It extends the shank and may assist in abducting the thigh away from the body's axis.

Extensor digitorum communis. This is the relatively large, medial muscle on the shank and pes. The origin is on the distal end of the femur, and it inserts proximally on the metatarsals. It extends the pes and digits.

Extensor fibularis. It is a small muscle on the posterodorsal part of the shank, posterior to the extensor digitorum communis. It extends from its origin on the femur to insert on the distal end of the fibula and tarsus. It extends the shank and foot.

Extensores breves. These small muscles lie on the dorsal surfaces of the metatarsals. They originate on the tarsus (ankle) and insert via tendons on the phalanges. They extend the digits.

CARL GANS and RONALD A. NUSSBAUM

20

The Mudpuppy: Visceral Organs

DIGESTIVE SYSTEM

As is the case with all salamanders, the mudpuppy is carnivorous and feeds on a wide variety of small aquatic animals. Worms, aquatic insects, and parts of crayfish are frequently found in the stomachs of mudpuppies; the contents discovered in stomachs vary with the seasons and with the length of time that the animals were held in captivity away from normal activity.

The morphology of the digestive system reflects the carnivorous habits of the mudpuppy. The mouth is terminal. Its sides have mechanically interlocking labial folds, which serve to seal the lips tightly when the mouth is closed and form a basket to trap food during the "gape" phase of "suck-and-gape" feeding. Unlike mammals, such as the cat, amphibians lack facial muscles along the upper jaw.

The numerous, needle-like teeth are grasping structures that can hold struggling prey. Such teeth do not cut or crush; rather, the food must be swallowed whole. Lack of cutting and crushing teeth limits the size of prey on which mudpuppies can feed. Both stomach and intestine of these carnivores are relatively short and simple, in contrast to the longer and more complicated digestive tracts of herbivorous animals.

DISSECTING INSTRUCTIONS

Make a longitudinal incision (about two and one-half centimeters, or one inch, long) through the muscular body wall, about two-thirds of the distance between the pectoral and pelvic girdles on the left-hand side of your specimen. (If your specimen was injected, there will already be an opening.) Place the tips of a pair of dull forceps through the opening and slide them, first anteriorly and then posteriorly, just deep to the body wall. Then spread the forceps and cut on top of them to extend the slit without slicing into the underlying viscera. Extend the incision transversely just posterior to the pectoral girdle and anterior to the pelvic one (use a mounted skeleton as a guide) and be careful not to destroy the ventral mesenteries. To see the heart you will have to cut through the pectoral girdle. To do so, carefully extend your cut anteriorly on the left side. In freeing the viscera, note the mesenteries, particularly the thin sheets extending from the ventral surface.

Attempt to separate the lips and pass a blunt probe through the esophagus, using the sagittally sectioned specimen to identify its position (see dissection instructions for muscles in Chapter 19). Next, move the visceral elements about carefully, shifting them from side to side for identification. Try not to rupture blood vessels and mesenteries. The major digestive and urogenital organs should now be exposed to view.

Finally, take one of the demonstration specimens and cut it laterally across the angle of the mouth on the right side so that the external nares, internal gill slits and similar structures may be probed. The cut should be extended to the level of the gular fold and then transversely toward the left before the mouth is bent back, or reflected. (Note the thick branchiohyoideus muscle in section.) It may be useful to use the more posterior portion of the same animal for a dorsal dissection by removing the body wall, including the vertebral column along the middle of the back. This will then expose the dorsal part of the coelomic cavity. More anteriorly, one can trace the arterial supply from the gills. This dissection has to be made very carefully as the animal then loses much of its stiffness. (What does this imply about the function of the axial column?) However, it does provide a very effective demonstration of the blood supply to lungs and gills.

DIGESTIVE AND VENTILATORY STRUCTURES

Buccal Cavity

The ***buccal cavity*** of the mouth is the anteriormost portion of the digestive tract (Fig. 20-1). It extends from the lips posteriorly to the ***pharynx***. The upper view contains the rows of premaxillary teeth, flanked medially by the ***vomerine-pterygoid tooth rows***. The two ***internal nares*** or ***choanae*** can be seen in the roof of the mouth just lateral to the middle of the vomerine tooth rows. The anterior portion of the ***tongue*** lies in the floor of the buccal cavity and has a free anterior edge. Note the rows of dentary teeth.

The pharynx, which continues the buccal cavity posteriorly, is roughly defined as the gill region. The lateral walls of the pharynx are defined by the gill arches and diagonal ***gill slits***. A tiny slit in the midline at the base of the tongue is the ***glottis***, or opening to the larynx, trachea, and bronchi that lead to the lungs. Carefully pass a blunt probe through each of these openings.

Posterior to the pharynx, the ***esophageal opening*** leads to the ***esophagus***, stomach, and lower digestive tract. Note that the esophagus lies dorsal to the air passage to the lungs.

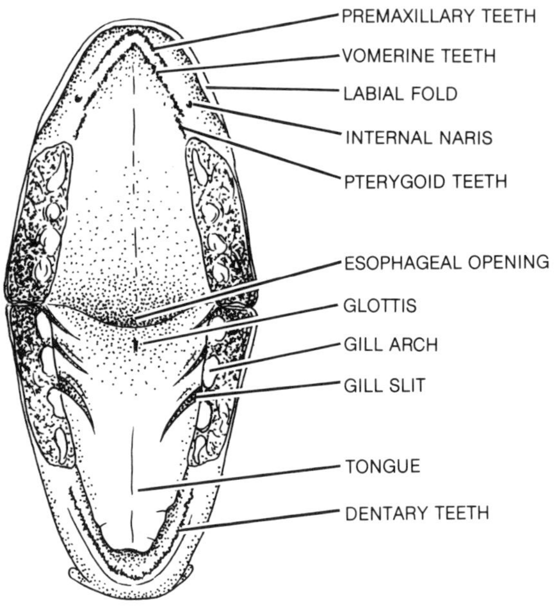

Figure 20-1. Anterior view of the buccal cavity and pharynx in a gaping mouth of a mudpuppy.

The Coelomic Cavity (Fig. 20-2)

This cavity contains the viscera. It consists of two divisions: (1) the ***pericardial cavity***, which contains the heart (discussed in Chapter 21), and (2) the ***pleuroperitoneal cavity***, which contains the lungs, alimentary tract and other organs. A continuous sheet of epithelium lines the visceral cavities and covers the organs within them. In the pericardial cavity this lining is called the ***parietal pericardium***, and in the pleuroperitoneal cavity the ***pleuroperitoneum.*** The two cavities are separated by a thin and transparent ***transverse septum*** that lies just at the level of the axilla (armpit). The covering of the internal organs, including the heart, is called the ***visceral peritoneum***.

Read over the following sections and identify major organs, such as the alimentary tract from esophagus to the cloaca, as well as the liver, lungs, pancreas and spleen; then attempt to identify the mesenteries, as these will rupture as you dissect the internal organs.

Mesenteries

The internal organs are supported by mesenteries; these are thin transparent sheets of connective tissue, which, in the mudpuppy, are evolutionarily primitive in that they pass from the dorsal wall to the viscera and hence to the ventral wall of the coelom. Upon opening the coelomic cavity, one first notes that the ***ventral mesentery*** is subdivided into the ***falciform ligament*** binding the liver to the body wall, the ***gastrohepatic ligament*** between the liver and stomach, the ***hepatoduodenal ligament*** between the liver and anterior small intestine, and the ***median ligament of the bladder*** running from the urinary bladder to the median, ventral body wall.

The other mesenteries generally attach to the middorsal wall of the coelom. The ***dorsal mesentery*** is subdivided into the ***mesogaster*** supporting the stomach, the ***mesentery*** supporting the coiled small intestine, and the ***mesorectum*** (or ***mesocolon***) supporting the large intestine. The spleen is bound to the stomach by a portion of the mesogaster named the ***gastrosplenic ligament***. ***Pulmonary ligaments*** attach the lungs to the mesogaster.

Alimentary Tract

The short esophagus leads into the ***stomach***, which normally lies on the left side of the pleuroperitoneal cavity and extends about two-thirds the length of the cavity. The stomach may be opened to examine food and to see the folds or ***rugae*** which line the stomach. (*Try to avoid cutting the arteries when opening the stomach*).

The stomach is separated from the ***small intestine*** by a muscular valve, the ***pyloric sphincter***, beyond which the internal architecture of the gut tube changes. The proximal portion of the small intestine, in contact with the pancreas (see below), is called the ***duodenum***. It curves transversely across the midline and leads into the coiled small intestine. If the small intestine is traced posteriorly, it is observed to enlarge and become thin-walled near the end of the body cavity. This short longitudinal segment, the straight ***large intestine***, is often filled with digestive waste. The large intestine empties into the ***cloaca***, which marks the end of the alimentary canal and is the common exit for sexual, excretory, and digestive products. *To see these structures you must pass a probe into the cloaca and cut anteriorly through the pelvic musculature. Try to make this incision to the left side of the midline in order to protect the ventral mesentery and the vessels traversing it.*

Figure 20-2. Ventral view of mesenteries and organs in the coelomic cavity of a mudpuppy.

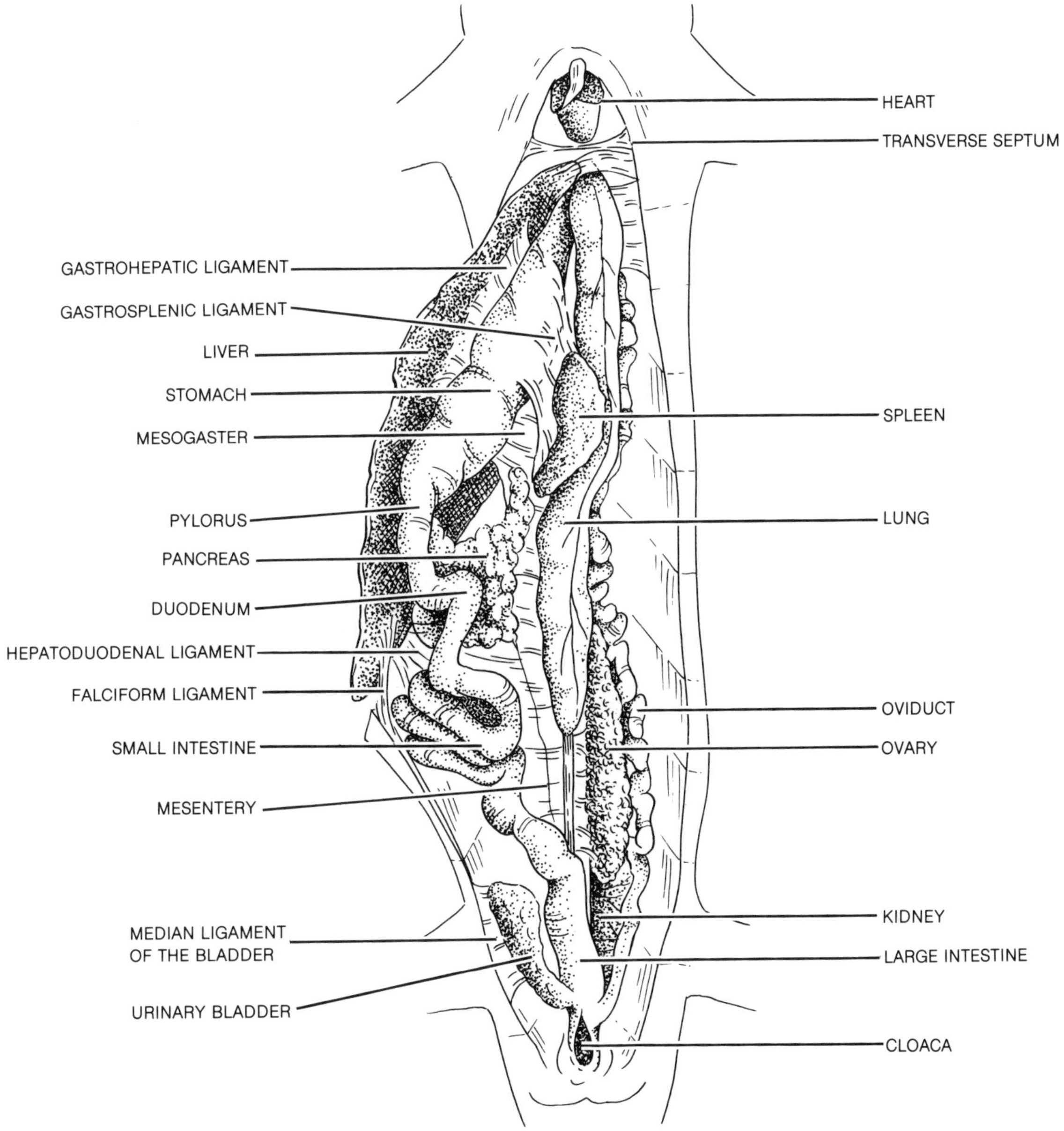

Other Organs of the Pleuroperitoneal Cavity

Digestive organs. The ***liver*** is the largest and most obvious organ in the body cavity. This solid organ has an irregular shape and generally lies parallel to the stomach. On the dorsal surface of the liver, near its terminus, lies the ***gall bladder***. The bladder stores bile, a substance secreted by the liver that digests and transports fat. The bile passes from the liver via several ***hepatic ducts*** that join the ***common bile duct*** and then enter the small intestine (duodenum). A ***cystic duct*** connects the gall bladder to the common bile duct. The various bile ducts are embedded in pancreatic tissue, which must be dissected away before the ducts can be fully exposed.

The ***pancreas*** is an irregularly-shaped organ found near the duodenum. Its ducts empty into the duodenum, but they are too small to see macroscopically

Figure 20-3. Ventral views of female and male urogenital systems in the mudpuppy.

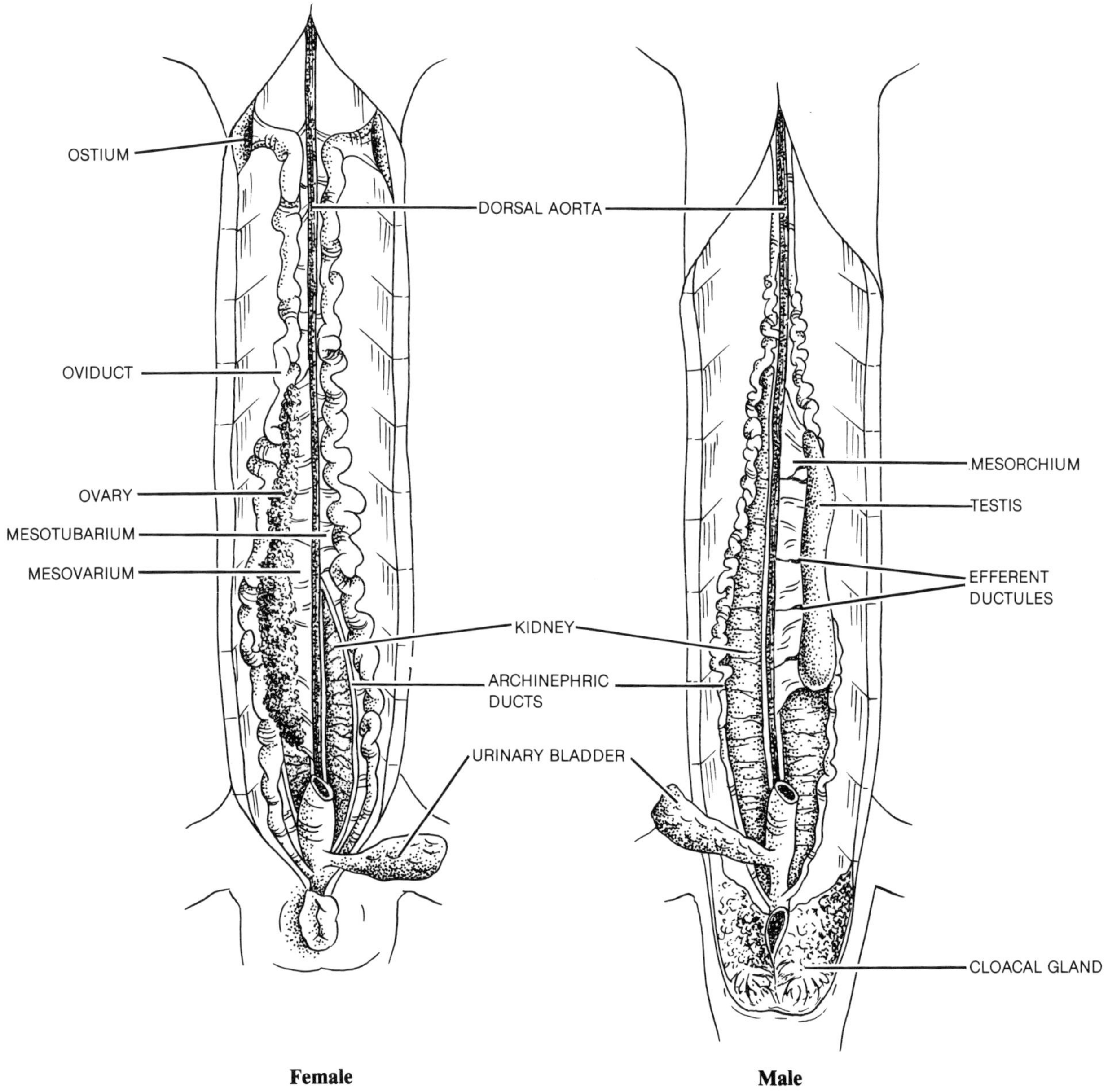

in most cases. The ***islets of Langerhans***, which are small patches of endocrine tissue, form part of the pancreas but are also too small to see.

The ***spleen*** is a reddish-brown, compact, elongate organ loosely attached to the dorsal surface of the stomach.

Ventilatory organs. The two ***lungs*** are thin-walled, simple sacs that lie in the anterior portion of the pleuroperitoneal cavity. They extend posteriorly for two-thirds the length of the coelom. The lungs can be filled with gas through anterior openings, the ***bronchi***, which lead to the short ***trachea***. The trachea opens into the ***larynx***, which opens dorsally to the pharynx by way of the glottis (mentioned above).

UROGENITAL SYSTEM (Fig. 20-3)

The urogenital system, though having two functionally separate parts (the two are anatomically connected only at the cloaca), is treated here in parallel with the digestive system; a single exposure will disclose both urogenital and digestive systems. As in most higher vertebrates, we find that excretory and genital functions have become anatomically interconnected. The genital products (eggs, sperm, and associated materials) are produced far anteriorly in the coelomic space, and their discharge proceeds via tubes that develop initially within or parallel to the primary nephric tissues.

After you have identified the sex of your specimen (see below), it will be necessary to find classmates with a specimen of the opposite sex to share observations. Most of the structures lie freely suspended in the visceral cavity; thus, no further dissection is required.

The Kidney and Urinary Ducts

The mudpuppy has ***opisthonephric kidneys*** that lie suspended from the posterodorsal wall of the pleuroperitoneal cavity, just adjacent to the mesentery on each side. They are elongate structures covered by visceral peritoneum. The kidneys of males and females differ somewhat in structure and function. The kidneys of both sexes are divided into a relatively narrow, anterior ***genital portion*** and a broader, posterior ***excretory portion***. The genital part of the female kidney is smaller than that of the male. In males this portion of the kidney transports spermatozoa to the archinephric duct.

The ***archinephric ducts*** extend down the lateral borders of the kidneys and drain them, carrying urine to the cloaca. As in the dogfish, these ducts are sexually dimorphic. Those of females are thin and straight, while those of males are wider and convoluted. The archinephric ducts are fed by small ***collecting tubules*** from the excretory parts of the kidneys. Urine passes from the cloaca into the thin-walled, ventrally-located ***urinary bladder***, where it is stored.

Male Reproductive Organs

The ***testis*** is an elongate, compact organ attached dorsally to the body wall by the ***mesorchium***, a mesentery that attaches to the base of the dorsal mesentery on each side. Spermatozoa leave the testis through small ***efferent ductules*** lying in the mesorchium and pass into the ducts of the genital part of the kidney. (The genital kidney is analogous with, and perhaps homologous with, the epididymis of the male mammal.) Spermatozoa are stored in the efferent ductules and genital kidney until needed, at which time they pass through the archinephric ducts to the cloaca where the ***spermatophore*** is formed. ***Cloacal glands*** secrete a gelatinous substance that encloses the spermatozoa (Fig. 20-4), and the entire structure is passed to the cloacal lips of the female. Normally, in other salamanders, this spermatophore is deposited on the substrate and is subsequently picked up by the female. But this process has not been observed in mudpuppies. Among vertebrates the indirect transfer of male gametes by means of spermatophores is unique to salamanders.

Figure 20-4. Spermatophore from a male mudpuppy (after Bishop); scale = 7 mm.

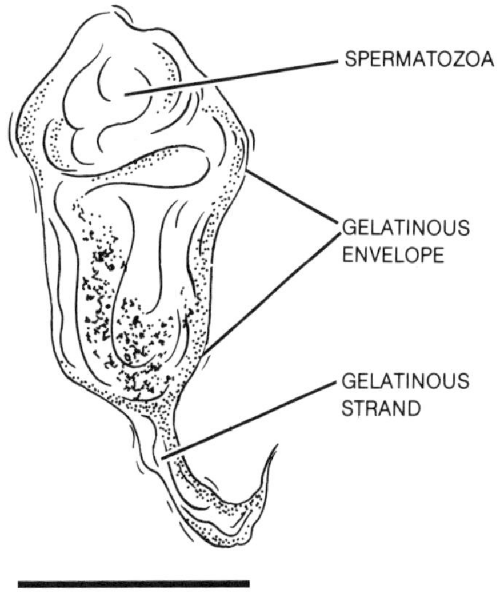

Male mudpuppies have a pair of ***vestigial oviducts*** along the outer edge of each archinephric duct. These thin, darkly-pigmented ducts extend anterior to the kidneys along the dorsal wall of the body parallel to the cardinal veins.

Female Reproductive Organs

The two ***ovaries*** vary in size depending on the age of the animal and the stage in the reproductive cycle when the animal was killed. If the ovary is filled with early oocytes, its surface will appear granular, but if mature or nearly mature eggs are present, these will be easily recognized. Each ovary is suspended by a mesentery, the ***mesovarium***, from the dorsal body wall.

The two ***oviducts*** are highly-convoluted, thick-walled tubes found lateral to the kidneys. The oviducts are suspended from the dorsal body wall by the ***mesotubarium***, a thin mesentery. Each oviduct has a large anterior opening, the ***ostium***, through which ovulated eggs enter the oviducts. The oviducts connect to the cloaca via a pair of ***genital papillae***.

CARL GANS and RONALD A. NUSSBAUM

21

The Mudpuppy: Respiratory and Circulatory Systems

Most amphibians spend different stages of their life in water and on land. Consequently, they obtain oxygen and give off carbon dioxide to water during one portion of their life cycle and to air during another portion. Because the change occurs at metamorphosis and influences both the larval and adult conditions, it has sometimes led to the mistaken impression that contemporary amphibians are somehow intermediate between fishes and reptiles.

Actually modern amphibians, such as *Necturus*, the mudpuppy, have a gas exchange pattern that is probably quite different from that of the ancestral forms that gave rise to reptiles. Furthermore, neotenous forms such as *Necturus* show various complications associated with their peculiar biology.

Our knowledge of the changes in the circulation between larval and metamorphosed amphibians is still most limited. Thus far, no one has traced blood flow to see how much of each pulse of blood leaving the heart passes through the gills and how much through the lungs. In mudpuppies, circulatory patterns suggest that little blood flows to the lungs, and more importantly, that this blood has already been oxygenated during passage through the gills. In this, mudpuppies differ substantially from post-metamorphic (non-neotenic) salamanders and from frogs in which the gill circulation is reduced or absent and in which blood passes through separate aortic arches, either to the lungs or to the body.

In the mudpuppy, as in all other animals, respiration takes place in the various tissues at the cellular level. Here oxygen is used to support metabolism and carbon dioxide is generated. Blood flowing through the arteries and veins of the circulatory system transports these respiratory gases to and from the tissues. Most gas exchange with the environment occurs in the gills and via the skin. (The air sacs referred to as lungs [Chapter 20] are mainly for buoyancy control.)

In certain lungfishes, metamorphosed salamanders, and frogs, the heart and system of arterial arches have become modified, so that blood returning from the lung spirals through the heart and is mainly directed to those aortic arches that lead to the body. The stream of blood draining the body performs a slightly displaced spiral, so that it mainly enters the pulmonary arteries and then the lungs. Thus, though the heart is three-chambered, oxygenated and unoxygenated blood mix only slightly in the ventricle and the efficiency of gas exchange is improved. In mudpuppies, all blood leaving the heart passes across the gills (where it is oxygenated), then proceeds to the tissues (where the oxygen is given up) and then enters the veins (which return it to the heart).

DISSECTION INSTRUCTIONS

The blood vessels in most animals used for dissection by students are injected with latex to make them visible. The color of the latex is added for convenience and does not automatically indicate whether a particular vessel observed is an artery or a vein, or whether the blood that passes through a given vessel is oxygenated or rich in carbon dioxide. Also, note that the latex fills the lumen (cavity) of a blood vessel. Whenever you see bare rubber, you know that the wall of the blood vessel has been destroyed. Many blood vessels pass into blocks of muscle and other tissues; their detailed dissection requires destruction of such tissues. Thus, follow the major vessels outward from the heart and trace them only as far as necessary to identify their location.

Begin by extending the opening on the left side of the pectoral region, just posterior to the girdle. Use a blunt scissors to pass just deep (on the ventral side) to the body wall and extend the incision anteriorly. You have previously noted that the heart lies in the pericardial cavity. Cut the walls of this sac midway along its height so that you can later reconstruct the shape of the cavity. Now identify the different parts of the heart. Then continue the dissection anteriorly along the large ventral aorta until you have traced the major blood vessels to the base of the three gills. Remember that these are arteries. Normally they are uncolored, as the latex is injected into the dorsal aorta and does not penetrate through the gill capillaries to the afferent side, where blood is carried to the gills.

Next dissect posteriorly, pulling away connective tissues until the major vessels, adjacent to the coelomic cavity, have become obvious. The ensuing dissection should expose the pectoral vessels leading from the pectoral region into the forelimb and those leading from the pelvic region into the hindlimb.

Always work out a flow diagram, asking yourself what branches are given off by an artery as it passes away from the heart or by the dorsal aorta as it passes away from the gills. Similarly, indicate which vessels are received by the major veins. Consider the association of arteries and veins so that you can identify the vein that receives the blood passed out through a particular artery. Note the capillary beds in the gills and the major portal systems.

Then, finally, open the heart to identify its internal chambers.

THE HEART (Fig. 21-1)

It is sometimes claimed that the three-chambered heart of the surviving amphibians is morphologically intermediate between the two-chambered heart of fishes and the four-chambered heart of the next higher tetrapod class, the Reptilia. However, all such claims of intermediacy should be viewed with suspicion, because the modern amphibians represent a side branch, far removed from the ancient amphibians that gave rise to reptiles. Indeed, some authorities argue that the three-chambered structure of modern amphibians is secondarily derived from a four-chambered heart, presumably present in ancient amphibians, by loss of the interventricular septum. Reduction of this septum may be related to the dependence of modern amphibians on cutaneous gas exchange, which requires a rerouting of oxygenated blood through the heart.

The heart of the mudpuppy consists of a single-chambered ***ventricle***, a partially divided ***atrium*** and a ***sinus venosus***. The ventricle is the large, solid, posteriorly-positioned ovoid structure that is most obvious and appears largest once the pericardial cavity has been opened from below. The ventricle receives blood from the two atria via the ***atrioventricular openings*** and pumps blood anteriorly into the ventral aorta through the more anteriorly positioned ***conus arteriosus*** and ***bulbus arteriosus***. Note that the ventricular exit lies on the right side of the ventricle.

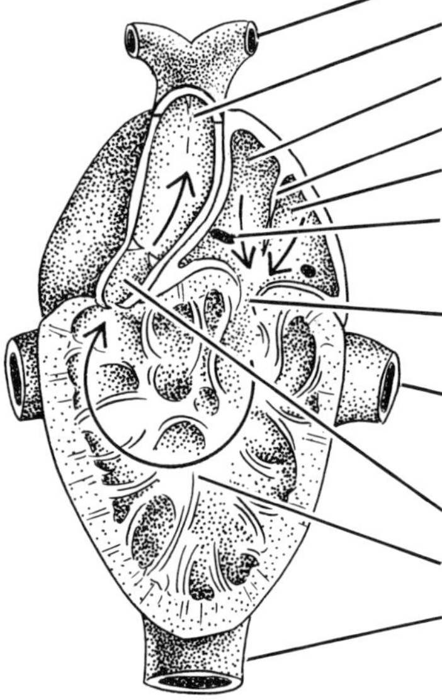

Figure 21-1. Ventral view of a sectioned mudpuppy heart; arrows indicate direction of blood flow.

The ***right*** and ***left atria*** lie anterior to the ventricle and appear, in ventral view, as dark, flabby, thin-walled sacs on either side of the conus and bulbus arteriosus. Their common connection may be seen once the heart is moved about in the pericardial cavity. The right atrium receives blood from various parts of the body via the ***sinus venosus***. The left atrium receives blood from the lungs through the small common pulmonary vein that passes dorsal to the ventricle. The right and left atria are divided internally by an incomplete ***interatrial septum***, which allows some mixing of oxygenated and unoxygenated blood even before the blood is discharged to the ventricle through the common atrioventricular opening.

Open the ventricle by a longitudinal incision, flush out the clotted blood and identify the internal connections and chambers. Also note the valves between the chambers and the muscular trabeculae that line the inside of the ventricle.

The sinus venosus is a short thin sac crossing the dorsal side of the heart. It receives blood laterally from the paired common cardinal veins and posteriorly from the paired hepatic sinuses. Blood from the sinus venosus enters the right atrium through the

sinatrial opening. The sinus venosus initiates each wave of contraction throughout the whole heart; accordingly, it is called the pacemaker.

VENOUS SYSTEM (Fig. 21-2)

The venous system is conveniently studied in five parts: (1) pulmonary veins, (2) ventral abdominal vein and its branches, (3) hepatic portal system (4) renal portal system, and (5) systemic veins. You may not be able to discover the smaller veins unless the venous and hepatic portal system have also been injected on your specimen.

Pulmonary Veins

There are two ***pulmonary veins***, one lying on the ventral surface of each lung. The right and left

Figure 21-2. Ventral view of major veins in the mudpuppy.

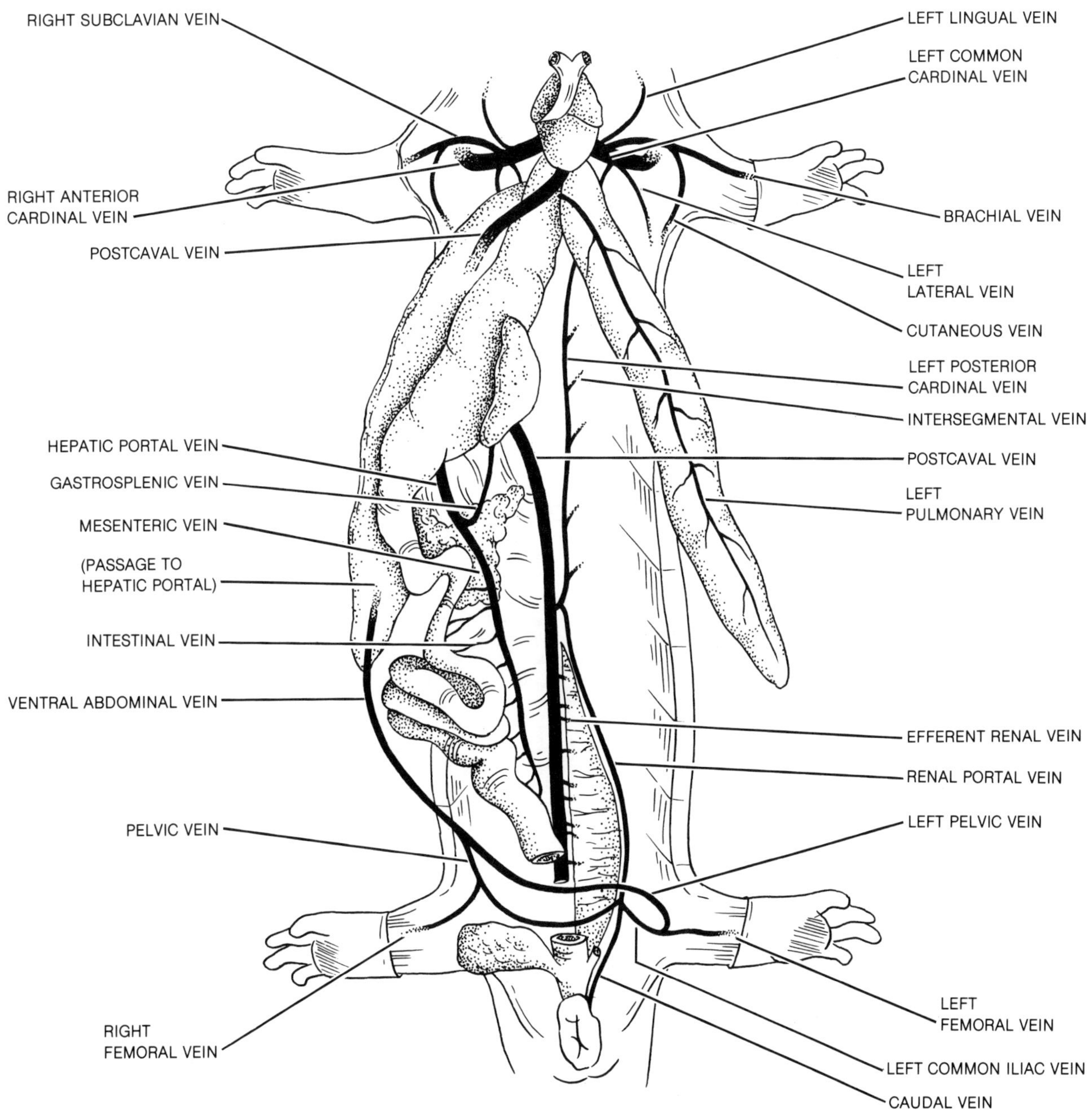

pulmonary veins join to form the ***common pulmonary vein***, which runs a short distance through the transverse septum to empty into the left atrium.

Ventral Abdominal Vein

The ***ventral abdominal vein*** extends along the ventral midline of the coelom, passing from the pelvic region to enter the liver and there joining the hepatic portal vein. Entry into the liver is through the falciform ligament from the posterior part of the coelom, where the ventral abdominal vein has received ***vesical veins*** from the urinary bladder. Still more posteriorly, the ventral abdominal vein branches into right and left ***pelvic veins***. Each pelvic vein bifurcates into a ***femoral vein*** that drains the hindlimb and a ***common iliac vein***, which is connected to the renal portal vein. Given these connections, blood from the hindlimb can return to the heart along two different routes. Either it passes dorsally through the common iliac veins to join the renal portal system, or it passes ventrally through the pelvic and ventral abdominal veins to join the hepatic portal system.

Hepatic Portal System

Besides the contribution from the ventral abdominal vein, which joins it in the liver, the ***hepatic portal system*** gathers blood from the digestive tract and associated structures and passes this blood into the capillary bed of the liver. The ***hepatic portal vein*** is formed by the union of the ***mesenteric*** and ***gastrosplenic veins***. The mesenteric vein is the major vein observed in the mesentery of the intestine; numerous ***intestinal veins*** can be seen radiating inward from the intestine to join the mesenteric vein. Note that the mesentery not only suspends the gut, but also serves as a tissue sheet in which these vessels run their course. Before joining the mesenteric vein near the pancreas, the gastrosplenic vein drains the stomach via the ***gastric vein*** and the spleen via the ***splenic vein*** (not illustrated).

Renal Portal System

The ***renal portal system*** gathers blood from the hindlimbs, tail, and posterior body region and carries it to the kidneys, where it is filtered. There are two ***renal portal veins***, one lying along the lateral border of each kidney. The renal portal veins receive blood from the single ***caudal vein*** (systemic vein) and from the two common iliac veins (described above). Blood passes from the renal portal veins into the kidney through numerous small ***afferent renal veins*** (these are only apparent when specially injected).

Systemic Veins

A series of veins returns blood to the heart from the hepatic and renal portal systems and from other parts of the body, exclusive of the lungs. The large ***common cardinal veins*** lie on each side of the heart and run directly into the sinus venosus. The common cardinal veins are joined by the ***posterior cardinal veins***, which form an anterior continuation of the renal portal veins from the kidney. Each lies parallel to and on either side of the dorsal aorta (see below). The body wall is drained by ***intersegmental*** or ***parietal veins***, and these can be seen joining the posterior cardinal veins at regular intervals.

Small ***efferent renal veins*** pass along the ventromedial surfaces of the kidneys. These veins drain filtered blood from the kidneys and pass it into the large ***posterior vena cava***, or ***postcaval vein***, which lies between the kidneys. The posterior vena cava also receives small ***gonadal veins*** from the ovaries (***ovarian veins***) or testes (***testicular veins***). The posterior vena cava passes anteriorly; here it may anastomose (fuse) with the posterior cardinal veins near the anterior ends of the kidneys. It continues anteriorly into the liver, where it receives ***hepatic veins***

Figure 21-3. Lateral view of arteries anterior to the heart on the left side of a mudpuppy.

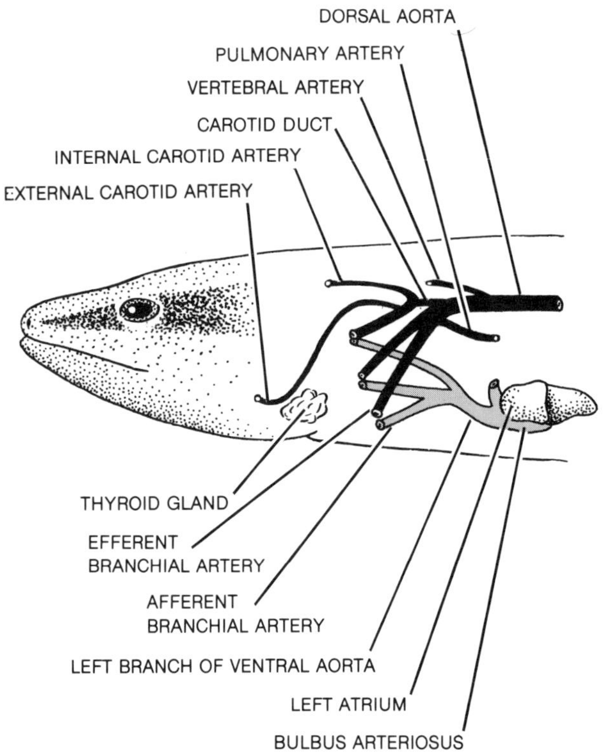

that have collected blood from the liver's capillary bed. The posterior vena cava passes anteriorly from the liver and through the transverse septum to divide into two ***hepatic sinuses*** that join the sinus venosus.

The anterior parts of the body are drained by veins that converge into the common cardinal veins, which were mentioned above. The tributaries of the common cardinal veins are the posterior cardinal veins (already described), the ***subclavian veins***, the ***common jugular*** or ***anterior cardinal veins***, the ***lingual veins***, and the ***lateral veins***. The large ***subclavian veins*** observed in the shoulder regions are formed by the union of the ***brachial veins*** from the forearms and the ***cutaneous veins*** from the skin. The common jugular veins, which drain the head, receive the larger ***external jugular veins*** from the gills and the smaller ***internal jugular veins*** from the roof of the mouth. The relatively small lingual veins drain the hypobranchial muscles. They pass posteriorly to join the common cardinal veins near the point where the subclavian veins enter the common cardinal veins. The lateral veins join the common cardinal veins from the posterior direction, lateral to the juncture of the posterior cardinal veins with the common cardinals. The lateral veins lie in the horizontal septum between the epaxial and hypaxial musculature.

ARTERIAL SYSTEM (Figs. 21-3, 21-4)

All arterial blood leaves the heart (ventricle) by way of the short conus arteriosus, seen lying between the atria in ventral view. The conus arteriosus enlarges to become the bulbus arteriosus, and the latter structure immediately bifurcates into the ***left*** and ***right branches*** of the ***ventral aorta***. This bifurcation occurs only a few millimeters anterior to the heart.

The later developmental stages of modern amphibians lack aortic arches I and II. In the mudpuppy, as in all salamanders, arches III through VI persist, but arch VI is modified as a pulmocutaneous arch.

Aortic Arches and their Major Arteries (Fig. 21-3)

Each branch of the ventral aorta provides three ***afferent branchial arteries*** that enter the three branchial arches and pass into the respective external gills where the blood is oxygenated. The anteriormost of these is the ***first***, the middle one is the ***second***, and the posteriormost is the ***third afferent branchial artery***. Trace these three branches into the gills by following along one of the branches of the ventral aorta, gently removing tissue as necessary.

Oxygenated blood from the capillaries of the external gills passes from the gills through the ***first***, ***second***, and ***third efferent branchial arteries***. The first efferent branchial artery gives rise to the ***external carotid artery***. It reverses direction, first passing ventrally parallel to the first efferent artery and then running anteriorly near the position of one of the ***thyroid glands***. The external carotid artery supplies numerous small arteries that serve the muscles of the pharyngeal region.

The efferent branchial arteries themselves and associated arteries are best observed by removing the mucous membrane from the roof of the mouth and pharynx. Trace the three efferent branchial arteries on one side to the point where they converge to form one of the two ***radices*** (roots), of the ***dorsal aorta***. The radices join in midline, dorsal to the heart, to form the dorsal aorta. Each radix gives rise to a ***vertebral artery*** that first extends anteriorly and then dorsally to penetrate the musculature at the base of the skull. Lateral to the vertebral artery, another small artery, the ***carotid duct***, arises from the radix. The carotid duct is very short. It immediately branches into the ***internal carotid artery***, which passes anteriorly into the base of the skull, and the first efferent branchial artery, which runs laterally into the first branchial arch. Further laterally along the radix, appears a posterior branch, the ***pulmonary artery***. (Remember that this artery carries blood already oxygenated in the gills *to* the lungs.) It should be traced to its respective lung, where it will be found lying on the opposite side of the lung relative to the pulmonary vein (Fig. 21-4). Lateral to the point of departure of the pulmonary artery, the radix bifurcates to become the second and third efferent branchial arteries. (*Note that much of this dissection proceeds in a direction opposed to that in which the blood flows. Also, remember the demonstration dissection recommended in Chapter 20. Removal of the dorsal body wall will clearly demonstrate the efferent arches.*)

Branches of the Dorsal Aorta (Fig. 21-4)

Return to the origin of the dorsal aorta, at the point of union of the two radices. Trace the dorsal aorta and its branches posteriorly, cutting away tissue along the right side of the animal until you have freed a path into the coelomic cavity. Just posterior to the juncture of the radices, two ***subclavian arteries*** arise and extend laterally. Near the base of the forelimb, each subclavian artery divides into a ***brachial artery***, which divides further to supply the forelimb, and a ***cutaneous artery*** that passes into the skin and adjoining tissues.

Posteriorly, the next major branch of the dorsal aorta is the single ***gastrosplenic artery*** which extends dorsolaterally and forms smaller branches to the

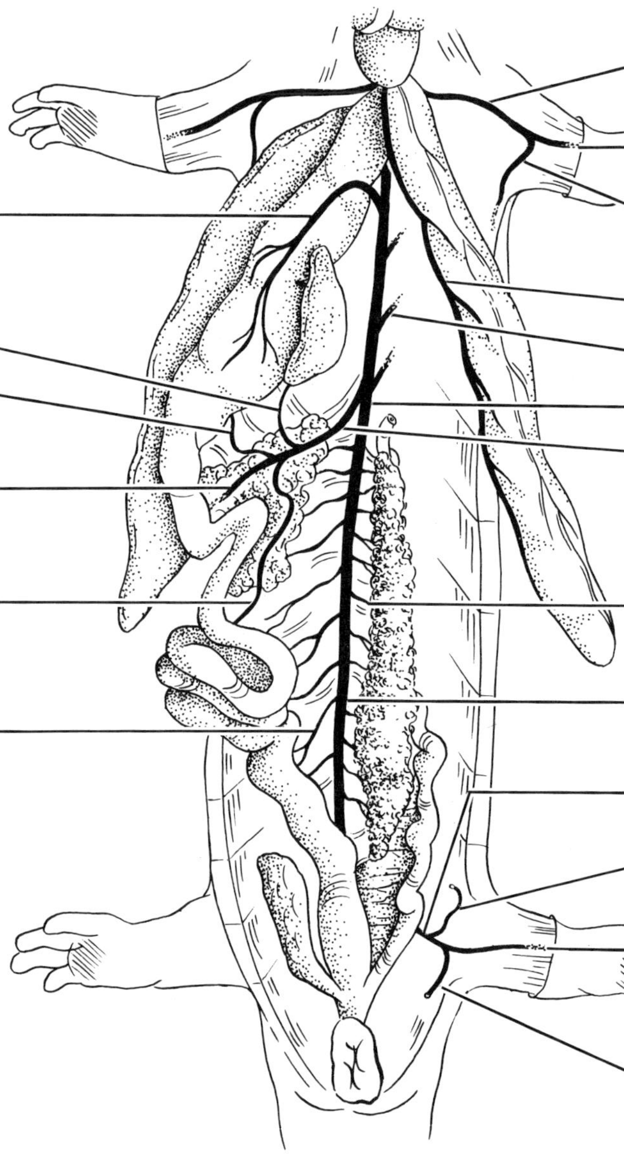

Figure 21-4. Ventral view of major arteries posterior to the heart in the mudpuppy.

stomach and spleen. At regular intervals along the length of the dorsal aorta, paired ***intersegmental (parietal) arteries*** arise from the dorsal aspect of the aorta to supply the body wall. Continuing posteriorly, the next major branch, the ***celiacomesenteric artery***, is also unpaired. It first gives rise to a ***splenic artery*** to the spleen and then a ***hepatic artery*** to the liver. Small ***mesenteric branches*** arise from the celiacomesenteric artery along its length. A major part of the celiacomesenteric artery continues as a ***pancreaticoduodenal artery*** to the pancreas and duodenum.

More posteriorly, the dorsal aorta gives rise to numerous ***mesenteric arteries*** that lie within the mesentery and pass to the intestine. Paired ***genital (ovarian*** to the ovaries, ***spermatic*** to the testis) ***arteries*** arise from the dorsal aorta and pass into the gonads; and small, short ***renal arteries*** pass from the dorsal aorta to the kidneys. Dorsal to the kidneys, the dorsal aorta gives rise to two large ***iliac arteries***. Close to its origin, each iliac artery has a branch, an ***epigastric artery***, which passes to the body wall musculature. Proceeding away from the dorsal aorta, each iliac artery branches into (1) a ***hypogastric artery*** to the cloaca and the urinary bladder and (2) a ***femoral artery*** to the hindlimb.

Caudally, the dorsal aorta gives off two ***cloacal arteries*** traceable to the cloaca. The aorta then passes posteriorly into the hemal canal of the caudal vertebrae, where it is named the ***caudal artery*** (not shown).

After tracing and reviewing these distributions of the dorsal aorta, consider your entire flow diagram for the mudpuppy. Try to visualize connections of the arterial system with the various veins studied earlier. Form a mental picture of the routes of blood flow from the heart (via arteries) and to the heart (through veins) so that a clear conception of the entire circulatory pattern emerges.

CARL GANS and RONALD A. NUSSBAUM

22

The Mudpuppy: Sensory and Nervous Systems

A few minimal comments on the sensory systems and brain of the mudpuppy are in order. Superficially, these appear to be highly specialized and limited structures that reveal little about the brain of amphibians in general, and even less about the transition between bony fishes and reptiles. As the mudpuppy is a neotenic animal, its brain, in many ways, reflects a semi-embryonic state, but so do the brains of most salamanders. The ratio of brain tissue to body mass is extremely small.

The organization of the brain of the mudpuppy is extremely simple, certainly simpler than that of the dogfish shark. Part of the simplicity reflects the rather limited motor capacity of the mudpuppy and the nature of its sensory systems. The most important senses are the anterior and posterior lateral line systems and the well-developed olfactory system, as well as the general sense of touch. The mudpuppy has very limited hearing ability and extremely reduced vision. (The number of retinal ganglion cells in each eye is between two and three thousand, perhaps ten percent of that in an eye of a dogfish.) The integumentary senses of mudpuppies, which detect vibrational changes, quite clearly play a very significant role in the recognition of other animals and objects in the surrounding environment.

SENSE ORGANS: DISSECTION AND IDENTIFICATION

Lateral line system. Lateral lines have already been traced during the dissection of the skin. In most salamanders, these organs disappear at metamorphosis; even in the paedomorphic mudpuppy, they are obscure. The sensory nerves (which are not assigned numbers) enter the central nervous system in association with the trigeminal and facial nerves and are located just anterior to, or as part of, the vagus.

Olfactory structures. The structures concerned with the sense of smell in amphibians differ from those of sharks and most fishes in being located in an ***olfactory sac*** or ***diverticulum*** that extends off of a continuous passage that runs from the external nostril to the internal naris. Place a probe through the passage and cut away part of the wall to expose the olfactory lamellae. Sensory cells situated along these surfaces provide the animal with information about compounds dissolved in the water in which it lives. Inspiration and expiration of water proceeds though these nasal passages. Mudpuppies apparently fill their lungs, as do lungfish, by gulping air into the buccal cavity and then pumping it into the lungs. The olfactory sac is normally water-filled and the nostrils are used only for irrigating the olfactory region.

Eye. The eye of the mudpuppy is very similar to that of the dogfish, though much smaller, both relatively and absolutely. Its general position has been described above (see Chapters 18 and 19). Note the optic nerve.

Ear. Salamanders lack an external ear opening, although a tympanic membrane is a common and very important feature in almost all frogs. Then again, salamanders produce only a few low sounds of uncertain origin and role, while sound production in frogs is extremely important and has complex roles in terms of mating, social spacing, and perhaps even defense.

The inner ear has already been described in the discussion of the skeleton (Chapter 19). The ***rod*** of the columellar bone reaches inward from the squamosal and the columellar footplate fills the fenestra ovalis. While there is no external auditory opening, the columella presumably serves as a transmitter of vibrations. Superficially, the basic arrangement of the ear is similar to that of fishes, although it has been suggested that the sensory organs of the ears of fishes, of amphibians, and of mammals may have arisen independently, in which case anatomical details would not be homologous. This possibility does not apply to the semicircular canals, which mirror those seen in the dogfish and most other fishes almost exactly.

Note the large ***otolith*** in the ***sacculus chamber*** of each inner ear. Such calcified "ear stones" help the mudpuppy sense the direction of gravity and detect linear acceleration.

THE BRAIN (Fig. 22-1)

The brain may be prepared either by standard dissection or by first soaking the specimen in 10% nitric acid for two to three days and then washing it well before dissection. This technique decalcifies the skull and facilitates dissection.

Remove the superficial tissues over the roof of the skull and then begin to chip away pieces of the cranial roof with the tip of a sharp scalpel. Twist the knife after each insertion and break off small pieces of cartilage and bone.

Once the cranial cavity has been penetrated you will see that the brain of the mudpuppy is surrounded by somewhat more obvious membranes than are found in the dogfish. These are the ***meninges***, consisting of an external ***dura mater*** and a much thinner, internal ***pia mater*** that adheres to the brain. They define the space occupied by the cerebrospinal fluid. The nerves leave the cranial capsule through foramina. Use the diagram (Fig. 22-1) to locate the major nerves and dissect carefully in the region where these structures are expected so that you do not break nerves when you remove the bones they perforate.

The brain of the mudpuppy can be divided into three major subdivisions: (1) ***forebrain*** (prosencephalon), (2) ***midbrain*** (mesencephalon), and (3) ***hindbrain*** (rhombencephalon). The first of these is further subdivided into the ***telencephalon*** and ***diencephalon***; the hindbrain is composed of the ***cerebellum*** (metencephalon) and ***medulla*** (myelencephalon).

The telencephalon comprises the large anterior ***cerebral hemispheres*** and the small ***paraphysis*** that rises dorsally between their posterior extent to form the ***pineal apparatus***. A faint ***lateral sulcus*** partitions off the anterior portion of the telencephalon; these are the ***olfactory bulbs*** or ***lobes*** from which the olfactory nerve passes to the wall of the olfactory sac. The cavities within the hemispheres are the ***first*** and

Figure 22-1. Dorsal and ventral views of the mudpuppy brain.

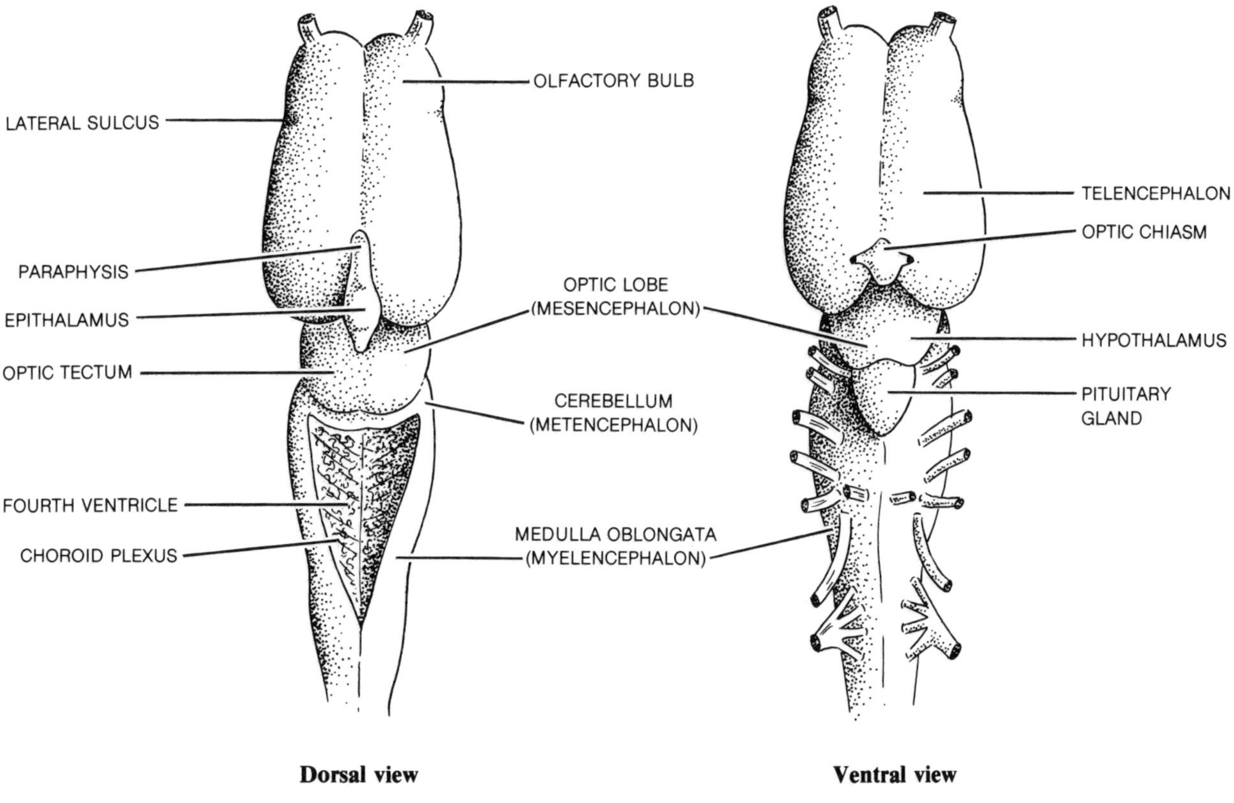

second ventricles. The telencephalon of amphibians is more poorly developed (its walls are thinner and less complex) than in other tetrapods, but its cell groupings (visible only in sectioned and stained specimens) are homologous to those in other vertebrates.

Immediately posterior to the cerebral hemispheres is the medial exposure of the roof of the ***third ventricle***, which is the cavity of the diencephalon or ***thalamus***. This roof is also known as the epithalamus, whereas the floor is the ***hypothalamus***; the ***pituitary gland*** descends from the hypothalamus. More anteriorly, the thalamus also contains the ***optic chiasma***, the site at which the two optic nerves enter the base of the brain.

The mesencephalon comprises the two relatively small, laterally-placed ***optic lobes.***

The narrow cerebellum is a structure that is much larger in most other vertebrates. The cerebellum is involved in motor control and assessment of information—for example body stance and locomotion. Ventral and posterior to the cerebellum lies the ***medulla oblongata*** or ***brain stem***, which, in some ways, represents an enlarged anterior prolongation of the spinal cord. The dorsal aspect of the medulla is open so that one can look directly into the fourth ventricle. Note that the roof is covered by a sheet, the ***choroid plexus***; it provides a site of vascular exchange with the cerebrospinal fluid filling the ventricles and the cerebrospinal canal.

Compare the general proportions of the brain of the mudpuppy with those of the brain of the dogfish. Which sections are relatively larger and which smaller? Bear in mind that the amount of tissue in some ways reflects the number of nerve cells and roughly the relative complexity of functions carried out by the various regions of the brain. Assume that vision and olfaction are processed primarily in the optic lobes and the cerebral hemispheres, respectively. What might this indicate about their relative importance in these two species?

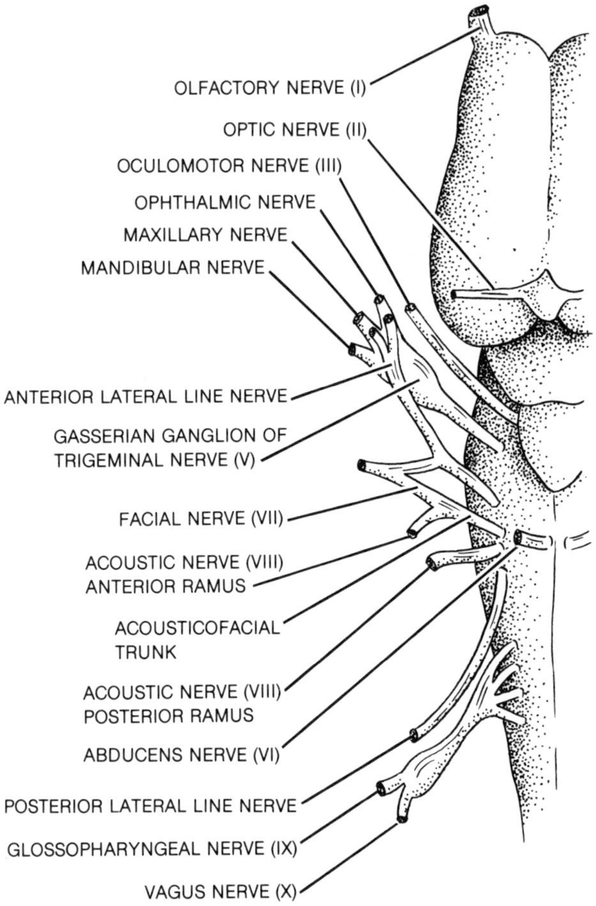

Figure 22-2. Ventral view of the right side of a mudpuppy brain, showing the roots of the cranial nerves.

THE CRANIAL NERVES (Fig. 22-2)

The ***olfactory nerve*** (I) has already been discussed. It carries information about chemicals dissolved in the surrounding water. The optic nerve (II) was demonstrated when the base of the brain (optic chiasma) was dissected. It transmits visual information. The nerves (III, IV and VI) that control the eye muscles are so small in the mudpuppy that a dissecting microscope and surface stains are needed to recognize them. They are much more easily studied in the dogfish.

The ***trigeminal nerve*** (V) leaves the side of the medulla and carries somatic sensory signals (pain, temperature and touch) from the anterior parts of the head. Also, its mandibular branch innervates the muscles of the jaws. Note its large ***Gasserian ganglion***, which contains the cell bodies of the sensory neurons but not of the visceral motor neurons. Identify the ***ophthalmic***, ***maxillary***, and ***mandibular branches***; their positions are very similar to those of the same nerves in the dogfish.

The ***anterior lateral line nerve*** (no number assigned) can be seen to leave the brain between the trigeminal and facial nerves and then to branch. One branch passes immediately anterior and then joins the path of the trigeminal, while the second continues laterally to join the facial nerve.

The ***facial nerve*** (VII) is complicated because it exits very closely posterior to the anterior lateral line nerve and is associated with the exit of the anterior ramus of the ***acoustic nerve*** (VIII)—forming the

acousticofacial trunk. The facial nerve carries taste information, as well as visceral motor and sensory information, mainly from the second branchiomere. These are the muscles of the first segment back of the jaws, as well as the skin and buccal lining of this region. Also, the ***superficial ophthalmic branch*** of the anterior lateral line nerve travels with the ophthalmic nerve branch of the trigeminal, and the ***buccal branch*** of the anterior lateral line nerve travels with the maxillary branch of the trigeminal. In neither case are the branches discrete; both are actually part of the anterior lateral line nerves, innervate the anterior lateral line organs, and represent independent sensory nerves. Both anterior and posterior rami of the acoustic nerve (VIII) transmit information from the ear and semicircular canals.

The ***posterior lateral line nerve*** leaves the medulla in a posterior direction, just anterior to the glossopharyngeal and vagus nerves. It passes within the dermis, just deep to the lateral line and transmits information from the mechanoreceptors.

The ***glossopharyngeal*** (IX) and ***vagal*** (X) ***nerves*** again leave the medulla from a closely adjacent position, just posterior to the acousticofacial trunk. The glossopharyngeal nerve innervates the third branchiomere, while the vagus includes the more caudal regions of the head. The vagus also contains a visceral component, later referred to as the parasympathetic trunk of the autonomic nervous system; it thus transmits information to the heart and gut derivatives.

THE SPINAL CORD AND ITS NERVES

The ***spinal cord*** may be seen by carefully opening the roof of the vertebral canal. At each intervertebral zone one sees the ***dorsal*** and the ***ventral ramus*** (root) of a ***spinal nerve***, and these join into a ***spinal ganglion*** on each side. The two anteriormost pairs of spinal nerves are extremely small.

The spinal nerves can also be approached from the periphery by tracing their ventral rami inward from their position between the oblique and transverse muscle at some level of the body. At this location one or more of these nerves may be freed along its path through the myoseptum and traced until it reaches the dorsal ganglion near the level of the vertebra. This nerve should join the dorsal root ganglion (and the rib).

The ***autonomic system*** of salamanders contains the ***sympathetic*** and ***parasympathetic systems***. Two ***sympathetic trunks*** can easily be dissected by noting that they pass alongside the dorsal aorta. As noted above for the vagus, these nerves carry visceromotor information to the various internal organs.

The Cat

23

Dissection of the Cat: General Instructions

SKINNING THE CAT (Fig. 23-1)

Using blunt forceps, grasp the loose skin along the midventral line of the chest region. Lift the skin and make a small superficial incision through it. Insert the blunt half of the large scissors into the opening. Being careful not to damage underlying muscles, cut cranially to the base of the neck and caudally to a point five cm from the genital area. Then cut laterally from the cranial limit of the incision on each side of the midline toward the forelimbs, and similarly from the caudal extreme of the midline toward the hind limbs. Extend these incisions along the medial surface of both sets of limbs to the wrists and ankles. Next cut away the skin around the legs above the paws. Using bone shears, remove the tail near its base.

To remove the skin, lift one open edge of the midventral incision. Separate the skin from underlying fatty tissue, using fingers to tease the layers apart. Large deposits of fat may be cut away with the scalpel later. Continue freeing the skin dorsally to the middorsal line. Repeat on the opposite side of the body and on each limb. (The skin of the head will be removed later, just prior to investigation of head musculature.) Remove the intact pelt, fold it, and set it aside. ***Do not discard the skin!***

Students with male cats should be careful to avoid damaging the spermatic cord and associated blood vessels. These course beneath the skin, superficial to the caudal abdominal musculature and extend caudally to the genital region. (See Fig. 24-29.)

STORING THE CAT

Cover all exposed surfaces with wet paper towels or a wet cloth. Wrap the cat's skin you previously removed around the towelling. Secure skin around body and limbs with rubber bands. Dress the wrapped body in a miniature jersey of your favorite N.F.L. (or any appropriate league) football team for identification. Place cat in a large plastic bag with a small amount of water. Seal tightly with string or wire.

COMMENTS

1. Adult female cats usually possess large pairs of lobular mammary glands, which extend over the ventral part of the abdomen and thorax. Examine these structures, then remove and discard them.

2. To identify members of superficial muscle groups, first remove the fatty connective tissue covering muscles. Individual muscles must be separated along their connective tissue boundaries. After scraping off external fat, locate the cleavage lines and carefully tease apart the neighboring muscles with the back of your scalpel or blunt probe. The direction in which component muscle fibers run is frequently a help in separating individual muscles.

3. When transecting a muscle, isolate it from ***all*** surrounding tissue. Cut the muscle perpendicular to its long axis, cleanly cutting through the thick belly of the muscle. To facilitate later identification, preserve the integrity of both ***origin*** and ***insertion*** of the muscle.

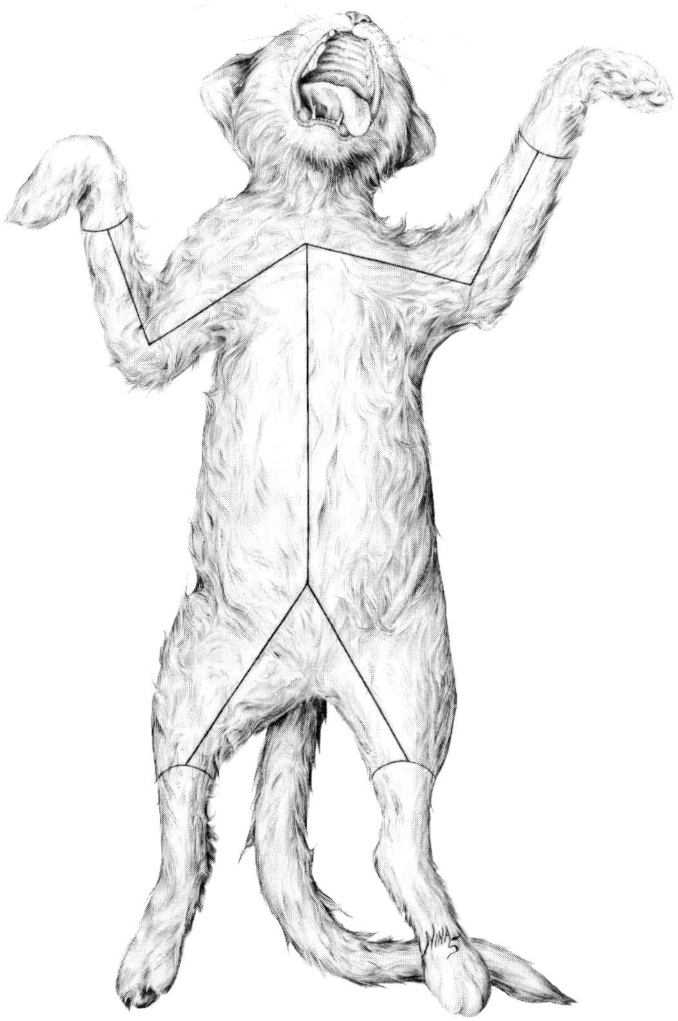

Figure 23-1. Ventral view of a cat, showing the incisions needed for proper skinning.

DAVID J. ANDERSON

24

The Cat: Musculoskeletal System

The move to land and the advent of tetrapody involved fundamental changes in the skeletal and muscular organization of the vertebrate body. Life on land was possible only for organisms that had developed new means of support for the body and new means of locomotion and respiration. The new niches available on land permitted different types of body activity to evolve. In this chapter, we will discuss some basic aspects of the evolutionary changes in the appendicular skeleton (limbs and limb girdles), the vertebral column, the skull, and the muscular system.

Without the support of surrounding water, a terrestrial vertebrate could either move on its belly or lift itself off the ground with its limbs. The limbs keep the body off the ground and move the body over the ground. The early tetrapod limb, like some fish fins, was a strong, jointed structure. In ancient amphibians, the limbs were oriented laterally, as in a salamander; they served to prop the body up off the ground and to augment the basic undulatory motion of the vertebral column during locomotion by pulling the body forward. In reptiles and all subsequent land forms, the appendages are reoriented under the body where they can serve as a more efficient means of support and locomotion.

As the limbs evolved in structure and function, so did the girdles that anchor them to the body. Because the "push" is a more efficient form of power than the "pull," the hindlimbs became the larger, more powerful set of limbs. In accordance, the pelvic girdle tends to be a large, heavy structure, generally possessing extensive surface areas for muscle attachment. In order that the thrust generated by the legs can be transferred efficiently to the body, the girdle has become fused directly to a specialized portion of the vertebral column called the ***sacrum***.

In contrast, the pectoral girdle at the front of the trunk has become a much lighter structure. In some cases, it is loosely connected to other trunk bones through the clavicle. But most often, especially in animals that are fast runners, the pectoral girdle is not directly attached to the vertebral column; the front of the body is suspended between the forelegs in a "sling" of muscle. (See Walker, 1965, p. 144.) This arrangement has vital functional significance. In the course of a stride or leap, the forelimbs accept the shock of landing. The jolt of impact is greatly mitigated by elastic muscular connections as it passes from the limbs to the trunk. This protects the bones and especially the skull from excessive shock.

With the development of a stance that raises the body off the ground, the vertebral column becomes a supporting girder. This is a more complex function than that found in fishes, where the vertebral column serves during locomotion as a compression strut. Because of the downward force imposed by gravity, the vertebrae of the land vertebrate tend to slip past one another. Interlocking articulating processes called *zygapophyses* on the anterior and posterior margins of each vertebra help to prevent such slippage.

In fishes, all caudal and trunk vertebrae have essentially the same shape. In tetrapods, vertebrae are differentiated into groups according to their location along the length of the body. This differentiation is related to four factors: (1) independent movement of the head (2) more complex trunk movements; (3) more sites for appendicular (limb) muscle attachment; (4) a new mode of respiration. These specializations involve such developments as unique articulation arrangements between vertebrae (as between the atlas and axis) and new bony processes acting as lever arms on which the muscles can act (such as the spinous processes).

The many new ways of life on land are reflected in numerous alterations in the vertebrate skull. The skull continues to encase the enlarging brain; it shows alterations that may be related to new modes of eating, which often require large jaw muscles. In general, the skull becomes "simpler" through fusion of certain bones and loss of others; the result is a compact, strong enclosure to house the brain. Another important change in the skull, which is seen in crocodiles and mammals, is associated with the chewing of food. The secondary palate, a bony surface found ventral to the primary palate, displaces the opening of the internal nostril toward the back of the mouth in many tetrapods. As a result, an organism can manipulate food with its mouth and breathe at the same time.

The skeletal system can be best studied as a functional entity when it is considered with the muscular system. Together they provide the vertebrate body's support and locomotion. Therefore, great changes in the muscular system occurred concomitantly with those in the skeleton. As noted in Chapter 8, the muscular system in fish consists basically of segmented axial muscle masses grouped as dorsal (epaxial) and ventral (hypaxial) muscles. Appendicular musculature in the fish consists only of rudimentary dorsal and ventral slips in the fins. In tetrapods, appendicular musculature increased in size and differentiated into a variety of discrete bundles designed for locomotive and supportive roles. In *Necturus* (the mudpuppy) we see such a pattern, but most of the musculature is still axial.

With the change from axial to appendicular locomotion, the hypaxial and epaxial muscles have been greatly reduced; they are often buried beneath the massive appendicular muscles. The hypaxials have lost their segmented character and form broad sheets that compress the visceral organs hanging beneath the vertebral column in animals that walk on four legs. The epaxials have also lost their segmented appearance; they join to form longitudinally oriented muscles which function in the varied movements of the vertebral column, the elevation of the head, and respiration. Epaxial muscles associated with the vertebral column are mostly flexors and extensors bending the body in the dorsoventral plane, a movement that aids running by extending the stride.

The extent to which the vertebrate muscular system has been modified can be seen by comparing the swimming modes of a fish and of a mammalian cetacean—the porpoise. The fish uses side-to-side flexion of the spine to propel itself, just as early vertebrates did; the dolphin swims with up-and-down undulations of its trunk and tail flukes.

The many muscles associated with the jaws and facial areas of tetrapods are a special category, derived from the branchial (gill) musculature. The muscles of the first gill arch have become the mandibular adductors and other smaller muscles. Those of the second arch have developed into jaw depressors, facial muscles, and such hyoid-associated muscles as the stylohyoid needed for swallowing.

The innumerable alterations in the bones and muscles of tetrapods can be viewed as meeting the functional needs of each animal's lifestyle. Understanding function is the key to appreciating structure. In the study of the skeletal and muscular systems, the student should strive to: (1) locate the element under study and consider its attachments with associated elements; (2) consider the function of the associated unit; and (3) attempt to understand how the function is carried out (see Table 24-1).

AXIAL SKELETON AND MUSCLES (Fig. 24-1)

The cat skeleton is composed basically of bone. In a few sites, however, cartilage can be found covering certain articular surfaces and connecting the ribs to the sternum. The bones are of two types—***appendicular***, the limbs and girdles; and ***axial***, the skull, mandible, hyoid, vertebral column, ribs, and sternum. The appendicular skeleton will be considered in association with the muscles, while the axial skeleton is considered separately in the following discussion.

Skull

Identify the various bones of the skull demarcated by sutures, using the dorsal, ventral, and lateral views. The ***foramina*** found in the skull and in other bones are channels allowing for the passage of nerves and blood vessels.

Dorsal view. **(Fig. 24-2)** In the anterior region the external nares are bounded by the ***premaxillaries*** and the ***nasals***. Lateral to the premaxillaries are the ***maxillaries***, which form the greater part of the upper jaw. Each maxillary contains five teeth. The teeth of cats consist of incisors, canines, premolars, and molars. The middorsal portion of the cranium is composed of the large ***frontal*** and ***parietal*** bones. Lateral to the parietal lies the ***squamosal portion*** of the ***temporal*** bone. Large processes extending posteriorly from the ***jugal***, or malar, and anteriorly from the squamosal form a great lateral arch called the ***zygomatic arch***, or zygomatic process, which serves as a site for muscle attachment. The arch bounds two spaces, the ***orbit*** anteriorly (for the eye), and the ***temporal fossa*** posteriorly (for the coronoid process of the lower jaw see Fig. 24-3).

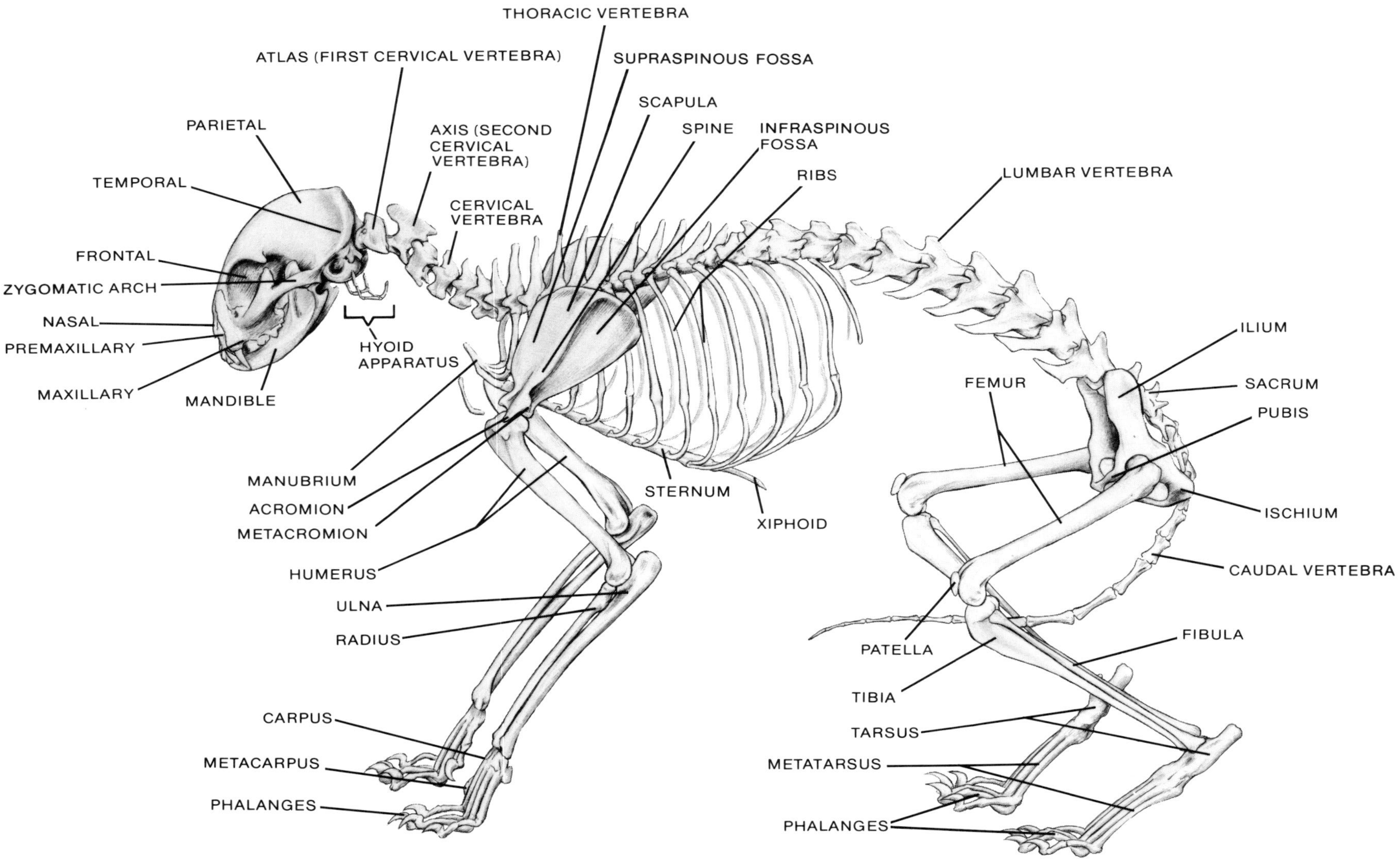

Figure 24-1. Lateral view of full cat skeleton.

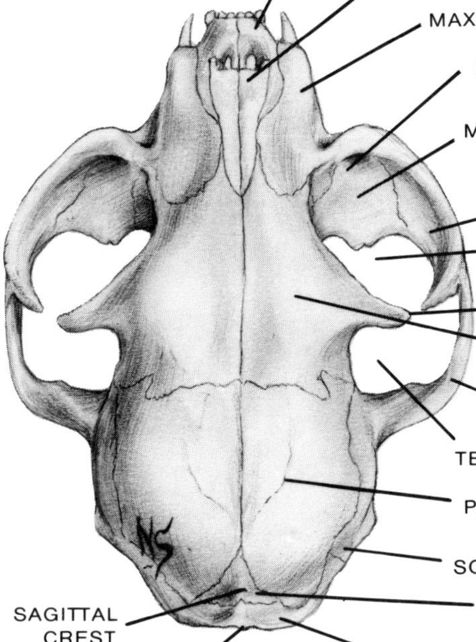

Figure 24-2. Dorsal view of a cat skull.

Posterior to the parietal is the small triangular ***interparietal***. Forming the posterior margin of the dorsal view is the ***supraoccipital***, which flares out prominently to the sides to form the ***nuchal crest*** (lambdoidal ridge). In an adult cat, the supraoccipital is one part of the single, large bone, the ***occipital***.

Lateral view. **(Fig. 24-3)** In this view the small bones forming the wall of the orbit, the ***lacrimal***, ***ethmoid***, ***palatine*** and ***orbitosphenoid***, can be seen, along with the large frontal bone. More posteriorly, the ***squamosal*** portion of the ***temporal*** bone can be clearly seen, as well as the ***mastoid process*** and the ***tympanic bulla*** which encloses the middle ear (see Figs. 24-2, 24-5 also).

Dentary. **(Mandible or lower jaw)** This is a structure composed of two halves united anteriorly in a symphysis. Note how the jaw operates with the skull. The articulation is made between the ***condyloid process*** of the dentary and the ***mandibular fossa*** of the zygomatic process (on each side of the skull, Fig. 24-5).

Hyoid. **(Fig. 24-4)** The hyoid is a thin, bony structure supporting the tongue and larynx. It is composed of one median bone, the ***body***, and two pairs of ***cornua***, horns, which articulate with the body. Each horn is made up of several small bones, homologous to the first two branchial arches of a fish.

Ventral view. **(Fig. 24-5)** Many of the bones seen in the dorsal view, such as the premaxillaries and maxillaries, are also visible in this ventral view. The large ***palatine*** bones are just posterior to the maxillaries, and the internal nares are found at the posterior margin of the palatines. Between the nares is the ***vomer***, a slender bone which forms part of the nasal septum.

The midventral surface of the skull is composed of the sphenoid group, the ***presphenoid***, ***orbitosphenoid***, and ***basisphenoid***. Posterior to the basisphenoid is the ***basioccipital***. Forming the most posterior portion of the ventral surface is the ***exoccipital***, which bears the ***occipital condyles***, the rounded projections that articulate with the vertebral column.

Vertebral Column, Ribs, and Sternum

The vertebral column consists of 51 to 53 bony subunits, the ***vertebrae***, which can be divided into five categories: 7 cervical, 13 thoracic, 7 lumbar, 3 sacral, and 21–23 caudal vertebrae. Although vertebrae in transitional regions between categories may be hard to identify, those in the middle of a region are

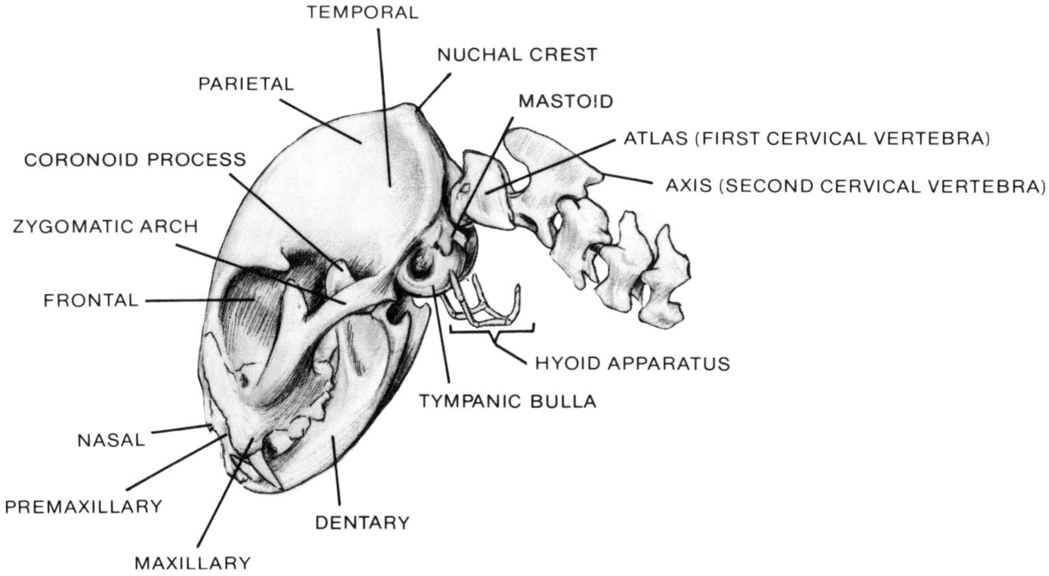

Figure 24-3. Lateral view of a cat skull.

easily recognizable, for each category has certain unique characteristics.

All vertebrae are made up of three major parts: the ***body***, or ***centrum***; the ***arch***, which surrounds the spinal cord; and the ***processes***. There are in turn three kinds of processes: ***spinous***, directed dorsally; ***transverse***, directed laterally; and ***articular*** or ***zygapophyseal*** processes.

Cervicals. The first two cervical vertebrae, the ***atlas*** (Fig. 24-6) and ***axis*** (Fig. 24-7), form a specialized composite unit allowing for the wide range of movement exhibited by the head. The large cup-like spaces on the cranial side of the atlas accommodate the occipital condyles of the skull, allowing for dorsoventral "nodding" movement of the head. Note that the body of the atlas is greatly reduced. Most of the centrum of the atlas has apparently been incorporated into the next vertebra, the axis, forming the protruding ***odontoid process***. The atlas-axis articulation, formed by the projection of the odontoid process into the atlas, allows for the rotational side-to-side movement of the head.

Four of the remaining five cervical vertebrae are characterized by ***transverse foramina*** (which house the vertebral arteries) lying along either side of the centrum (Fig. 24-8).

Figure 24-4. Hyoid apparatus of a cat.

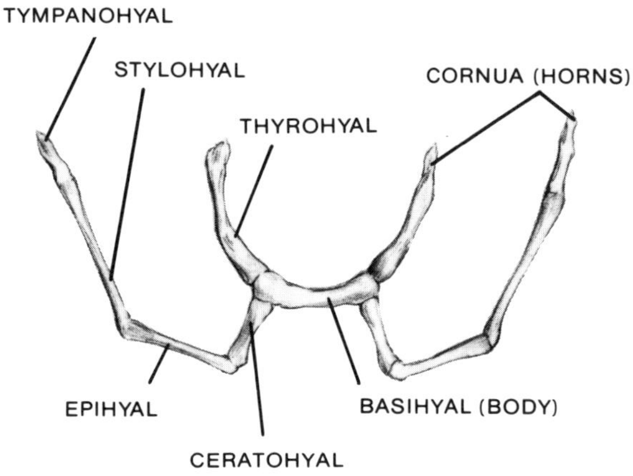

Thoracic vertebrae. **(Fig. 24-9)** These are the vertebrae with which the ribs articulate. The sites of rib articulation, the ***costal facets*** on the transverse processes, and the ***costal demifacets*** on the lateral margins of the centra, are the distinguishing features of the thoracic vertebrae. The costal facets articulate with the tubercles of the ribs and the costal demifacets with the heads of the ribs (Fig. 24-10).

Ribs. **(Figs. 24-1, 24-10)** There are thirteen pair of ribs. The first nine are "true" ribs, because they are attached directly to the sternum through their own costal cartilages. The next three are "false" ribs, having a ***costal cartilage*** joining that of the ninth rib rather than the sternum. The last is a "floating" rib which has no ventral skeletal articulation.

Most of the ribs articulate with the vertebral column at two points. The ***head*** articulates with two

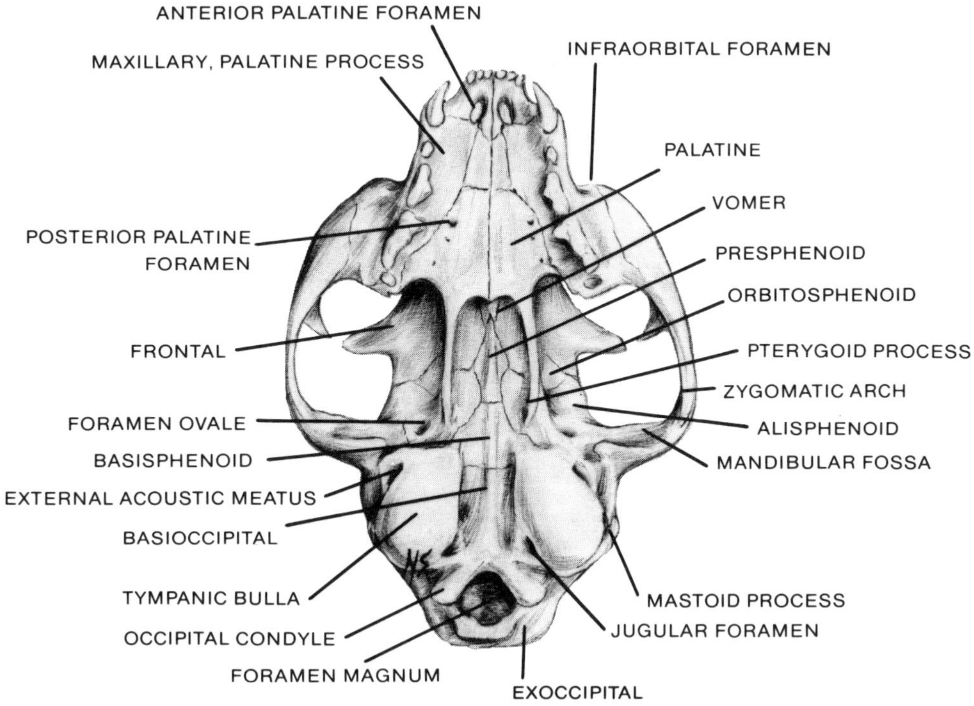

Figure 24-5. Ventral view of a cat skull.

demifacets of adjacent vertebrae and the ***tubercle*** articulates with the costal facet on the transverse process of one thoracic vertebra.

Sternum. (**Figs. 24-1, 24-11**) This is a segmented bony structure to which the ***costal cartilages*** and many muscles attach.

Lumbar vertebrae. (**Fig. 24-12**) These are easily recognized by their large transverse processes and cranially oriented spinous processes. They lack transverse foramina and costal facets.

Sacrum. (**Fig. 24-13**) This represents the fusion of the three sacral vertebrae, a modification for articulation with the pelvic girdle.

Caudal vertebrae. (**Fig. 24-14**) These can be distinguished by their small size and simple form. In the cat, the small chevron bones articulate with the

Figures 24-6, 24-7. Left lateral view of the first two cervical vertebrae of the cat, the atlas and the axis.

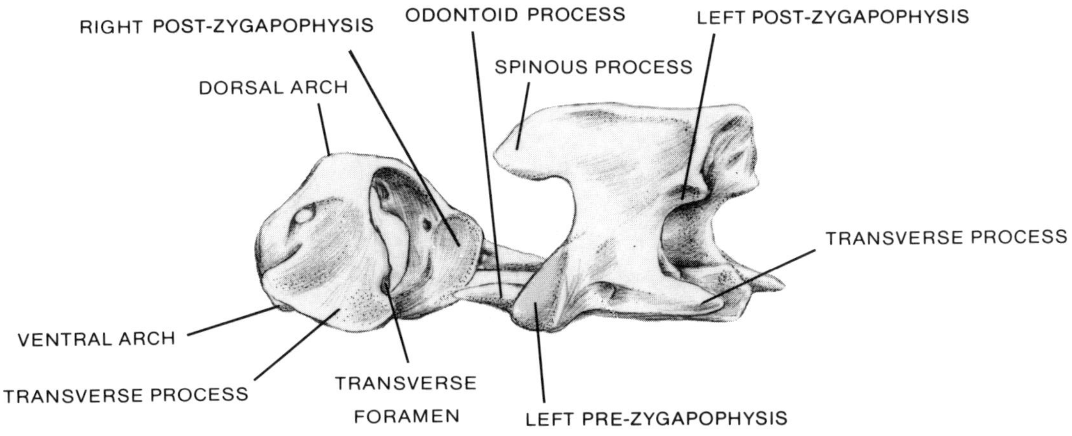

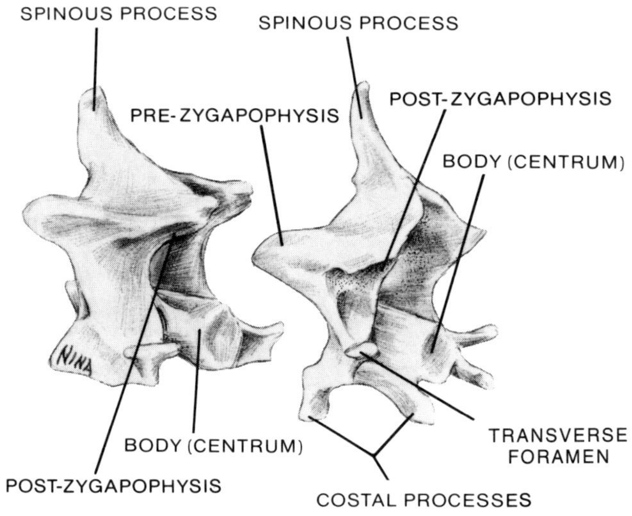

Figure 24-8. Left lateral view of typical cervical vertebrae of the cat.

hemal processes of the proximal caudal vertebrae and function to protect the caudal artery and vein.

MUSCLES: DISSECTION AND IDENTIFICATION

The following listing of muscles is intended as a study guide. You will find it very helpful to read the section "Muscle Attachment to the Skeleton" at the end of this chapter *prior* to beginning muscle dissection. Also, remember to locate bone landmarks that will aid in determining positions and actions of a muscle group (see Figs. 24-26, 24-32, 24-34). Trace muscles to their attachments (origin and insertion) and use the orientation (direction) of muscle fibers to identify specific muscles.

Figure 24-9. Left lateral view of typical thoracic vertebrae of the cat.

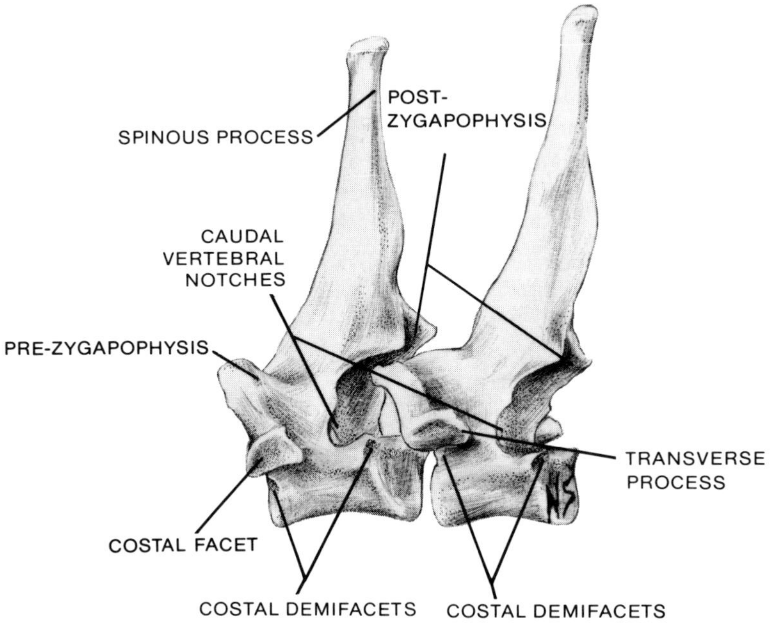

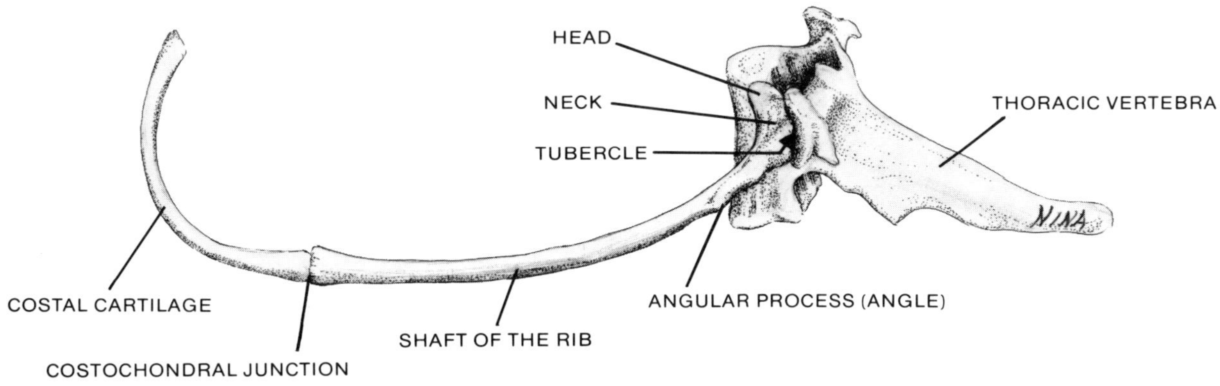

Figure 24-10. Left lateral view of a typical rib articulating with a thoracic vertebra.

Cutaneous Muscles (Fig. 24-15)

Identify the cutaneous muscles described below. Remember to keep muscles moist during dissection.

Cutaneous maximus. This is a very thin layer of superficial muscle that lies in proximity to the skin over the chest, shoulder, thoracic, and abdominal regions. Because the muscle lies so close to the skin, and because it is generally separated from the main body of muscles by a layer of fat, it is usually removed with the skin. It will appear as light striations on the skin's undersurface.

Origin: linea alba; ventral cranial surface of latissimus dorsi; ventral axillary region of forearm.

Insertion: on the skin.

Action: movement of the skin.

Platysma. This is another thin superficial muscle covering the neck and face regions.

Origin: fascia of the middorsal line.

Insertion: on the skin.

Action: movement of the skin of the neck and face.

Superficial Head and Neck Muscles (Fig. 24-16)

Remove the tough skin of the head and neck just prior to examining the musculature. Continue the midventral incision to the chin, then cut laterally along the lower lip. Make a middorsal incision from the neck to the forehead. Tease the skin from each incision laterally until skin is separated from musculature. Remove skin and large fatty deposits, avoiding the large glands of the lateral neck. Except for the sternohyoid, these muscles are classified as branchiomeric (or branchial).

Sternomastoid. This is a strap of muscle proceeding diagonally from a medial position just cranial to the pectoantebrachialis to a lateral position on the mastoid region of the skull.

Origin: manubrium of sternum.

Insertion: nuchal crest, mastoid region of skull.

Action: acting unilaterally, turns the head; acting bilaterally, flexes the neck.

Figure 24-11. Ventral view of the sternum.

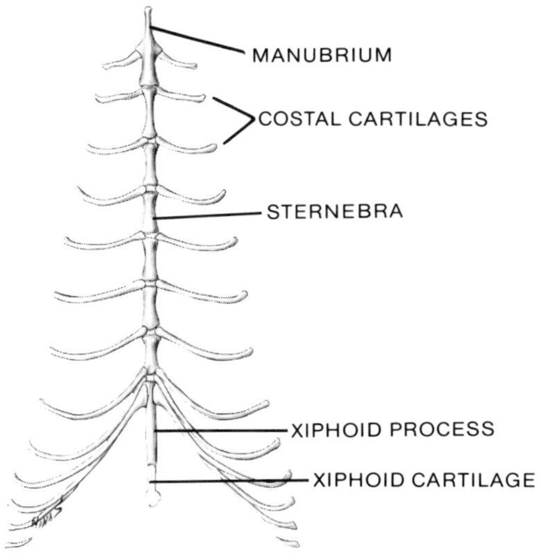

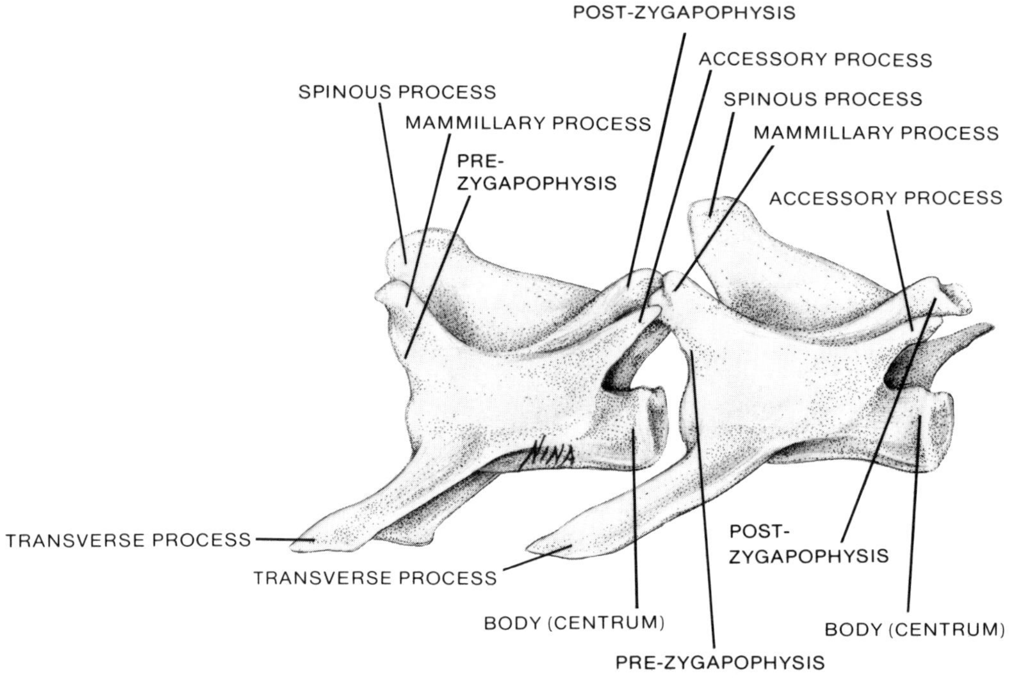

Figure 24-12. Left lateral view of typical lumbar vertebrae of the cat.

Cleidomastoid. To see this muscle clearly, the sternomastoid must be cut and retracted. The cleidomastoid lies directly beneath the sternomastoid; it extends from the clavicle cranially to the mastoid region.

Origin: mastoid process.

Insertion: clavicle.

Action: if clavicle is fixed, the head is pulled down; if the head is fixed, the clavicle is drawn forward.

Sternohyoid. This is a thin band of muscle originating at the sternum and becoming superficial where the sternomastoids come together medially. From this point, the muscle proceeds cranially to the hyoid.

Figure 24-13. Dorsal view of the sacrum of the cat.

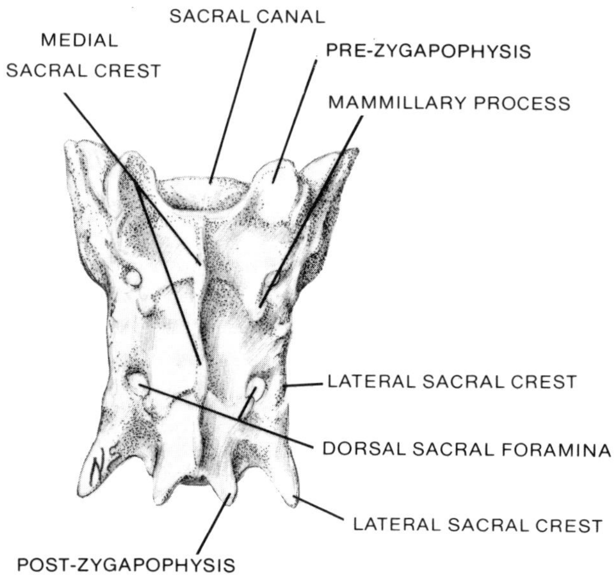

Figure 24-14. Dorsal view of a typical caudal vertebra of the cat.

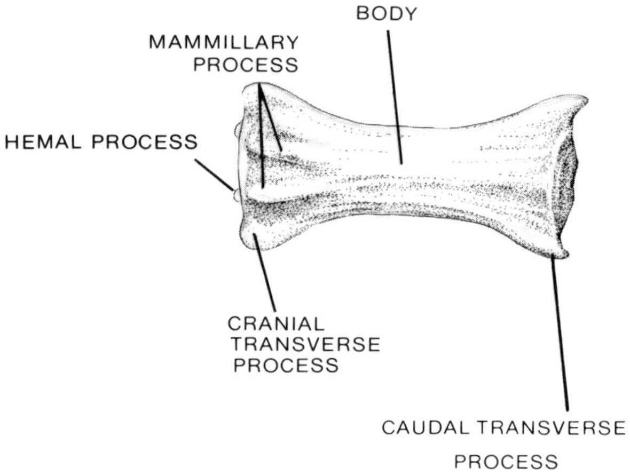

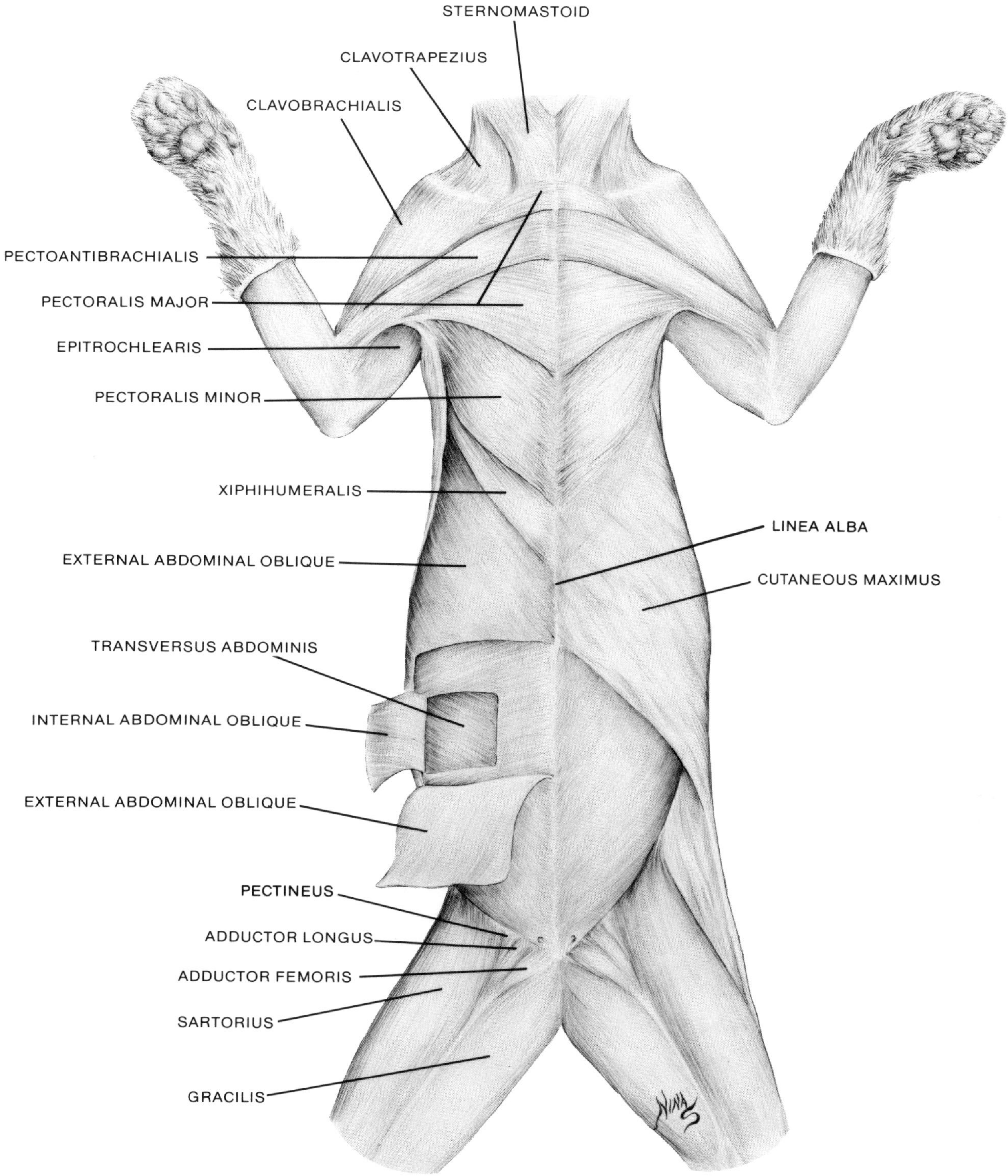

Figure 24-15. Superficial muscles of the ventral surface of the cat, excluding those of the head. Flaps indicate layered abdominal musculature.

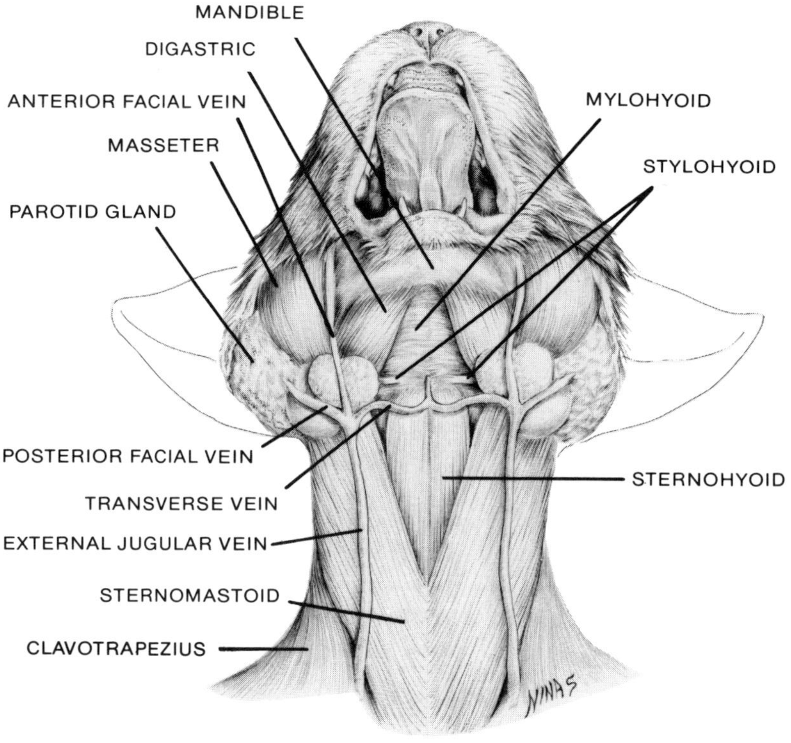

Figure 24-16. Ventral view of the head and neck of the cat, showing superficial muscles, blood vessels, and glands.

Origin: anterior margin of first costal cartilage.

Insertion: body of hyoid.

Action: if hyoid is fixed, the sternum is drawn forward; if sternum is fixed, the hyoid is drawn posteriorly.

Mylohyoid. This is a thin muscle. Its striations are oriented transversely. It extends between the mandible and the midventral raphe, visible between the digastrics.

Origin: median surface of mandible.

Insertion: midventral raphe.

Action: raises the floor of the mouth.

Stylohyoid. This is a small strand of muscle lying transversely along the caudal margin of the mylohyoid.

Origin: stylohyal of hyoid.

Insertion: body of hyoid.

Action: raises the hyoid.

Digastric. This is a band of muscle extending from the mastoid region of the skull past the salivary glands to the ventromedial surface of the mandible.

Origin: mastoid process of skull.

Insertion: ventromedial margin of mandible.

Action: depresses mandible to open mouth.

Masseter. This is a thick wall of muscle lying just cranial to the salivary glands.

Origin: zygomatic arch.

Insertion: coronoid fossa and adjacent areas.

Action: elevates jaw to close mouth.

DEEP MUSCLES OF THE HEAD AND NECK (Fig. 24-17)

Transect and reflect the mylohyoid and sternomastoid on the right side. Deflect the geniohyoid medially to observe the genioglossus muscle. Cut the digastric at the level of the mandible and reflect it to observe the styloglossus muscle.

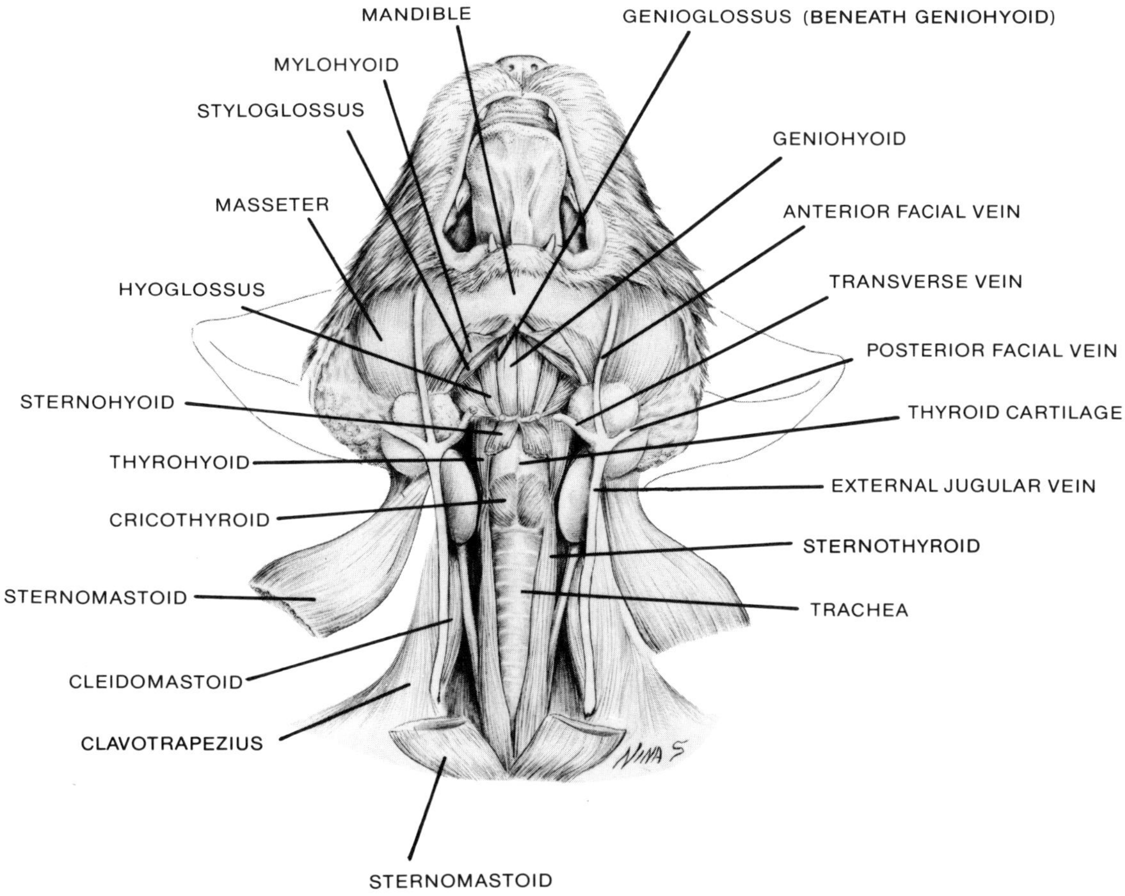

Figure 24-17. Ventral view of the deep muscles of the throat, larynx, and hyoid apparatus, after reflection and partial removal of the sternomastoid, sternohyoid, and mylohyoid muscles.

Geniohyoid. This is a thin, longitudinally oriented muscle lying below the mylohyoid.

Origin: symphysis of mandible.

Insertion: basihyal of hyoid.

Action: pulls hyoid forward.

Hyoglossus. This is a band of muscle originating just lateral to the insertion of the geniohyoid and proceeding cranially, at a slight lateral angle, to the tongue.

Origin: body of hyoid.

Insertion: tongue.

Action: retracts tongue.

Genioglossus. This lies directly beneath the geniohyoid and medial to the hyoglossus.

Origin: medial surface of mandible.

Insertion: tongue.

Action: draws tongue forward.

Styloglossus. Reflection of the digastric will reveal this small muscle located below and running parallel with the line of the mandible.

Origin: mastoid process of skull.

Insertion: tongue.

Action: retracts tongue.

Sternothyroid. It is necessary to transect and reflect the sternohyoid to see this muscle lying below, extending from the sternum to the thyroid.

Origin: costal cartilage of first rib.

Insertion: thyroid cartilage of larynx.

Action: draws larynx caudad.

Thyrohyoid. This also lies beneath and lateral to the sternohyoid. It extends from the thyroid region of the larynx to the hyoid.

Origin: thyroid cartilage of larynx.

Insertion: thyrohyal of hyoid.

Action: elevates larynx.

Cricothyroid. This appears as a small patch of muscle on the ventral surface of the trachea at a point near the thyroid.

Origin: cricoid cartilage.

Insertion: thyroid cartilage.

Action: tenses vocal cords.

Abdominal Muscles (Figs. 24-15, 24-18)

Beginning about 2 cm lateral to the linea alba in midabdomen and proceeding toward the cat's right, cut superficially three sides of a square through the thin outermost layer of muscle, as in Fig. 24-15. Reflect this layer caudally, exposing the first layers of underlying muscle. Cut three sides of a second square from the exposed part of the underlying muscle, then reflect the patch laterally. This reveals a third muscle layer. Reflect the covering white fibrous sheath to observe the rectus abdominus (Fig. 24-18). Direction of fiber orientation is crucial to correct identification of these muscles. In the outermost muscle, the fibers run in an anteriodorsal direction.

External oblique. This is a broad sheet of muscle visible as anteriodorsal striations superficially covering the abdominal region.

Origin: last nine or ten ribs; lumbodorsal fascia.

Insertion: via an aponeurosis—a sheet of white, shiny connective tissue—to the linea alba.

Action: compresses the abdominal region.

Internal oblique. This is another broad sheet of muscle; its striations run perpendicular to those of the external oblique. The internal oblique is located directly beneath the external oblique and can be seen by cutting carefully through that superficial muscle layer and reflecting it back.

Origin: dorsolumbar fascia and ventral portion of the iliac crest.

Insertion: via an aponeurosis to the linea alba.

Action: compresses the abdomen.

Transversus abdominis. This is the third muscle layer of the abdomen, lying deep to the internal oblique and consisting of transverse striations. The internal oblique must be cut and reflected back if this muscle is to be seen.

Origin: costal cartilages; transverse processes of lumbar vertebrae; ventral border of ilium.

Insertion: linea alba.

Action: compresses the abdomen.

Rectus abdominis. (Fig. 24-18) This is a narrow band of muscle lying on either side of the linea alba, extending from the pubis to the first rib.

Origin: pubis.

Insertion: costal cartilage of the first rib; cranial portion of the sternum.

Action: compresses abdomen, draws ribs and sternum caudad.

Superficial Thoracic Muscles (Fig. 24-15)

Separate and identify individual overlapping members of the superficial pectoralis group.

Pectoantibrachialis. This is a very thin superficial muscle extending transversely across the thoracic area.

Origin: manubrium of the sternum.

Insertion: proximal part of the ulna.

Action: adducts forelimb.

Pectoralis major. This is a broad triangular sheet of muscle, extending from the sternum to its insertion along the humerus. This muscle is composed of two parts: a smaller superficial anterior portion, located for the most part beneath the pectoantibrachialis and clavotrapezius; and a larger, deeper portion.

Origin: midventral thoracic raphe and manubrium of the sternum.

Insertion: pectoral ridge of humerus.

Action: adducts and retracts forelimb.

Pectoralis minor. This is another broad sheet of muscle found just caudal to and partially overlapped by the pectoralis major. Separation between the major and minor is difficult, requiring a close exam-

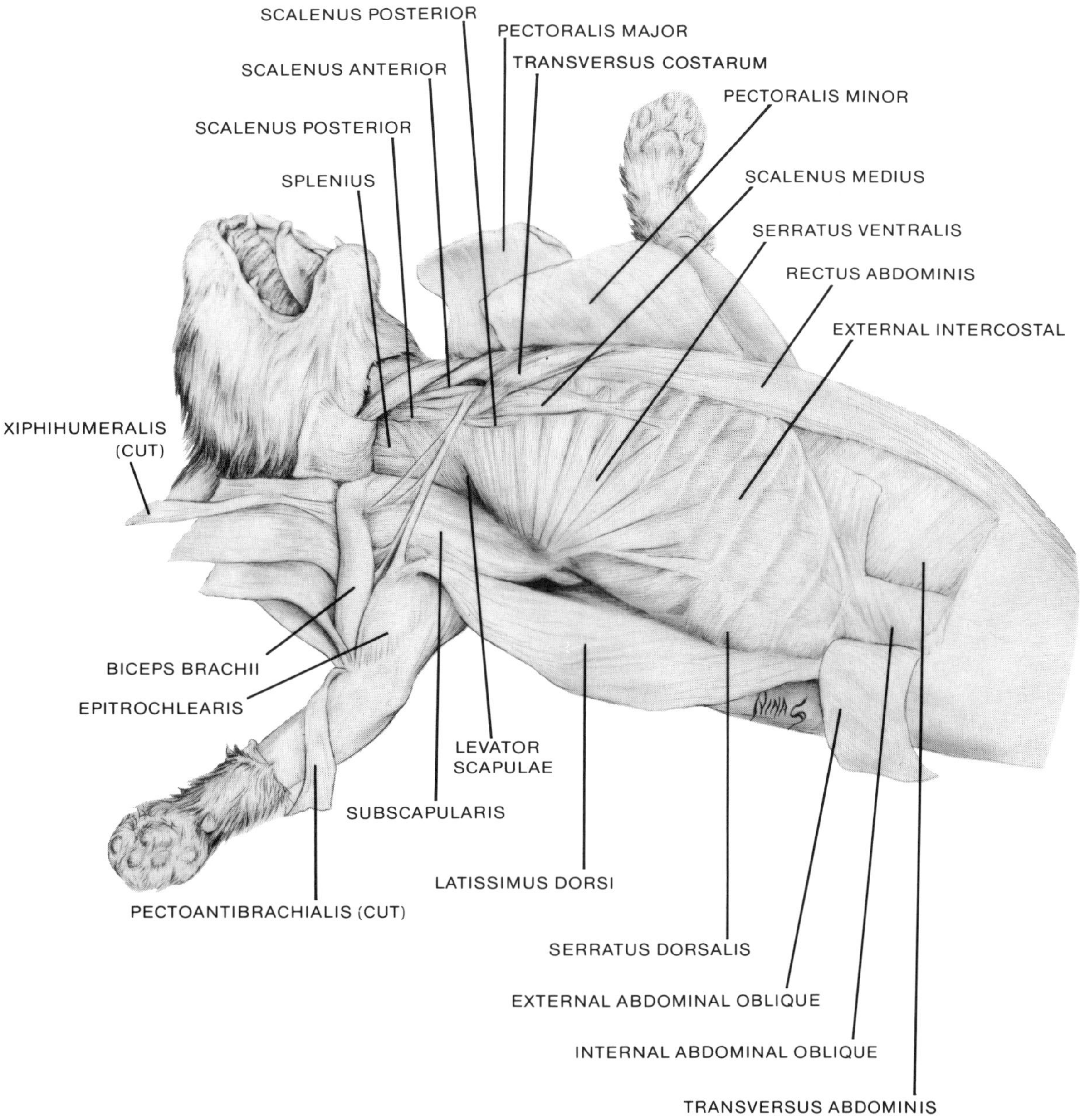

Figure 24-18. Ventral view of deep thoracic muscles of the cat, after reflection of pectoralis group muscles and removal of the cranial portion of the external abdominal oblique. Muscular suspension of scapula and forelimb is included. (Do not cut blood vessels in the arm.)

ination of the striations. Those of the minor run more in a craniocaudal direction.

Origin: the sternebrae.

Insertion: just beneath the pectoralis major, the pectoralis minor splits to form a dual insertion on the pectoral ridge of the humerus and the bicipital groove.

Action: adducts and retracts forelimb.

Xiphihumeralis. This most caudal member of the pectoralis group is a thin strip of muscle. Ventrally it lies adjacent to the pectoralis minor. Moving dorsally it tapers and is found beneath the pectoralis minor as it proceeds to its insertion on the humerus.

Origin: midventral raphe in region of xiphoid process.

Insertion: bicipital groove of humerus.

Action: adducts forelimb.

Deep Thoracic Muscles (Fig. 24-18)

Transect and reflect the xiphihumeralis, pectoralis minor, pectoralis major, and pectoantibrachialis muscles of the right side. Cut each muscle at a different position along its length so that the sections can be matched accurately in later study. Isolate the anterior rectus abdominis and divert it medially, keeping it intact. Lift and reflect the remaining anterior external oblique muscle, which inserts on the caudal edge of the ribcage. Cut in a medial direction along the anterior margin of the external oblique, then cut in a caudal direction along the anterior-medial insertion of this muscle to the posterior edge of the ribcage. Reflect the muscle sheet. Be very careful not to cut arteries and veins in the arm.

Scalenus. This is a long ribbon muscle running ventrolaterally along the neck and thoracic cavity. Although much more finely divided, the muscle generally appears as three bands. The posterior (dorsal) and anterior (ventral) are shorter; the medius is longer and broader.

Origin: the ribs.

Insertion: transverse process of cervical vertebrae.

Action: acting bilaterally, flexes the neck; acting unilaterally, turns the head.

Transversus costarum. This is a small, thin, diagonally oriented muscle lying ventral to the insertion of the rectus abdominis.

Origin: lateral margin of the sternum between the third and sixth ribs.

Insertion: first rib.

Action: draws the sternum forward in respiration.

Levator scapulae. This is a broad triangular sheet of muscle lying laterally along the most cranial portion of the thoracic area, just beneath the scapula.

Origin: the tubercles of the transverse processes of the last five cervical vertebrae.

Insertion: medial surface of scapula.

Action: draws scapula cranioventrad.

Serratus ventralis. This appears as a series of bands just caudal to the levator scapulae, fanning out from an insertion on the medial portion of the scapula to multiple origins on the ribs. The separation between the levator scapulae and serratus ventralis can most easily be seen by lifting the large medial band of the scalenus.

Origin: the most anterior nine or ten ribs.

Insertion: medial surface of scapula.

Action: pulls scapula toward the thorax; supports trunk.

External intercostals. These can most easily be seen by removing the external oblique. They appear as short muscles interconnecting adjacent ribs.

Origin: caudal edge of ribs.

Insertion: cranial edge of adjacent ribs.

Action: pull ribs forward.

Internal intercostals. These are short muscles interconnecting adjacent ribs; they have striations perpendicular to those of the external intercostals. For the most part, the internal intercostals lie directly beneath the externals, but they are visible ventrally if the rectus abdominis is retracted.

Origin: cranial edge of ribs.

Insertion: caudal edge of adjacent ribs.

Action: pull ribs backward.

Superficial Muscles of the Back (Fig. 24-19)

Separate and identify individual muscles of the trapezius group. These muscles are nonsegmental. Use fiber orientation information as well as attachments to identify the parts of the trapezius.

Clavotrapezius. This is a broad strip of muscle running from the dorsal area of the neck to the clavicle. The position of the clavicle is demarcated by a thin line of connective tissue transversing the distal end of the muscle.

Origin: lambdoidal crest and middorsal raphe over the axis.

Insertion: clavicle and laterally extending raphe above the clavicle.

Action: draws scapula craniodorsad.

Clavobrachialis. This appears as a continuation of the clavotrapezius, extending from the clavicle to the proximal portion of the ulna.

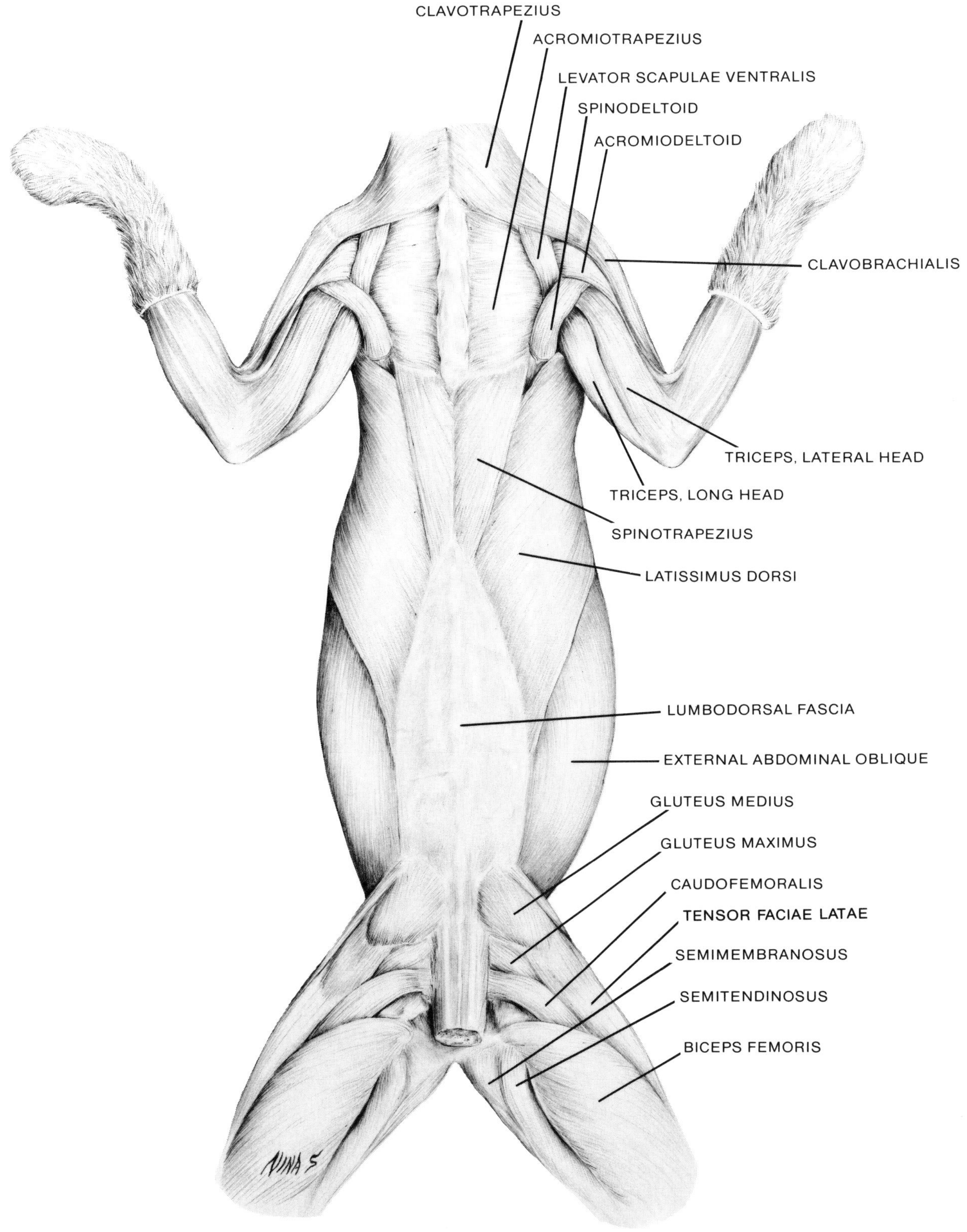

Figure 24-19. Superficial muscles of the dorsal surface of the cat, excluding those of the head.

Origin: clavicle and laterally extending raphe above the clavicle.

Insertion: joins the distal end of the brachialis; together they insert on the ulna, distal to the trochlear notch.

Action: flexes forearm and rotates limb anteriorly.

Acromiotrapezius. This is a broad, thin triangular sheet of muscle caudal to the clavotrapezius. It extends from the middorsal line to the spinodeltoid, at which point it proceeds deeply to insert on the dorsal surface of the scapula.

Origin: extends from the spinous process of the axis to the spinous processes of the first four thoracic vertebrae.

Insertion: spine and metacromion of scapula.

Action: pulls scapula anteriorly and upward; together the right and left muscles hold the scapulae together.

Spinotrapezius. This is the most caudal member of the trapezius group. It extends in a more longitudinal direction from the middorsal line to the scapula.

Origin: the spinous processes of the fourth to the twelfth thoracic vertebrae.

Insertion: fascia of supraspinatus and infraspinatus muscles; tuberosity of the spine of the scapula.

Action: fixes dorsal part of scapula as rest of trapezius rotates ventral part anteriorly.

Latissimus dorsi. This is the broad triangular sheet of muscle posterior to the spinotrapezius, originating along the dorsal midline and converging to insert on the humerus.

Origin: the spinous processes of the fifth thoracic to the sixth lumbar vertebrae; lumbodorsal fascia.

Insertion: medial surface of the proximal end of the shaft of the humerus.

Action: pulls arm caudodorsad.

Deep Muscles of the Back (Fig. 24-20)

Transect and reflect the latissimus dorsi, the spinotrapezius, acromiotrapezius, and clavotrapezius on the right side. Manipulate the forelimb and scapula to identify associated musculature. Remove the lumbodorsal fascia of the cat's right side, beginning at its caudal boundary. The fascia fuses cranially with the spinalis dorsi and cannot be separated there.

Splenius. This is a broad sheet of muscle covering the dorsal and lateral surfaces of the neck.

Origin: middorsal cervical fascia.

Insertion: lambdoidal ridge.

Action: acting bilaterally, elevates the head; acting unilaterally, turns the head.

Rhomboideus. This is a thick, transversely oriented bunching of muscle extending between the middorsal line and the dorsal and caudal margins of the scapula.

Origin: spinous processes of the first four thoracic vertebrae, and from interspinous ligaments caudal to this.

Insertion: vertebral margin of scapula.

Action: pulls scapula dorsad and fixes the dorsal part of the scapula during rotation.

Rhomboideus capitis. This appears as a distinct thin band of muscle lateral to the midline. It arises from the anterior portion of the rhomboideus and proceeds cranially to the lambdoidal ridge.

Origin: lambdoidal ridge.

Insertion: cranial angle of the scapula.

Action: rotates and draws scapula craniad.

Serratus dorsalis (anterior and posterior). These originate from a broad aponeurosis on the middorsal line and insert as slips on the outer surfaces of the ribs. The anterior slips, which for the most part lie beneath the serratus ventralis, slant caudally; the posterior slips run in a more transverse direction.

Anterior, origin: from a broad aponeurosis at the middorsal raphe, extending from the axial spinous process to that of the tenth thoracic vertebra.

Insertion: lateral surfaces of first nine ribs.

Action: draws ribs craniad for inspiration.

Posterior, origin: lumbar spinous processes.

Insertion: last four ribs.

Action: draws last four ribs caudad for expiration.

Longissimus dorsi. To see this muscle clearly, remove the superficial lumbodorsal fascia and deeper fibrous thoracolumbar fascia that cover the caudal middorsal area. The longissimus dorsi is composed of three longitudinally oriented bundles in the lumbar region. The most lateral mass continues beyond the lumbar region and forward to the last cervical vertebra. The muscle generally occupies the area be-

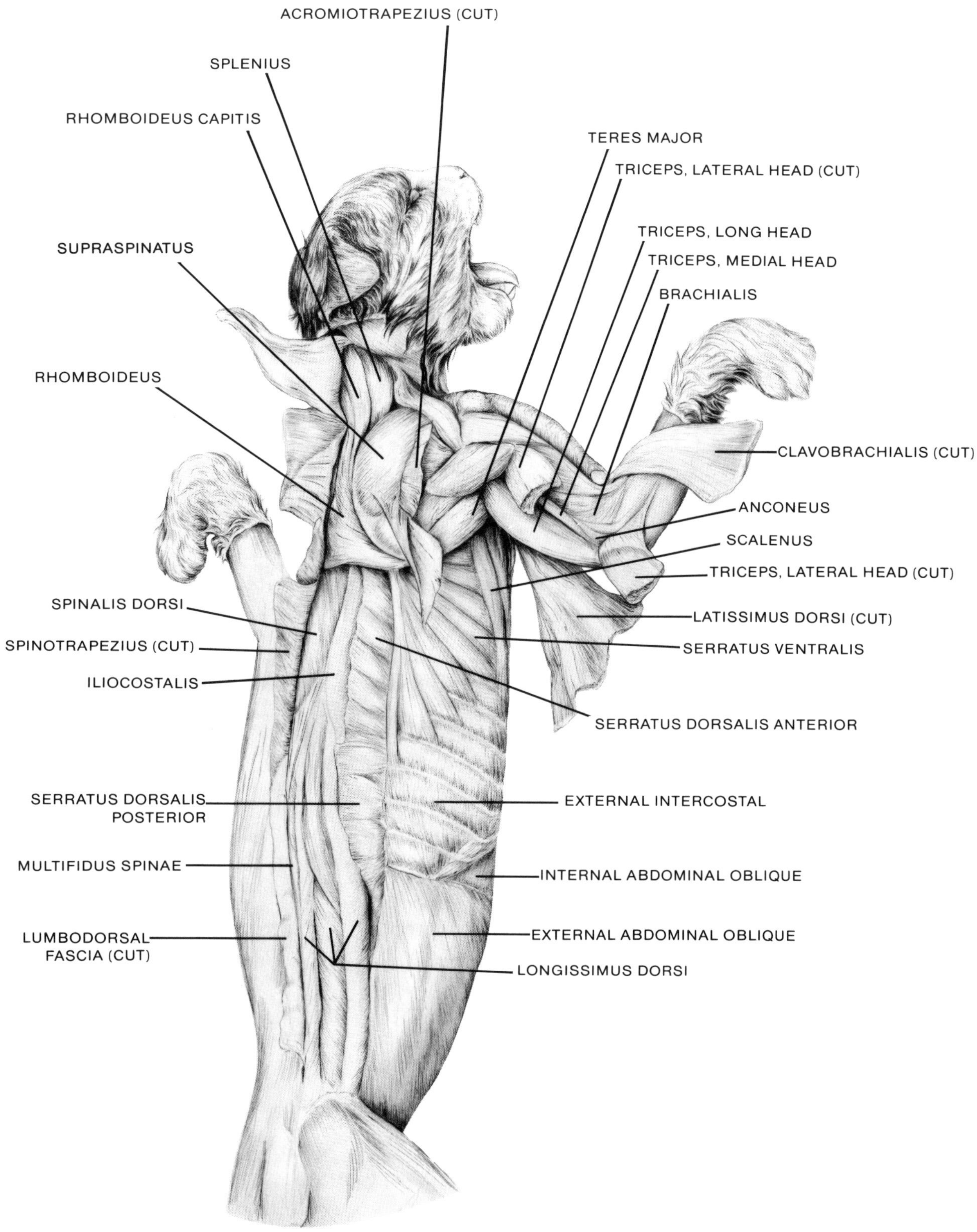

Figure 24-20. Dorsolateral view of deep musculature of the scapula, forelimb, and back, after reflection of trapezius group muscles and removal of the cranial portion of the external abdominal oblique.

tween the spinous processes and transverse processes of the vertebrae.

Origin: this muscle is made up of a mass of complicated connections. Fibers originate from all the vertebrae, the sacrum, and the ilium.

Insertion: the insertions likewise occur on many parts of all vertebrae.

Action: extends vertebral column.

Multifidus spinae. This is closely associated with the vertebral column. It is composed of many fibers extending from the transverse processes of one vertebra to the spinous processes of a more anterior vertebra. This muscle can be most easily seen in the lumbar region, adjacent medially to the longissimus dorsi.

Origin: transverse processes of vertebrae.

Insertion: spinous processes of vertebrae.

Action: extends or bends vertebral column.

Spinalis dorsi. To see this muscle clearly, fascia attaching the slips of the posterior serratus dorsalis to the midline must be removed. The spinalis dorsi lies superficially along the midline medial to the longissimus dorsi at the level of the serratus dorsalis.

Origin: spinous processes of the four most caudal thoracic vertebrae.

Insertion: spinous processes of anterior thoracic and most of the cervical vertebrae.

Action: extends vertebral column.

Iliocostalis. Retraction of the serratus dorsalis will reveal this muscle as it originates lateral to the longissimus dorsi and attaches to ribs by fine slips.

Origin: ribs.

Insertion: third or fourth rib cranial from the rib of origin.

Action: draws ribs together.

APPENDICULAR SKELETON AND MUSCLES

Bones of Shoulder and Forelimb (Figs. 24-21, 24-24, 24-26)

The clavicle, scapula, humerus, radius, ulna, and forefoot bones comprise the skeleton of the shoulder and forelimb. The scapula is a thin blade of bone held in place by muscle. The head of the humerus articulates with the scapula at the glenoid fossa, the greater tubercle of the humerus projects laterally, and the lesser tubercle projects anteromedially. The humerus articulates with the trochlear notch of the ulna and with the radius at the radial notch. The radius is in a lateral position at this point; it articulates with the radial notch of the ulna, but as it moves distally, it crosses dorsally over the ulna to a medial position. The radius and the ulna articulate at this distal point by way of articulating facets.

The clavicles and scapulae are the bony elements of the pectoral girdle. In the cat, the clavicles are reduced to small bones held in place by surrounding muscle and connective tissue attaching them to the scapulae, which are suspended along the sides of the ribcage.

Bones of Pelvis and Hindlimb (Figs. 24-1 24-22, 24-32)

The pelvic girdle, femur, tibia, fibula, and hindfoot bones make up the skeleton of the hindlimb and its support.

The pelvic girdle is composed of two sections, the ***innominate bones***, which are anchored to the sacrum dorsally and fused in the pubic symphysis ventrally. Each innominate bone has three skeletal components, the ilium, ischium, and pubis. The ilium projects upward from the acetabulum, its medial surface articulates with the sacrum. The ischium is the posterior portion of the innominate; it and the anterior pubis surround the obturator foramen.

The femur head articulates with the pelvic girdle at the acetabulum. Distally the lateral and medial condyles of the femur articulate with the lateral and medial condyles of the tibia. The patella is a sesamoid bone; that is, it develops, and is suspended in tendon tissue. Proximal and distal facets on the tibia allow for articulation with the fibula, a slender bone lying lateral to the tibia.

Muscles of the Scapula and Proximal Forelimb: Lateral View (Figs. 24-20, 24-21, 24-23, 24-24)

Separate and identify the musculature of the scapula, the deltoid group, and of the lateral surface of the right forelimb. Isolate superficial portions of the triceps muscle. Transect and reflect the lateral head of the triceps to observe the medial head of the triceps and anconeus muscles.

Supraspinatus. This muscle covers the supraspinous fossa of the scapula.

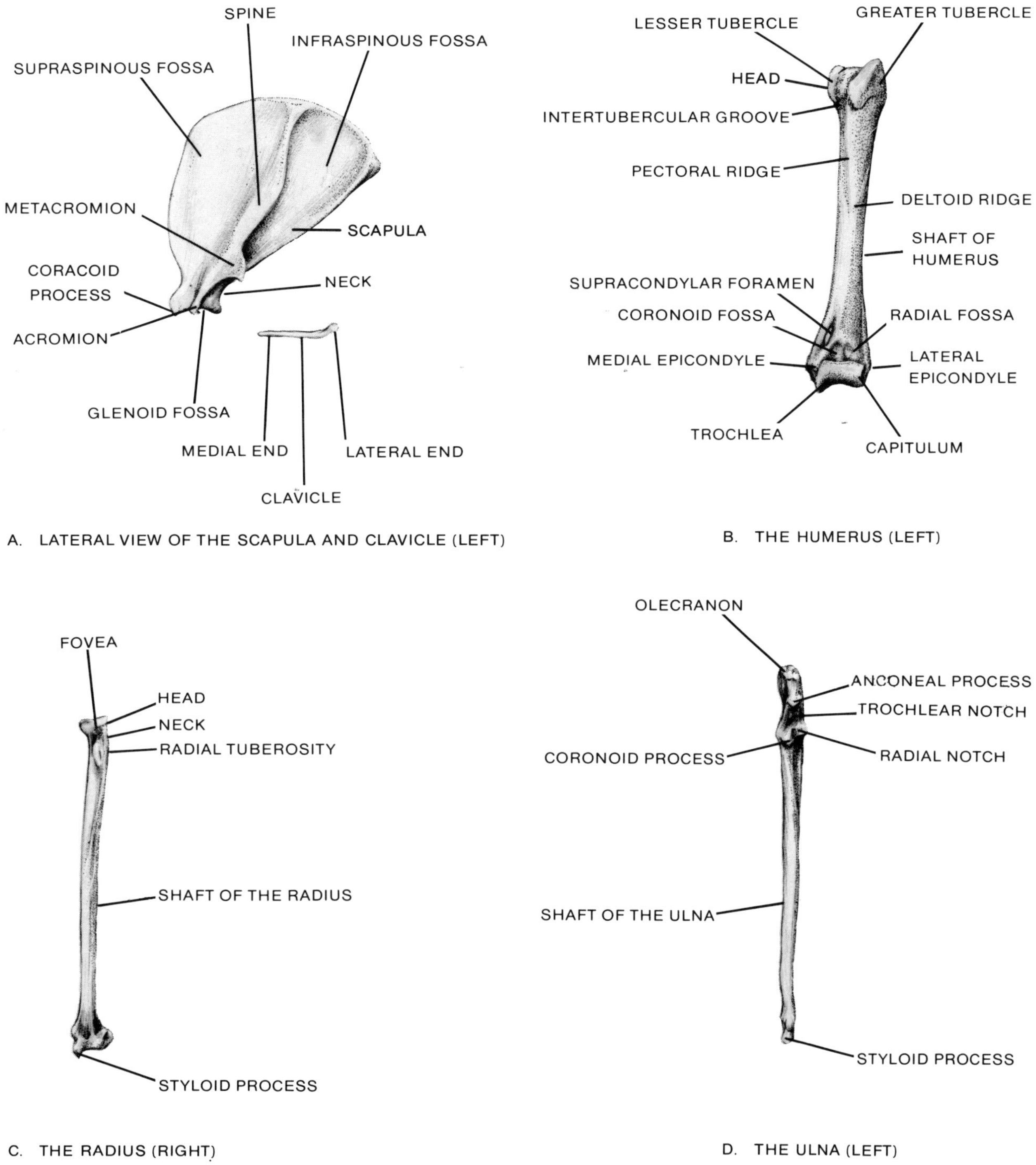

Figure 24-21. Bones of the shoulder, forelimb, and clavicle of the cat.

Origin: supraspinous fossa.

Insertion: greater tubercle of humerus.

Action: extends shoulder to advance forelimb forward.

Infraspinatus. This covers the infraspinous fossa of the scapula.

Origin: infraspinous fossa.

Insertion: greater tubercle of humerus.

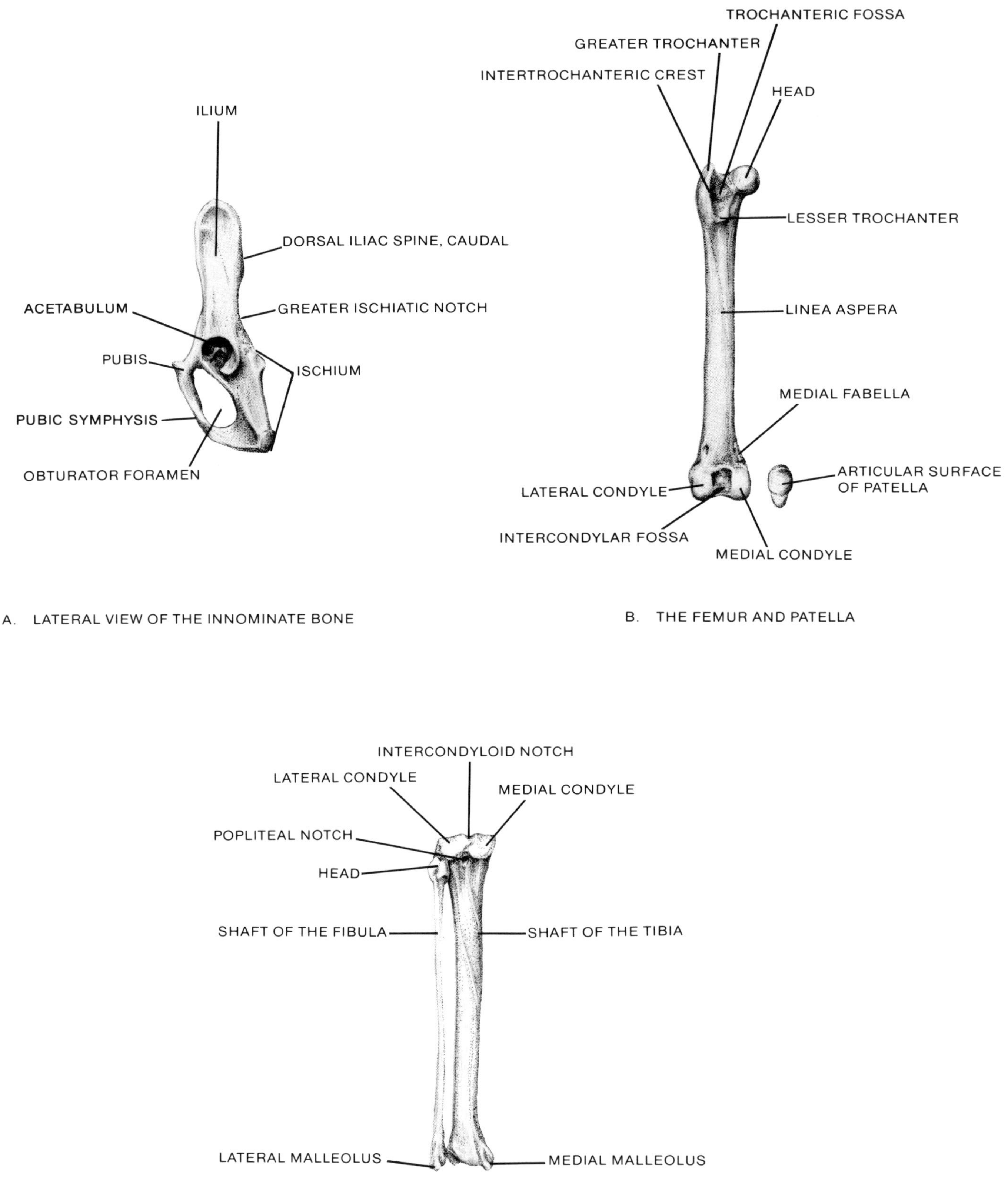

A. LATERAL VIEW OF THE INNOMINATE BONE

B. THE FEMUR AND PATELLA

Figure 24-22. Bones of the left side of the pelvic girdle and the left hindlimb of the cat.

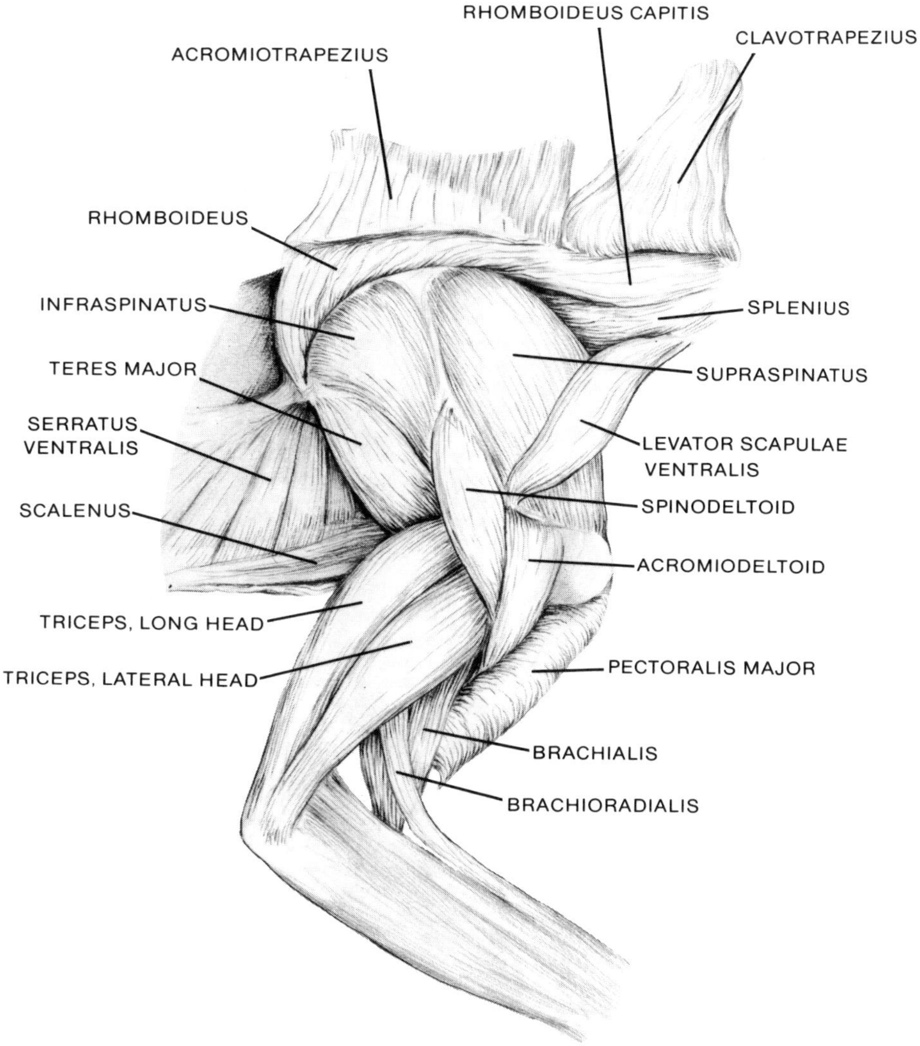

Figure 24-23. Lateral view of the muscles of the scapula and forelimb of the cat.

Action: abducts and rotates humerus outward.

Teres major. This is a short, thick muscle that runs along the caudoventral margin of the scapula.

Origin: caudal margin of scapula; fascia of the infraspinatus.

Insertion: medial surface of the proximal portion of the shaft of humerus.

Action: flexes and rotates the humerus.

Teres minor. (Locate this muscle after identification of the deltoids.) Upon transecting and reflecting the spinodeltoid, this small muscle appears beneath it, arising from the caudal border of the scapula, and often from the proximal area of the long head of the triceps. It proceeds anteriorly to the humerus.

Origin: an area of the caudal border of the scapula.

Insertion: greater tubercle of humerus.

Action: acts with infraspinatus to rotate humerus.

Spinodeltoid. This is a short band of muscle lying over the spine of the scapula and the head of the humerus.

Origin: lower portion of the spine of the scapula.

Insertion: deltoid ridge of the humerus.

Action: flexes humerus and rotates it outward.

Acromiodeltoid. This is a band of muscle running over the proximal portion of the humerus and across the distal end of the spinodeltoid.

Origin: acromion of the scapula.

Insertion: the fascia of the spinodeltoid. Together the two muscles insert on the deltoid ridge of the humerus.

Action: flexes humerus and rotates it outward.

Levator scapulae ventralis. This is a long ribbon of muscle extending from the atlas to the metacromion of the scapula. It can be seen superficially only in the angle between the acromiotrapezius and the clavotrapezius.

Origin: transverse process of atlas and basioccipital bone in the skull.

Insertion: metacromion of the scapula.

Action: draws scapula dorsad.

Triceps brachii. This is the most obvious muscle of the upper forelimb. It consists of three parts, or ***heads***, which all share a common insertion and action.

Insertion: olecranon process of ulna.

Action: extends forearm.

The three heads of the triceps are:

Long head. This is a very thick, large muscle comprising the caudal margin of the arm.

Origin: border of glenoid fossa of scapula.

Lateral head. This is also a large muscle, adjacent to the long head, covering the medial area of the upper arm.

Origin: deltoid ridge of humerus.

Figure 24-24. Lateral view of the muscles of the scapula and forelimb of the cat, with associated skeletal structures.

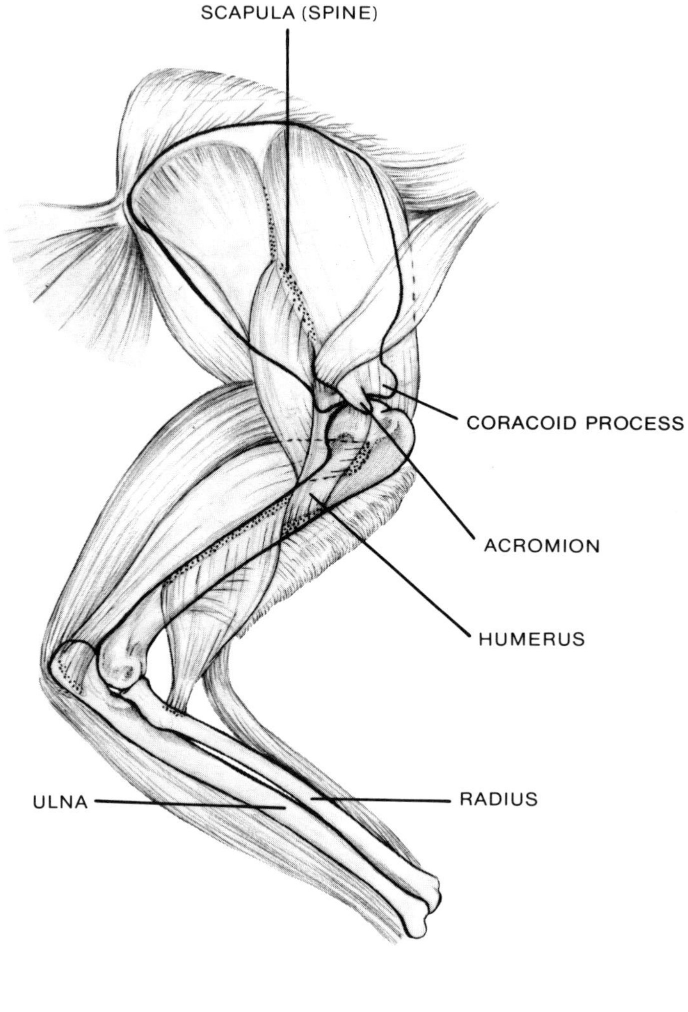

Medial head. Transecting and reflecting the lateral head reveals the narrow medial head just beneath it.

Origin: proximal end of dorsal side of humerus.

Anconeus. This is a triangular muscle lying at the elbow below the distal end of the lateral head of the triceps. Some consider it to be a fourth head of the 'triceps.'

Origin: dorsal side of the distal end of humerus.

Insertion: olecranon process of ulna.

Action: tenses elbow joint capsule.

Brachialis. With the lateral head reflected, it is easy to see the brachialis, a long band of muscle running along the humerus, cranially, adjacent to the medial head of the triceps.

Origin: the distal lateral surface of the humerus.

Insertion: just distal to the trochlear notch.

Action: flexes forearm.

Muscles of the Scapula and Arm: Medial View (Figs. 24-18, 24-25, 24-26)

Dissect the ventral chest region and the medial surface of the forelimb. This dissection provides an opportunity to study the architecture of a joint and the action of muscles surrounding the joint.

Subscapularis. (See Fig. 24-25) This large muscle covers the medial surface of the scapula. It can be partially seen in a dorsal view between the teres major and the infraspinatus.

Origin: entire surface of subscapular fossa.

Insertion: lesser tubercle of humerus.

Action: adducts humerus.

Figure 24-25. Medial view of the superficial muscles of the scapula and forelimb of the cat.

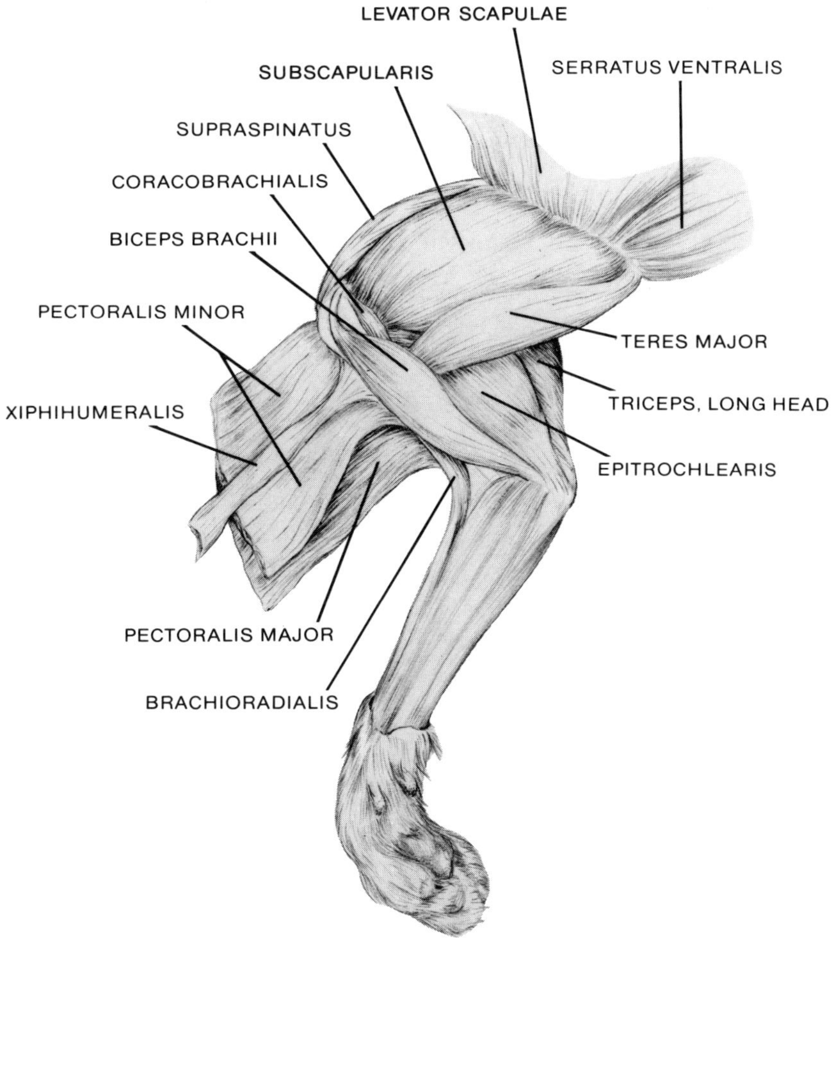

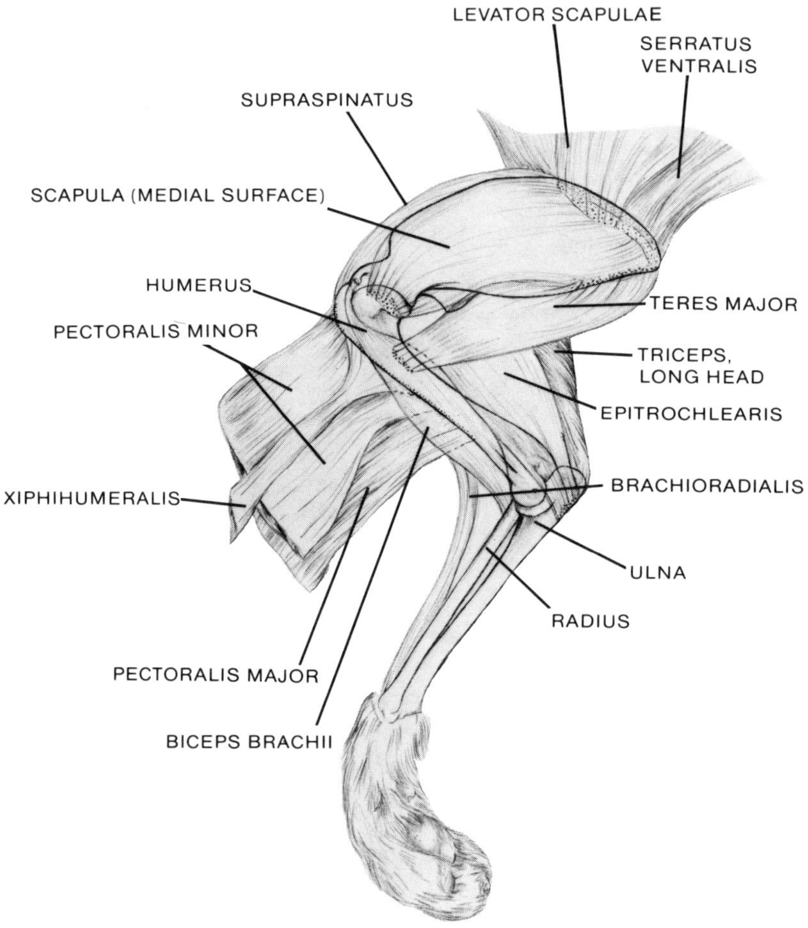

Figure 24-26. Ventral view of the superficial muscles of the scapula and forelimb of the cat, with associated skeletal structures.

The Shoulder Joint

Cut the small coracobranchialis muscle (Fig. 24-25) to see the subscapularis insertion clearly. Cut the subscapularis fibers away from their origin by passing a scalpel along the medial surface of the scapula. Leave the serratus, teres major, and supraspinatus muscles intact. The subscapularis origin is extensive, so your scalpel must be passed along the scapula several times. Reflect the muscle toward its insertion. The pinnate fiber architecture is obvious on the deep surface of the muscle; in order to study its actions, locate the shoulder joint, itself. Its membranous capsule lies just deep to the muscle (portions of the capsule may even adhere to the muscle). If you have difficulty identifying the position of the joint, rotate the humerus slightly. Note the position of the subscapularis muscle relative to the joint. It lies so close to the medial portion of the capsule that it has relatively little leverage in humeral adduction. It can rotate the humerus against the glenoid fossa and has weak leverage in humeral retraction or flexion. Put the muscle back into position and note that most of the force would be directed through the joint axis and would act to lock the surfaces together. The major function of the subscapularis (and the infra- and supraspinatus muscles) is to maintain the joint and control movement of the joint surfaces while muscles, such as the teres major retract or protract the limb.

Reflect the muscle again and examine the shoulder joint capsule with forceps. There are no obvious ligaments, and it tears rather easily. The shoulder is an unusual joint because it is not maintained by ligaments; rather, the subscapularis, infraspinatus, and supraspinatus muscles are largely responsible for maintaining the joint and permit a wider range of movement than would ligaments. *Open the joint capsule and examine the smooth, cartilage-covered joint surfaces.* This is a synovial joint (see section on arthrology following this rundown of the muscles).

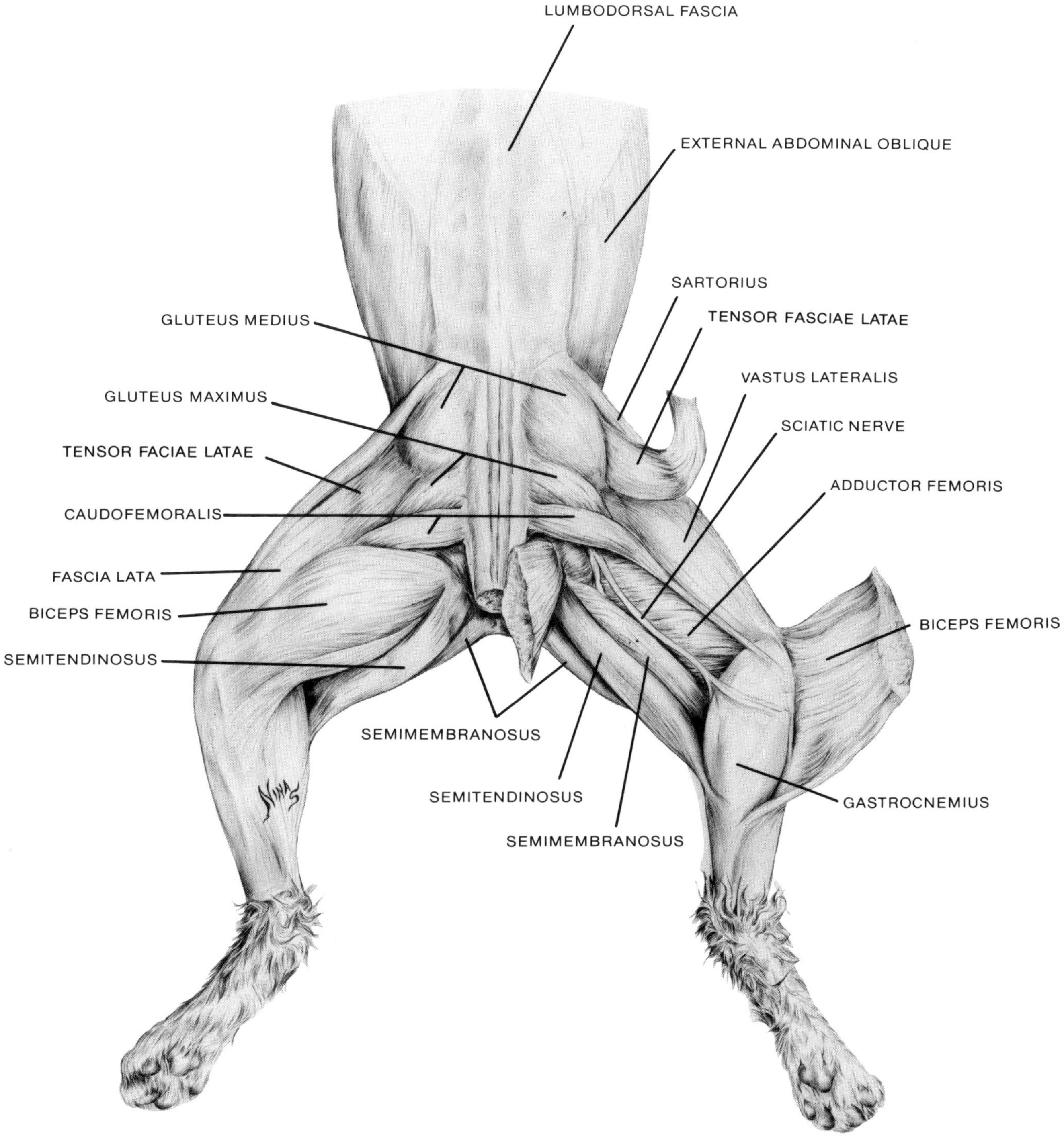

Figure 24-27. A comparison of superficial (left side) and deep (right side) muscles of the hindlimb's lateral surface. Flaps indicate transected and reflected tensor fasciae latae and biceps femoris muscles.

Biceps brachii. (**Fig. 24-25**) This appears as a bar of muscle lying along the ventral surface of the humerus beneath the pectoralis major and minor. In the cat it has only one head, though two are found in humans.

Origin: coracoid process of scapula.

Insertion: radial tuberosity of radius.

Action: flexes forearm.

Epitrochlearis. (**Fig. 24-25**) This is a very thin sheet of superficially visible muscle lying over the ventral surface of the long head of the triceps.

Origin: ventral margin of latissimus dorsi.

Insertion: olecranon process of ulna.

Action: extends forearm.

Superficial Muscles of the Hindlimb: Lateral View (Figs. 24-19, 24-27)

Separate and identify the superficial muscles of the lateral surface of the hindlimb.

Tensor fasciae latae. This is a wedge of muscle on the cranioproximal portion of the hindlimb. It extends from the fascia covering the gluteus medius to the fascia lata, the great sheet of tissue covering the lateral surface of the thigh. The main mass of the muscle gives off a short strap from the cranial margin that extends down into the fascia; this should not be confused with the adjacent sartorius.

Origin: ventral border of ilium.

Insertion: fascia lata.

Action: pulls on fascia; extends leg.

Biceps femoris. This is a very large mass of muscle covering the greater portion of the lateral surface of the leg.

Origin: ischium.

Insertion: patella and proximal third of the tibia.

Action: retracts the limb; flexes shank.

Deep Muscles of the Hindlimb: Lateral View (Fig. 24-28)

Transect and reflect the fascia lata and the biceps femoris. Avoid disturbing translucent tendonous material linking the caudofemoralis to the knee. Avoid the thin tenuissimus muscle located beneath the caudofemoralis.

Figure 24-28. Deep muscles of the hindlimb's lateral surface, showing the gluteus medius and caudofemoris muscles transected and reflected to show underlying structures.

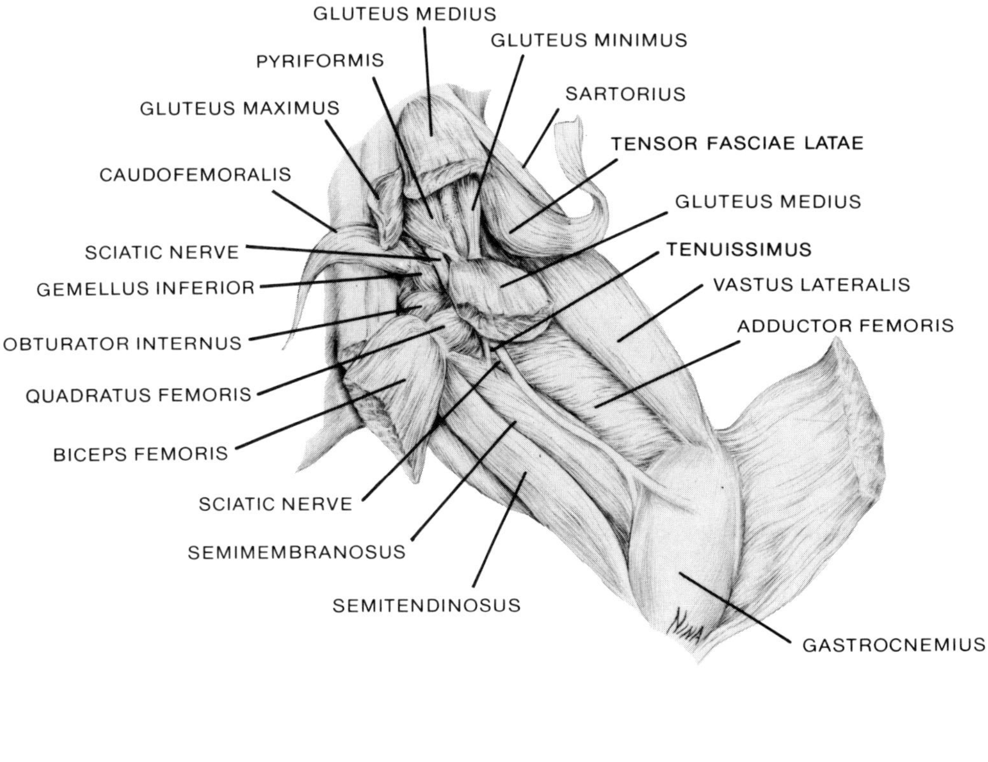

Gluteus medius. This is a thick mass of muscle found medial to and partly beneath the tensor fasciae latae.

Origin: transverse processes of the last sacral and first caudal vertebrae.

Insertion: greater trochanter of femur.

Action: abducts thigh.

Caudofemoralis. This strap of muscle extends laterally from the midline through the medial portion of the leg. It is superficial near the middorsal origin, but it is covered for the most part by the biceps femoris.

Origin: transverse processes of second and third caudal vertebrae.

Insertion: patella.

Action: abducts thigh; extends shank.

Gluteus maximus. This is a small muscle lying near the middorsal line just cranial to the caudofemoralis and partially overlapped by it.

Origin: transverse processes of last sacral and first caudal vertebra.

Insertion: greater trochanter of femur.

Action: abducts thigh.

Vastus lateralis. This is a large muscle lying beneath the fascia lata; it comprises the mass of the anterior portion of the lateral surface of the leg.

Origin: greater trochanter and shaft of femur.

Insertion: patella.

Action: extends shank.

Tenuissimus. This is a thin string of muscle lying just beneath the biceps femoris, running from the midline to the lower leg.

Origin: transverse process of second caudal vertebra.

Insertion: fascia of biceps femoris.

Action: abducts thigh.

Semitendinosus. This is a thick strap of muscle forming, along with the gracilis, the caudal margin of the upper leg.

Origin: ischium.

Insertion: proximal end of tibia.

Action: flexes shank; extends thigh.

Gastrocnemius. This is a large mass of muscle on the posterior portion of the lower leg. It consists of two heads, the ***lateral head*** and the ***medial head***.

Origin: the lateral head from a complex of origins on the patella and fascia of shank; the ***medial head*** from the distal portion of the femur.

Insertion: on the heel (calcaneus) via the tendon of Achilles.

Action: extends foot.

Superficial Muscles of the Hindlimb: Medial View (Figs. 24-15, 24-29, 24-31)

Separate and identify the superficial muscles of the medial surface. Students with male specimens must avoid disturbing the spermatic cords and related blood vessels.

Sartorius. This is a large flat sheet of muscle covering the cranial half of the medial surface of the leg. The sartorius may also be seen in the lateral view along the anterior margin of the leg.

Origin: ilium.

Insertion: proximal end of the tibia and patella.

Action: adducts and rotates thigh; extends tibia.

Gracilis. Another large flat sheet of muscle covering the other half of the medial surface of the hindlimb.

Origin: ischium and pubic symphysis.

Insertion: proximal end of tibia.

Action: adducts thigh.

Deep Muscles of the Hindlimb: Medial View (Figs. 24-29, 24-30, 24-31, 24-32)

Transect and reflect the sartorius and gracilis muscles. Once these are reflected, identify, transect, and reflect the vastus medialis muscle to observe the deeper vastus intermedius muscle.

Quadriceps. This is a group of four muscles, all inserting on the patella. We have already discussed the ***vastus lateralis***, which appears in this view as a thick band of muscle running along the anterior margin of the leg beneath the sartorius. The other three muscles of the group are: the ***rectus femoris***, ***vastus medialis***, and ***vastus intermedius***.

Rectus femoris. This is found beneath the sartorius, running down the anterior portion of the leg, adjacent to the vastus lateralis.

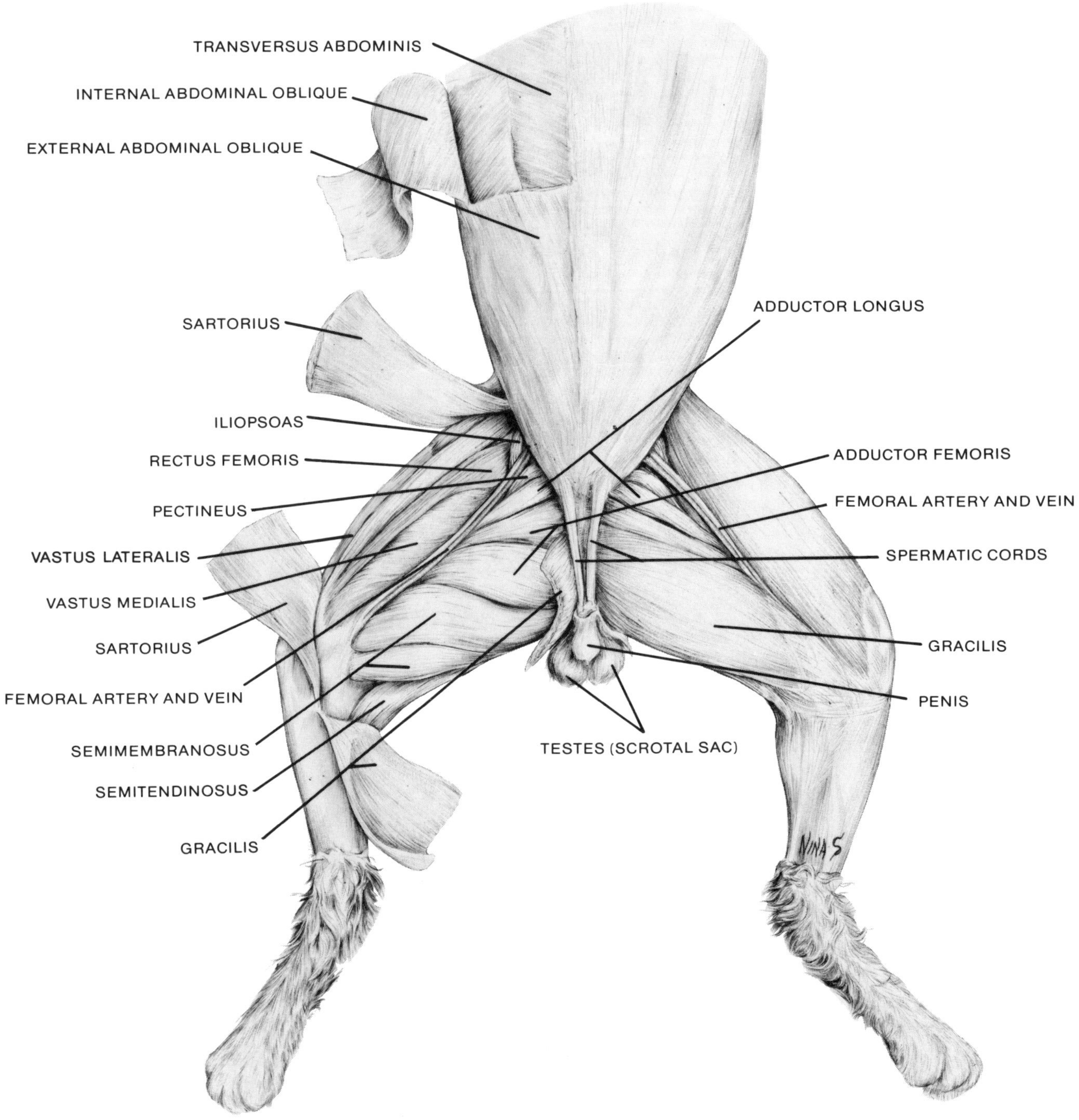

Figure 24-29. A comparison of superficial (right side) and deep (left side) muscles of the hindlimb's medial surface. Flaps indicate transected and reflected sartorius and gracilis muscles.

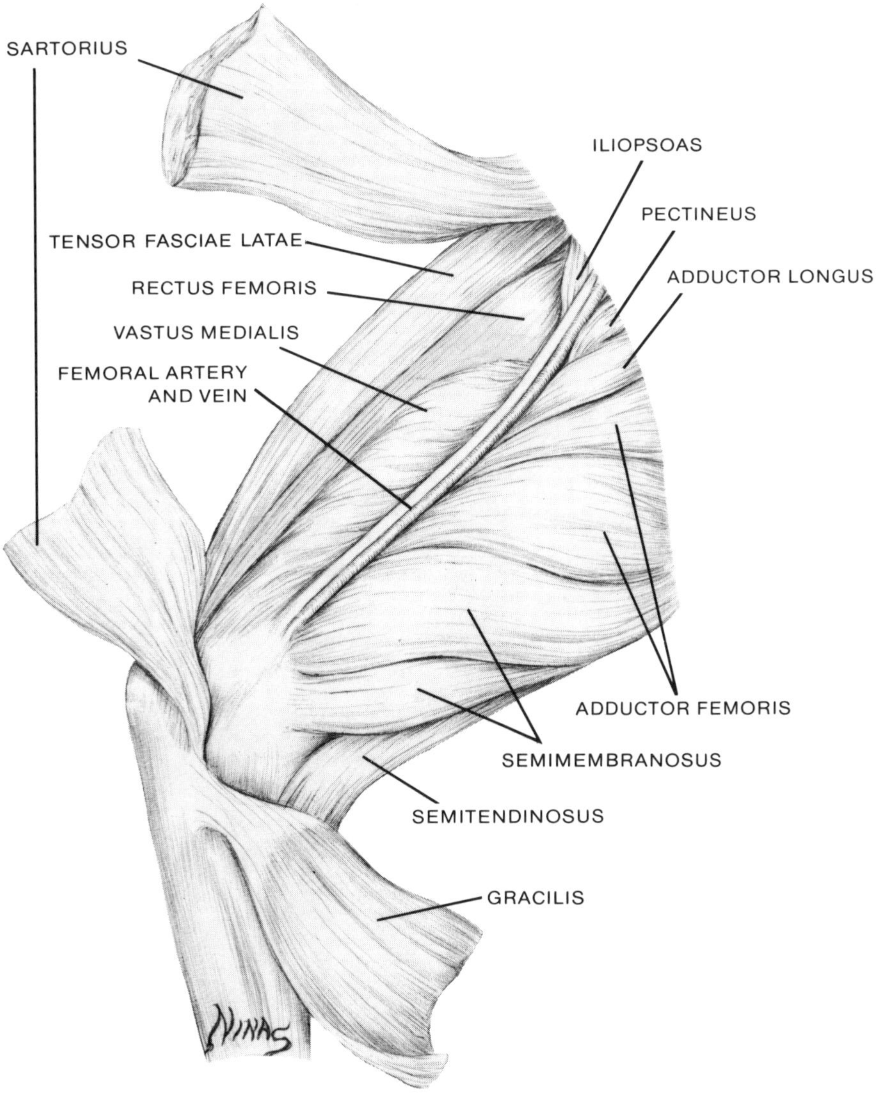

Figure 24-30. Medial view of the muscles of the hindlimb. Flaps indicate the transection and reflection of the sartorius and gracilis muscles.

Origin: ilium.

Action: flexes hip; extends knee.

Vastus medialis. This runs parallel with and medial to the rectus femoris. Distally it overlaps the rectus femoris.

Origin: shaft of femur.

Action: extends leg.

Vastus intermedius. Transecting and reflecting the vastus medialis reveals the vastus intermedius below, running adjacent to the rectus femoris, along the femur.

Origin: shaft of femur.

Action: extends leg.

Iliopsoas. This muscle runs along the ventral side of the vertebral column. Here it is possible to see the terminal portion of the muscle extending out of the abdominal cavity to a proximal medial position on the leg.

Origin: ilium; centra and transverse processes of the lumbar and the most posterior thoracic vertebrae.

Insertion: lesser trochanter of femur.

Action: rotates and flexes thigh.

Adductor femoris. This is a thick, triangular muscle extending from the midventral line to the femur.

Origin: ischium and pubis.

Insertion: shaft of femur.

Action: adducts and extends thigh.

Adductor longus. This is a thin ribbon of muscle running along the cranial margin of the adductor femoris from the midventral line to a distal point on the shaft of the femur.

Origin: pubis.

Insertion: midportion to distal portion of shaft of femur.

Action: adducts thigh.

Pectineus. This is a thinner, shorter muscle running parallel to the adductor longus, along its cranial margin.

Origin: pubis.

Insertion: proximal portion of shaft of femur.

Action: adducts thigh.

Semimembranosus. This is a large, thick mass of muscle caudally adjacent to and partially overlapped by the adductor femoris. The muscle can be separated into two almost equal parts.

Origin: ischium.

Insertion: distal end of femur; proximal end of tibia.

Action: extends thigh, flexes knee.

FUNCTION

Although we have been considering the musculoskeletal system as a single functional unit, it is obvious that this system has two distinctive elements, each with a separate role. The skeleton is the framework on which the muscles operate to control posture and movement; and the skeleton resists compression, bending, and torsion while the muscles generate tension.

Figure 24-31. Deep muscles of the hindlimb's medial surface. The vastus medialis muscle is transected and reflected to show the vastus intermedius.

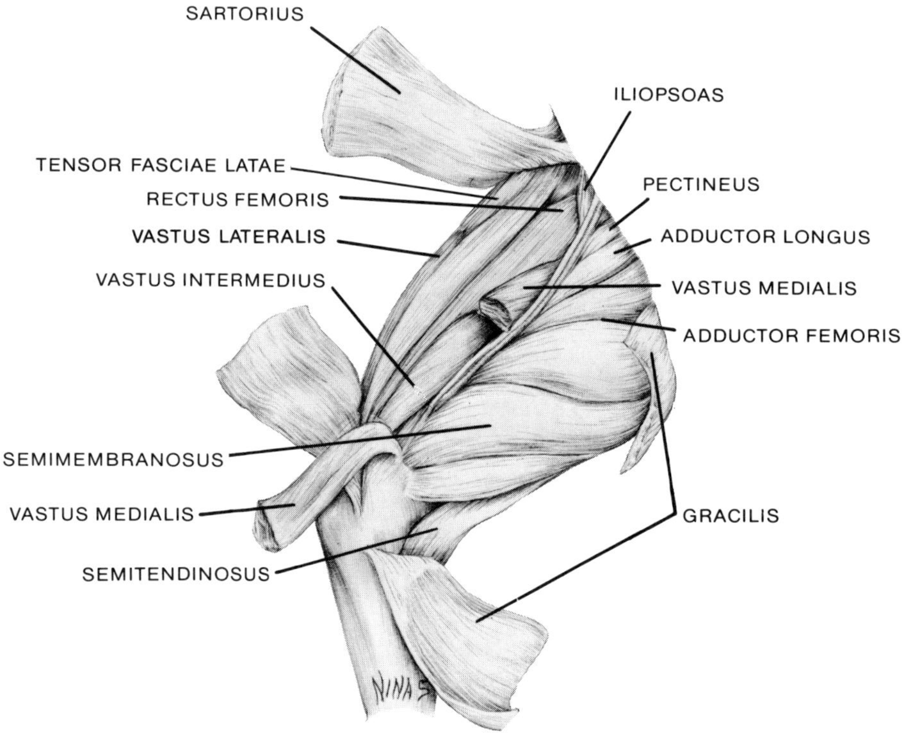

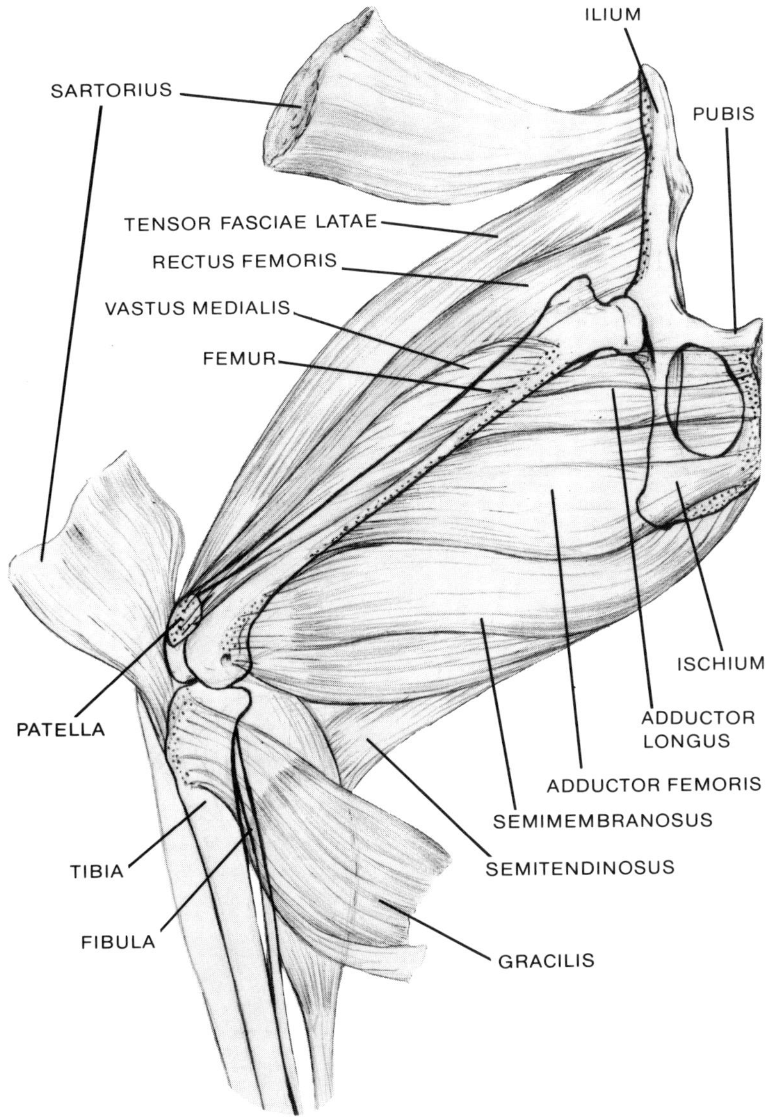

Figure 24-32. Medial view of the muscles of the hindlimb, with associated skeletal structures. Flaps indicate the transection and reflection of the sartorius and gracilis muscles.

Arthrology

Some bones are joined into units that resist mechanical displacement, whereas others move relative to one another. The points of fixed or flexible contact are called ***articulations***. These articulations determine the kinds of rotations and translations that bony elements may perform with respect to each other.

1. ***Fibrous articulations*** are those in which bones are held together by fibrous connective tissue. These may involve immovable apposition, as in the sutures of the skull of some vertebrates, or a freer connection using longer strands of connective tissue, as in the distal articulation of the tibia and fibula in certain mammals.

2. ***Fibrocartilaginous articulations*** are formed by dense fibrous connective tissue and cartilage. They are designed for situations requiring a great deal of stability yet with the possibility of some movement. Fibrocartilaginous articulations are found between the vertebrae and the sternebrae in many vertebrates and at the pubic symphysis. This type of connection at the pubic symphysis is of great importance to mammals since, under hormonal influence, the fibrous connection may be loosened at the time of birth to allow for enlargement of the birth canal.

Table 24-1. Summary of major muscles of the cat and their action.

Muscle	*Action*
Abdominal Muscles	
External oblique	Compresses abdomen
Internal oblique	Compresses abdomen
Transversus abdominis	Compresses abdomen
Rectus abdominis	Compresses abdomen, draws ribs & sternum caudad
Thoracic Muscles	
Superficial:	
Pectoantebrachialis	Adducts forelimb
Pectoralis major	Adducts and retracts forelimb
Pectoralis minor	Adducts and retracts forelimb
Xiphihumeralis	Adducts forelimb
Deep:	
Scalenus	Bends head to one side; pulls rib cage forward
Transversus costarum	Draws sternum craniad
Levator scapulae	Draws scapula cranioventrad
Serratus ventralis	Draws scapula to thorax, supports trunk
External intercostal	Pulls ribs forward
Internal intercostal	Pulls ribs backward
Muscles of the Back	
Superficial:	
Clavotrapezius	Draws scapula craniodorsad
Clavobrachialis	Flexes forearm
Acromiotrapezius	Draws scapula dorsad
Spinotrapezius	Draws scapula caudodorsad
Latissimus dorsi	Pulls arm caudodorsad
Deep:	
Splenius	Turns and elevates head
Rhomboideus	Draws scapula dorsad
Rhomboideus capitis	Draws scapula craniad
Serratus dorsalis	Draws last four ribs caudad
Longissimus dorsi	Extends vertebral column
Multifidus spinae	Extends or bends vertebral column
Spinalis dorsi	Extends vertebral column
Iliocostalis	Draws ribs together
Muscles of the Head and Neck	
Superficial:	
Sternomastoid	Turns head
Cleidomastoid	Pulls head down or clavicle forward
Sternohyoid	Draws hyoid posteriorly or sternum forward
Mylohyoid	Raises floor of mouth
Stylohyoid	Raises hyoid
Digastric	Lowers mandible
Masseter	Elevates mandible

Muscle	Action
Deep:	
Geniohyoid	Pulls hyoid craniad
Hyoglossus	Retracts tongue
Styloglossus	Retracts tongue
Genioglossus	Draws tongue forward
Sternothyroid	Draws larynx caudad
Thyrohyoid	Raises larynx
Cricothyroid	Tenses vocal cords
Muscles of the Scapula and Proximal Forelimb	
Lateral:	
Supraspinatus	Extends shoulder
Infraspinatus	Rotates humerus outward
Teres major	Flexes and rotates humerus
Teres minor	Rotates humerus
Acromiodeltoid	Flexes and rotates humerus
Spinodeltoid	Flexes and rotates humerus
Levator scapulae ventralis	Draws scapula dorsad
Triceps brachii	Extends forearm
Brachialis	Flexes forearm
Anconeus	Tenses elbow joint capsule
Medial:	
Subscapularis	Adducts humerus
Biceps brachii	Flexes forearm
Epitrochlearis	Extends forearm
Muscles of the Proximal Hindlimb	
Lateral:	
Tensor fasciae latae	Pulls on fascia latae, extends shank
Biceps femoris	Retracts leg
Gluteus medius	Abducts thigh
Caudofemoralis	Abducts thigh & extends leg
Gluteus maximus	Abducts thigh
Vastus lateralis	Extends shank
Tenuissimus	Abducts thigh
Semitendinosus	Flexes shank
Medial:	
Sartorius	Adducts and rotates thigh
Gracilis	Adducts thigh
Rectus femoris	Flexes hip and extends knee
Vastus medialis	Extends leg
Vastus intermedius	Extends leg
Iliopsoas	Rotates and flexes thigh
Adductor femoris	Adducts and extends thigh
Adductor longus	Adducts and extends thigh
Pectineus	Adducts thigh
Semimembranosus	Extends thigh

3. ***Synovial articulations*** offer the widest range of mobility and are the most complex. In the synovial joint (Fig. 24-33) the articulating surface of one bone is usually convex, that of the other concave. Each surface is covered with cartilage, which cushions the opposing elements. The whole joint is enclosed by a fibrous capsule which contains fluid that functions to lubricate and supply nutrients to the articular cartilages. Occasionally, as in the knee and the jaw, the space within the capsule is traversed by a disc or ***meniscus*** of connective tissue, which allows extra independent motion for both of the articulating bones.

There are many types of synovial joints, each designed to fulfill particular functional needs. The atlas-axis joint provides for rotation; the humerus-ulna hinge joint provides for movement around a single axis (Fig. 24-34); and the femur-acetabular ball-and-socket joint provides rotation on two axes. However, even within a specific type of joint, there is variation. The scapula-humerus articulation (Fig. 24-34) is also a ball-and-socket joint, but because of the wider range of movement required by the forelimbs of the tetrapods, the bony components of this joint have been reduced. Stability has been provided by the increase in elastic fibrous capsule and surrounding muscle. In the cat, there are considerably more muscles extending from the trunk over the girdle-limb joint in the pectoral region than over the joint in the pelvic region. Of course, these extra muscles require more complex innervation and yield greater variability in use of the forelimb.

Muscle Attachment to the Skeleton

Muscles in the vertebrate body can be divided into three categories according to type: (1) striated muscles, which are controlled voluntarily and closely associated with the skeleton; (2) cardiac muscle, which is also striated but of a type unique to the heart; and (3) smooth, involuntarily controlled muscles. Smooth muscles invest the walls of many of the organs and, through contraction, aid the functioning of those organs. Striated muscle, which moves the parts of the skeletal system, is the type treated below,

Muscles always attach to bone by connective tissue and not by the muscle fibers themselves. The attachment can be accomplished in either of two ways: (1) through a ***tendon***, which is a narrow band of connective tissue extending between muscle and bone; or (2) through an ***aponeurosis***, which is a broad sheet of connective tissue extending between bone and muscle.

Muscles operate by generating forces. In some cases they pull a distal element attached to one end of the muscle toward the other end, which is anchored. The anchored, relatively immovable, point of attachment is called the ***origin***. The bony element at this point is often modified for its role of anchoring the muscle. For example, the spinous processes of the thoracic vertebrae, the origin of the splenius muscle that supports the weight of the head, are enlarged and oriented anteriorly to provide a maximum lever advantage.

The point of bone attachment for the other end of the muscle is called the ***insertion***. The location and means of insertion can have great effect on the action produced. A good example of this is the insertion of the spinodeltoid muscle on the humerus (Fig. 24-23). The spinodeltoid originates on the lower portion of the scapular spine; it inserts on the upper portion of the humerus. Basically, it acts to pull back the forelimb. But the insertion of this muscle is along the deltoid ridge, a special attachment process on the humerus that slants from a proximal, lateral point down to a more distal point on the middorsal surface. Because this insertion is at an angle, contraction not only pulls the humerus back, but rotates it outward as

Figure 24-33. Idealized view of the synovial joint. (After Crouch.)

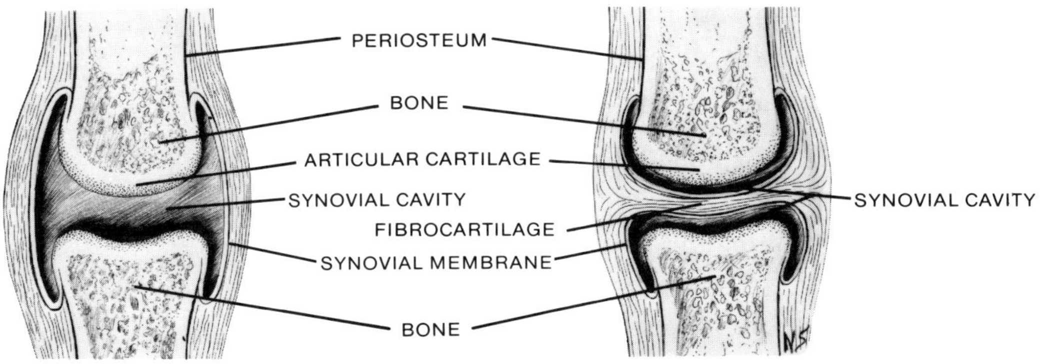

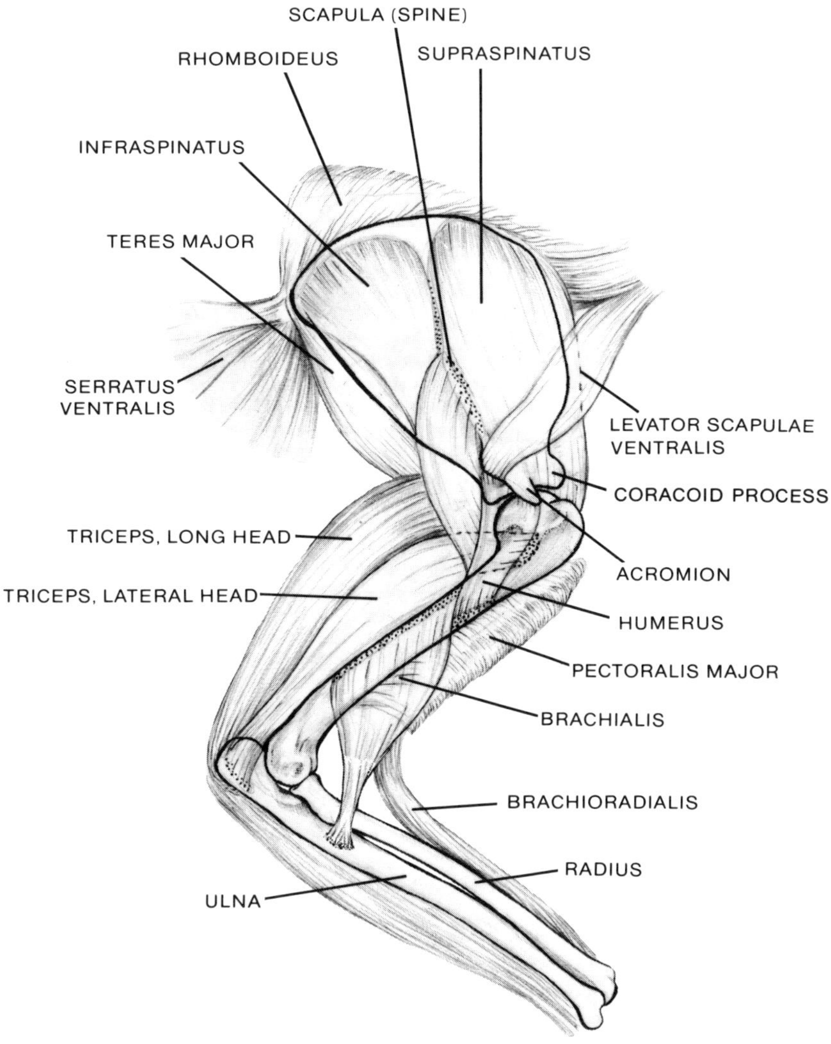

Figure 24-34. Right lateral view of the joints of the scapula-humerus and of the humerus-radius-ulna of the cat.

well. Again, the morphology of the bone is modified to function effectively with the muscular system.

The term ***flexion*** refers to the muscle action that moves a joint or extremity toward a bent position in which the distal part is closer to the proximal part of a given structure. ***Extension*** is the reverse action and results in an increase in the angle between the bones involved. Muscles that ***protract*** move a structure away, and those that ***retract*** pull it back toward its foundation. In a few cases vertebrates use skeletal elements as lever arms on which muscles act. In Fig. 24-35, it is clear that the ***power arm*** is much shorter than the ***load arm***. Muscles generate the most power when they contract slowly. The shortness of the power arm represents a sacrifice of power for amplification of speed. The proximity of the insertion point to the axis of movement—the fulcrum—also helps to keep the bones meeting at a joint together, thus strengthening the joint.

The basic short-power-arm system is modified, of course, to serve the particular needs of each animal. Both origin and insertion sites vary among species. Figure 24-36 shows that the insertion of the teres major of the badger is distal to that of the cheetah. The resultant longer power arm yields a significant strength advantage in the badger's toes. But the price is loss of speed. The cheetah achieves a speed advantage with its shorter power arm, but it sacrifices power. Each arrangement is effective for the animal's particular way of life.

All body movements are very complex processes; whole groups of muscles cooperate to produce the

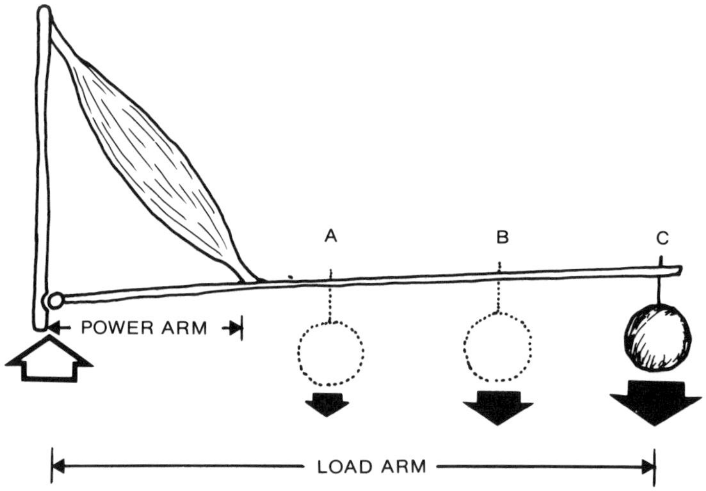

Figure 24-35. Diagram of a lever system involving a muscle. The joint acts as a hinge and fulcrum. If the power arm remains constant, greater muscular contraction is required to suspend or raise the weight as the load arm is increased by moving the weight from A to B to C.

Figure 24-36. An idealized view of the power-arm system of the cheetah and badger drawn so that z is to the same scale. Distance y is longer than x, indicating that the badger has a relatively longer power arm than the cheetah. Distance z is the length of the leg. (After Hildebrand.)

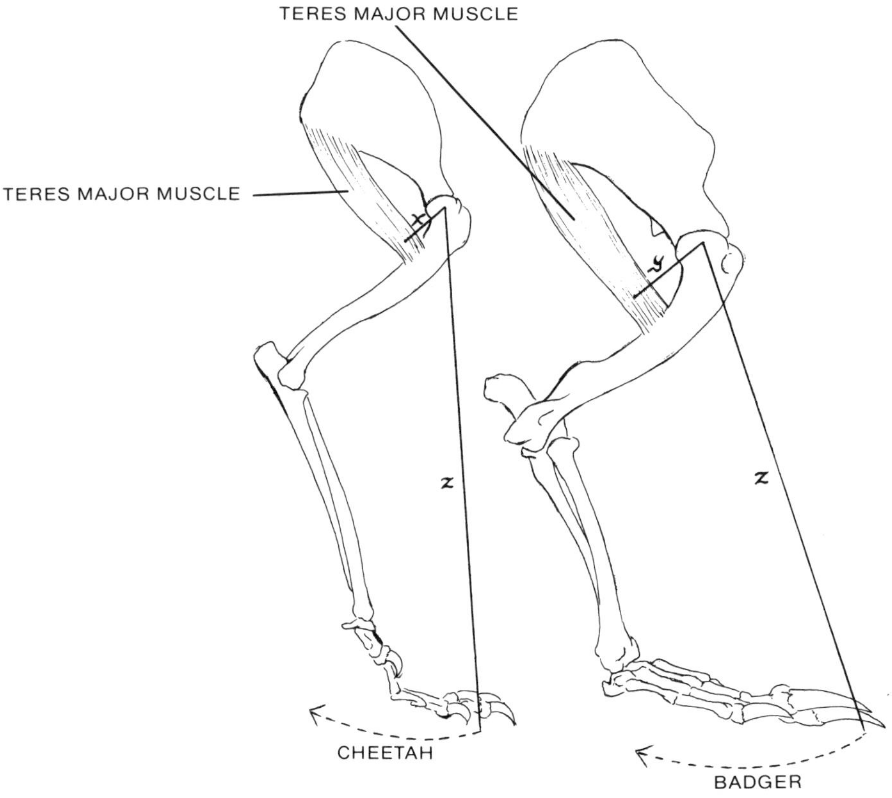

desired action. To see how sets of muscles act together and how such muscle actions are described, it is helpful to use the human hand as an example.

In some kinds of movement there is one particular muscle called the ***prime mover***. As the name implies, this muscle effects the basic action. In the hand, there are digital extenders which operate as prime movers. As they carry out their action, the wrist is also extended. So, in order to extend only the fingers and not the wrist, other muscles must be called into play—in this case the wrist flexor. Muscles that fill such stabilizing roles are called ***fixators***. To complete the extension of the terminal elements of the fingers, other muscles of the hand and finger must also be brought into action. These muscles are termed ***synergists***. To obtain the precise movement required in the use of the hand, a final group, the ***antagonists***, must operate. In quick movements, the antagonists relax completely, operating only to protect the joint from hyperextension. But in slow, metered manipulations, the balanced effect of both prime movers and antagonists is required. Actual electromyographical measurements of complex muscle systems, such as in the hand or jaw, suggest that the prime mover-fixator-synergist-antagonist hypothesis is an oversimplification. The exquisite complexity of the nervous control of muscle contraction apparently permits much finer tuning than the hypothesis implies.

The cleidomastoid in the cat offers a good example of how cooperative effort between muscles may control an action. The cleidomastoid can operate either to turn the head or to pull the clavicle forward. The action depends on associated fixators. If the clavicle is fixed, it becomes the origin, and the head turns. If the head is fixed, it becomes the origin, and the clavicle moves forward.

Another operation that requires cooperative muscular effort is respiration. In the cat, inspiration is brought about through contraction of the diaphragm and forward movement of the ribcage. This occurs through action of the *scalenus* and anterior slips of the *serratus dorsalis*. Expiration is brought about by a constriction of the abdominal muscles that pushes the viscera anteriorly against the relaxed diaphragm. At the same time, the natural elasticity of the ribcage and action of the *tranversus thoracic* and posterior slips of the *serratus dorsalis* move the ribcage posteriorly. On the basis of the anatomy of the cat, it would seem that the intercostal muscles should have a fundamental role in respiration. Those muscles may maintain inter-rib distances as well as help expand or contract the cage.

NATHAN SCHAFER

25

The Cat: Digestive and Respiratory Systems

As noted in the discussion of the shark, the digestive system consists of the alimentary canal, or "gut," and certain associated glands. In a mammal such as the cat, the alimentary canal includes the mouth, oral cavity, pharynx, esophagus, stomach, small intestine, large intestine, and anus. Glands of the digestive system include the salivary glands, liver, and pancreas, all of which are physically separate from the gut, though connected with it by ducts. The respiratory system, as defined by routes of air flow, consists of the nasal cavities, pharynx, larynx, trachea, and lungs.

There were many problems attendant to ingesting large pieces of food rather than filtering fine particles from the water as was done by the first chordates. Various vertebrates grind their food in pharyngeal mills, rasp it to pieces, or grind it by masticating. Chemical breakdown of food follows the processing in the mouth, pharynx, or gizzard (birds). Because chemical digestion is time-consuming, storage organs are necessary in which the digestion can occur. Two major trends in vertebrate evolution are elongation of the digestive tract and increase in its internal surface area. By tracing the digestive tract of the cat, we can note some of the important changes that have occurred.

In mammals, and land vertebrates in general, the portion of the digestive tract anterior to the pyloric sphincter takes on greater importance. Specifically, the esophagus is greatly elongated and the stomach becomes more highly developed. In most vertebrates, the stomach functions both as a storage and as a treatment center for food prior to entry into the intestine. The development of the neck, a distinctive structure in the reptiles and their descendants, contributes to even greater lengthening in the esophageal region.

In many mammals the intestine is substantially longer and more highly developed on its internal surface than it is in lower vertebrates. As discussed in Chapter 9, the spiral valve in the shark intestine slows the movement of food through the intestine and increases its absorptive surface area. The cat accomplishes these ends with the macrovilli and microvilli which line the inner surface of its intestine. The macrovilli are finger-like projections of the intestinal epithelium that greatly increase the absorptive surface area. The microvilli are minute projections on the luminal surface of individual epithelial cells that cover the macrovilli.

The digestive process continues beyond the small intestine in the colon. Here, foods that have resisted digestion are treated further. In the mammalian large intestine, food breakdown is facilitated by the abundant intestinal flora, which contributes to the digestive process as well as supplying the host with certain nutrients. In the large intestine, water is resorbed and fecal formation occurs.

The cat digestive tract terminates at the anus; in the shark, the tract ends at the cloaca. The reduction of the cloaca evolved concomitantly with separation of channels for waste products and gametes. Why this alteration is advantageous is not clear (though some have suggested aesthetics as a cause!).

The digestive tract would not be able to perform its function without the assistance of two important parts of the system: the liver and the pancreas.

Amphioxus, a cephalochordate, probably does not possess a liver. All true vertebrates, however, possess a liver which develops as an outpocketing of the embryonic digestive tract. In mammals, the liver produces bile, which it secretes into the intestine to aid digestion. In addition, the liver has important functions as an organ of metabolism involved in the storage and processing of nutrients.

The pancreas is responsible for the production of many digestive enzymes. It too develops as an evagination of the embryonic gut. Pancreas-like cells are seen in amphioxus and in cyclostomes, and a discrete pancreas is present in the shark. The pancreas consists of two distinct—and seemingly unrelated—functional units: the endocrine pancreas, which produces insulin and glucagon, and the enzyme-producing exocrine pancreas.

As vertebrates moved onto land, problems resulted from the absence of a liquid environment. Food was not ingested in a volume of water; it may have been more difficult to handle during its early stages in the alimentary canal. Salivary glands evolved in aquatic vertebrates to lubricate the food and assist in swallowing. Secondarily, the salivary glands came to have a digestive function, secreting enzymes into the saliva. Some think this may have been related to an initial use of the saliva as "toothpaste." The muscular tongue evolved as an aid to manipulating food during chewing. And, in fact, tongues show important modifications among reptiles, birds, and mammals that permit them to be used for many specialized functions besides simple food manipulation.

Respiration also underwent many changes in the course of vertebrate evolution. Respiratory capacity depends on many parameters, one of the most important being the amount of surface area available for gas exchange. Thus, one of the changes associated with increased oxygen demand is a larger respiratory surface area. The surface area may be increased either by increasing the size of the respiratory organ or by increasing the amount of surface area only within the gas exchange tissues, maintaining organ size as a constant. The latter seems to have occurred in both mammals and birds. Mammalian bronchi, for instance, are more branched than are those in earlier vertebrates; the number of terminal branchings and alveolar sacs involved in gas exchange is greatly expanded. This increases the surface area over which diffusion of oxygen and carbon dioxide can occur.

The movement of air in and out of the lungs is caused by the compression and expansion of the chest cavity. In mammals, this movement is effected by the diaphragm and the ribs with their associated muscles. The reptiles, predecessors of mammals, have an expandable ribcage, but not a diaphragm. Because chest muscles can cause expansion of the cavity in which the lungs reside, the pressure on the outside of the lungs is lowered below atmospheric pressure. As a result, air moves through the nasal passages and into the lungs from outside the organism.

In mammals, the lungs lie in separate pleural cavities. Posterior to the lungs is the diaphragm. This muscular sheet moves posteriorly and flattens during inspiration; at the same time the ribcage expands. Both actions increase the volume of the pleural cavities, and, as in reptiles, air enters the lungs due to the difference between external and internal pressure.

The anterior end of the respiratory tract in mammals has also undergone evolutionary modification. In particular, the food and air passages are separated in all regions except the pharynx, where the two passages cross. There, the epiglottis guards the trachea from the entry of food particles. This region of the larynx and its associated sound-producing elements is, in turn, the sight of crucial alterations.

All the changes seen in the modern mammalian digestive and respiratory systems may, in some sense, make these systems more "efficient." These changes make it possible to maintain a high, constant body temperature, high activity levels, and the generally complex biology of mammals becomes possible.

BODY CAVITIES, PLEURA, AND PERITONEUM (Figs. 25-1, 25-2)

The following discussion of cavities and their associated membranes is presented here as an aid for subsequent dissections.

In the mammal, the bulk of the digestive system, the lungs, and the heart, are all suspended in the coelomic cavities. The mammalian coelom is composed of four main parts: the ***abdominal cavity***, the ***pericardial cavity***, and two ***pleural cavities***. The main coelomic cavities, such as the pleural cavities housing the lungs, are lined with sheets of coleomic epithelial tissue, which are designated as ***pleura***, ***peritoneum***, or ***pericardium***.

Thoracic cavity. The mediastinum, the pericardial, and the two pleural cavities comprise the thoracic cavity. This cavity is bounded by the body wall containing the ribs and sternum, the vertebral column, and the diaphragm. Thus it comprises the coelom anterior to the diaphragm.

Mediastinum. This is a space between the medial walls of the parietal pleura of the left and right pleural cavities. The mediastinum contains the ***pericardial cavity***, which surrounds the heart, and the trachea, esophagus, thymus, and associated blood vessels.

Parietal pleura. The parietal pleura is a sheet of coelomic epithelial tissue that lines the pleural cavities.

Visceral pleura. The visceral pleura is a continuation of the parietal pleura that covers the surface of the lungs. During respiration, when the lungs expand, the visceral pleura comes into contact with the parietal pleura. The two sheets can slide over each other because of the pleural fluid, which acts as a lubricant.

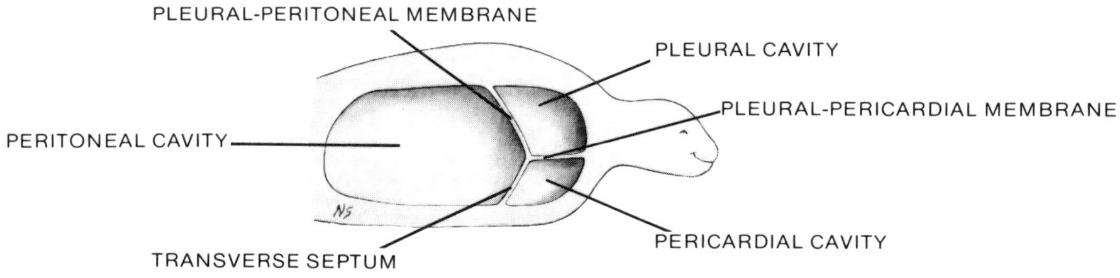

Figure 25-1. Diagram of mammalian body cavities, lateral view. (After Romer.)

Abdominal cavity. **(Peritoneal cavity)** The abdominal cavity is the portion of the coelomic cavity posterior to the diaphragm.

Parietal peritoneum. This is the coelomic epithelium which lines the peritoneal cavity.

Visceral peritoneum. This membrane connects both dorsally and ventrally to the parietal peritoneum. It covers the viscera.

Mesenteries. The mesenteries, which are double layers of peritoneum, connect organs of the digestive tract to the body wall. For a full discussion of the mesenteries see Chapter 9. The arrangement of mesenteries in the cat is similar to that in the dogfish.

STRUCTURES OF THE HEAD AND NECK

In order to expose the salivary glands and ducts, remove the skin and underlying connective tissue from the left side of the head (Fig. 25-3).

Note: Much of the following dissection is done in close proximity to the blood vessels of the head. Great care should be taken to maintain the integrity of these vessels so they will be intact for later study of the circulatory system (Chapter 27).

Salivary Glands (Fig. 25-3)

Locate and identify the salivary glands by examining the lateral head and neck regions.

Figure 25-2. Diagram of mammalian body cavities, transverse view. This diagram is oriented with the chest upward, corresponding to the position of a cat specimen lying on its back. (After Crouch.)

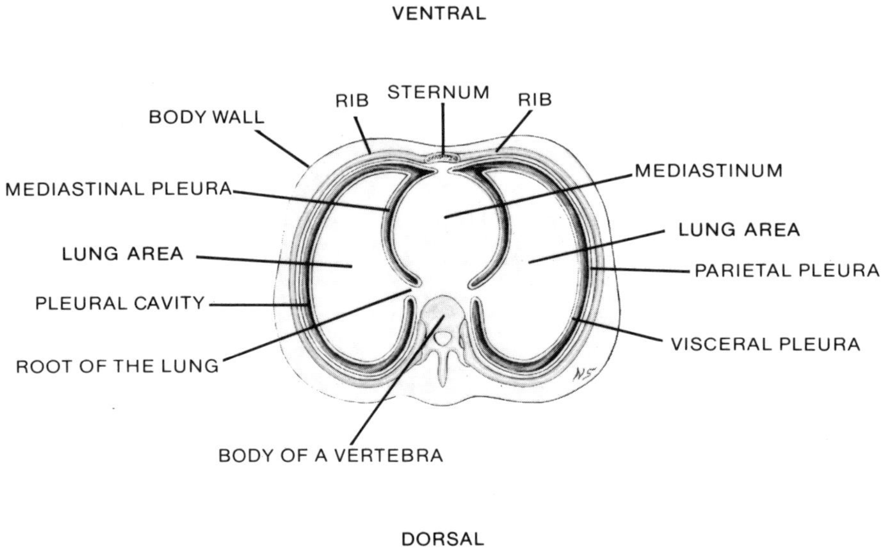

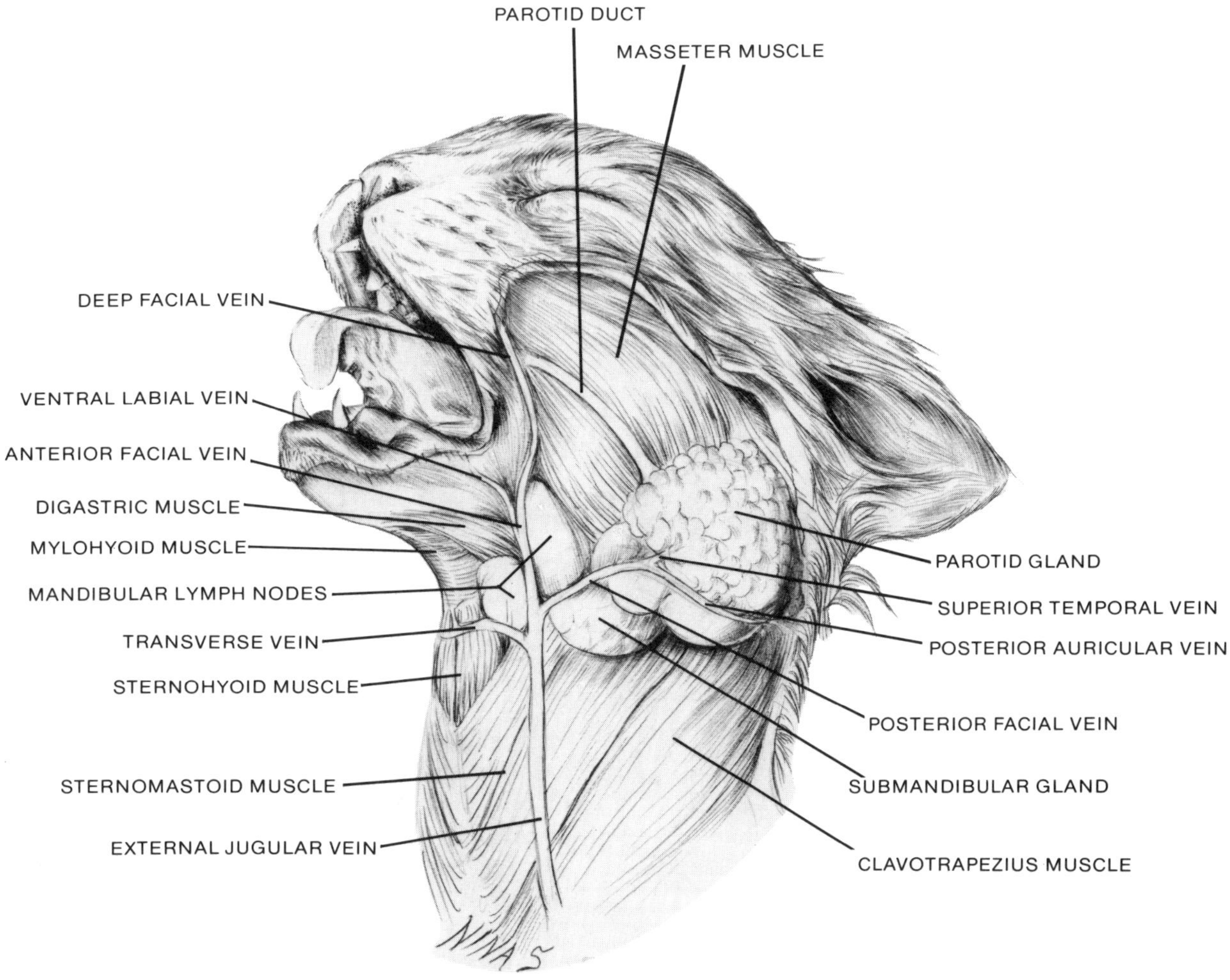

Figure 25-3. Ventrolateral view of a cat head and neck, with specific reference to salivary glands.

Parotid gland. A parotid gland is found in the cheek region on each side of the head. This oval-shaped salivary gland is lobulated in texture.

Parotid duct. The duct carrying the products of the parotid gland extends superficial to the masseter muscle and opens into the mouth inside the upper lip. The facial nerve is also found in the vicinity of the duct. This nerve may be found under the parotid gland; it passes over the masseter on its way to the facial muscles. Be careful not to confuse the nerve with the parotid duct. Usually, branches of the facial nerve are found both dorsal and ventral to the parotid duct.

Submandibular gland. (Mandibular gland) Oval and lobular, the submandibular glands lie ventral to the parotid glands. (Do not mistakenly identify the two smooth bodies in this region as salivary glands; they are lymph nodes.)

Submandibular duct. The submandibular duct proceeds deep to the digastric muscles from the anterior end of each submandibular gland. It terminates in the mouth in front of the midventral septum of the tongue. The termination may be difficult to find.

Sublingual gland. This salivary gland on each side is elongate in shape. It lies beside the duct of the submandibular gland and touches the anterior portion of this gland. The sublingual gland duct is difficult to see, but it runs parallel to the duct of the submandibular gland.

Molar gland. (Not in Fig. 25-3) This gland lies in the posterior part of the lower lip, between mucous membrane and skin.

Molar duct. (Not in Fig. 25-3) The duct of the molar gland empties on the inside of the lower lip. Both the duct and its opening are difficult to find.

The Oral Cavity and Pharyngeal Region (Fig. 25-4)

On the right side of the cat's head, cut dorsolaterally from the corner of the mouth toward the angle of the jaw. With bone shears, cut through the mandible near the angle of the jaw. Then, with a small scalpel, cut caudally along the right lateral oropharynx and laryngopharynx to a point just cranial to the thyroid gland.

Free the jaw on the left side by cutting through the body of the mandible midway between the symphysis and the angle of the jaw on that side. Reflect the mandible and observe the structures of the mouth and pharynx.

Identify as many of the following structures as possible.

Oral cavity. The oral cavity includes all of the mouth region ventral to the palate and posterior to the teeth.

Vestibule. The vestibule is the space between the lips and the teeth.

Teeth. Note the different types of teeth. ***Incisors*** are small and spike-like; ***canines*** are larger spikes; and ***premolars*** and ***molars*** are broad, cusped teeth. In each ***half*** of the upper jaw, there are three incisors, one canine, three premolars, and, most posteriorly, one small molar. Each side of the lower jaw holds three incisors, one canine, two premolars, and one large molar; this latter large lower molar and the last premolar in the upper jaw are known as the ***carnassials*** and are characteristic only of carnivores.

Hard palate. The hard palate is the more anterior portion of the palate which forms the roof of the mouth. It is supported by the palatine parts of the maxillae, processes on the premaxillae, and the palatine bones.

Soft palate. The soft palate, posterior to the hard palate, is not supported by bone as is its hard counterpart. It separates the oral cavity from the nasopharynx.

Nasopalatine ducts. (Not shown on Fig. 25-4) The nasopalatine ducts open into the mouth via two small openings in the hard palate, immediately behind the incisors of the upper jaw. The ducts extend through the premaxilla bone to Jacobson's organ in the nasal cavity. Jacobson's organ is a patch of what appears to be olfactory epithelium in the ventral part of the nasal cavity. Its function is unknown (some researchers suggest that it is used in smelling food in the mouth).

Choanae. (Internal nostrils) (Not visible in this dissection) The choanae are two openings in the skull just dorsal to the posterior edge of the hard palate. They connect the nasopharynx with the nasal cavity. The choana can be considered as the posterior opening of a cylinder which begins anteriorly at the external nostril. The choanae are among the structures which permit the separation of digestive and respiratory pathways. They also enable the nose to serve as both a respiratory organ and an organ of olfaction by permitting air to continue its flow to the lungs after passing over the olfactory epithelium.

Tongue. The tongue is a muscular organ found in the floor of the oral cavity. It is attached to the floor by a structure called the ***lingual frenulum***. The most posterior portion of the tongue lies outside the oral cavity in the pharynx.

Papillae of the tongue. The papillae are small raised bodies found on the dorsal surface and the sides of the tongue. In anterior-posterior sequence, the papillae in the cat are of fungiform, filiform, foliate, circumvallate, and conical types. Taste buds are found on all except the filiform and conical papillae. Differences between the papillae have to do with their taste sensitivities (sweet, sour, salt, and bitter).

Glossopalatine arches. (Not shown in Fig. 25-4) These are lateral folds of skin extending dorsally from the posterior portion of the tongue to the soft palate. They can best be seen by pulling the tongue forward in the mouth. The glossopalatine arches separate the oral cavity from the pharynx. The passage between these arches is called the ***isthmus of the fauces***.

Pharynx. The pharynx is the chamber of the "throat" region which is common to both the digestive and respiratory systems. It is divided somewhat arbitrarily into nasal, oral, and laryngeal portions called the ***nasopharynx***, ***oropharynx***, and ***laryngopharynx***. The pharynx has great evolutionary significance because of the important structures derived from the pharynx and its walls. A few of these pharyngeal derivatives are the hyoid apparatus, the visceral arch muscles, the thyroid tissues, thymus tissue, and the parathyroid glands.

Nasopharynx. The nasopharynx is the upper portion of the pharynx, posterior to the openings of the nasal cavities (the choanae) and dorsal to the soft palate.

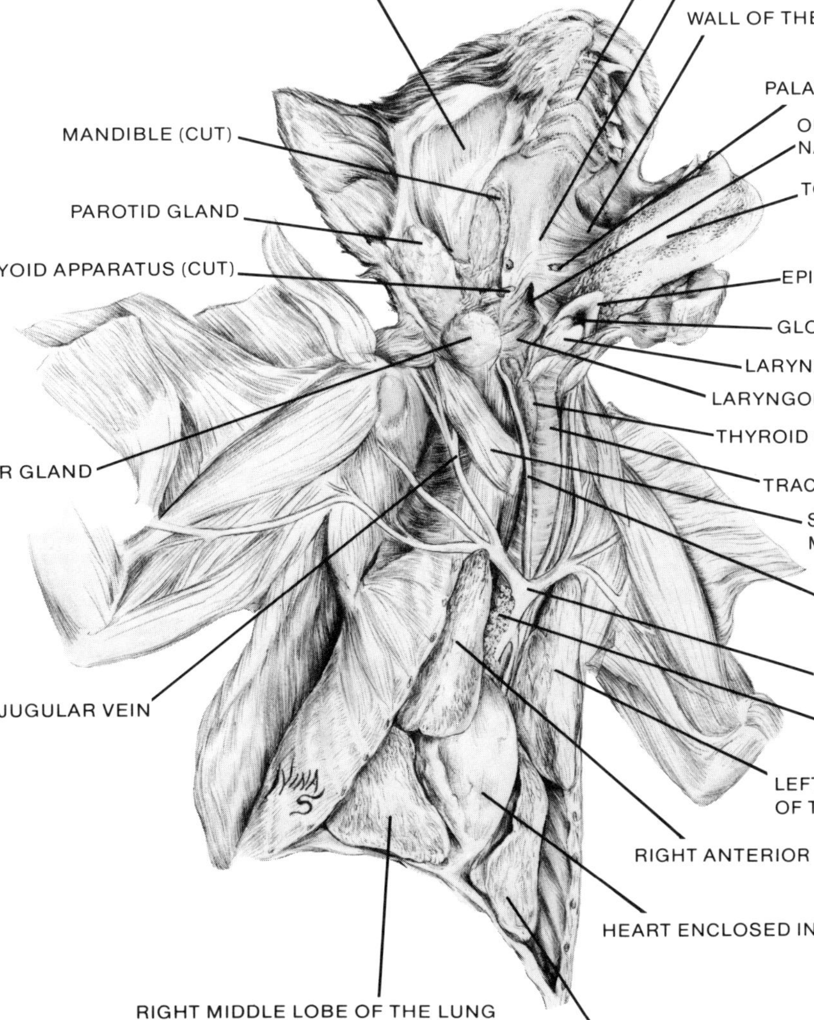

Figure 25-4. Ventrolateral view of the head, mouth, oral cavity, and pharynx after reflection of the mandible.

Auditory or eustachian tubes. (Not shown in Fig. 25-4, but revealed by cutting through the ventral surface of the nasopharynx) The auditory tubes connect the middle ear with the pharynx by slit-like openings in each lateral wall of the nasopharynx. They serve to equalize pressure on the inner and outer surfaces of the tympanic membrane by allowing air to enter the middle ear. This pressure equilibration is essential to the proper mechanics of sound conduction.

Oropharynx. The oropharynx, or middle segment of the pharynx, is located behind the oral cavity and ventral to the nasopharynx. It is considered as part of the digestive pathway. Its anterior boundary is the glossopalatine arches; its posterior boundary is the free posterior margin of the soft palate.

Palatine tonsils. The palatine tonsils are masses of lymphoid tissue located in the lateral walls of the oropharynx, near the posterior end of the soft palate.

Tonsillar fossae. These are indentations in the oropharyngeal wall in which the palatine tonsils are embedded.

Laryngopharynx. The laryngopharynx is the most posterior portion of the pharynx. It is the space immediately dorsal to the larynx and posterior to the edge of the soft palate.

Epiglottis. The epiglottis is the flap of tissue supported by cartilage. It guards the opening of the larynx (the glottis). It lies anterior and ventral to the glottis.

Glottis. The glottis is the opening between the vocal cords of the larynx by which the pharynx and larynx are connected.

Larynx. **(Voice box)** The larynx, a structure in the floor of the posterior laryngopharynx, is also considered as the anterior continuation of the trachea. Air must flow through the larynx in order to enter or leave the trachea. *To expose the true and false vocal cords, make a longitudinal incision in the exposed surface of the larynx.*

Ventricular folds. **(False vocal cords)** The ventricular folds are a pair of membranous folds on either side of the glottis, accessory to the true vocal cords. They are anterior to the real vocal cords and stretch vertically from the epiglottis to the larynx. They have no function in sound production.

Vocal folds. The vocal folds are the true vocal cords. They are located just posterior to the false vocal cords, on both sides of the glottis; they project medially into the larynx.

Esophagus. **(Fig. 25-5)** The esophagus is the continuation of the digestive pathway between the posterior laryngopharynx and the stomach. It is located dorsal to the trachea. There is a short "abdominal" segment of the esophagus after it pierces the diaphragm, just prior to its entrance into the stomach. The walls of the esophagus are muscular.

Trachea. The trachea is the respiratory pathway from the larynx to the lungs. The walls of the trachea are supported by characteristic "rings" of cartilage, which are visible externally along its length. They prevent collapse of the air tube.

Glands of the Pharyngeal Region

Thyroid. The thyroid consists of right and left lobes located on either side of the trachea at its anterior end, just caudal to the larynx. The two lobes of the thyroid are sometimes connected by a narrow piece of thyroid tissue (the ***isthmus***) which crosses the ventral surface of the trachea.

Parathyroid. The parathyroid glands consist of two pairs of glands embedded in the thyroid. The glands are embedded in both the anterior dorsal and posterior dorsal portions of each lobe of the thyroid. The four parathyroid glands are lighter in color than the thyroid gland, but they are hard to observe on gross examination.

THORACIC CAVITY (Fig. 25-4)

From the most posterior rib on each side, cut cranially with a large scalpel and bone shears on a straight line through the ribs about 1 cm lateral to the sternum. Then make a transverse incision posterior to the sternum, through the two most posterior ribs on each side. The lateral thoracic walls can then be reflected. (If necessary, ribs may be broken to do this.)

Locate and examine the lungs and heart in the thoracic cavity. Portions of the posterior trachea, and adjacent areas of the heart may be obscured by the loose, grey-brown thymus gland present in immature specimens. If so, carefully remove the thymus and discard it. Avoid damaging the trachea and blood vessels. (These procedures preserve blood vessels for later dissection. If desired, the entire sternum can be removed and discarded.)

Identify the following structures shown in Figure 25-4:

Thymus. The thymus is dark, irregularly shaped, and of variable size. It is found in the anterior segment of the mediastinum, ventral to the heart and trachea. The thymus may extend from the caudal half of the trachea as far as the diaphragm, or it may be virtually absent in an old animal. The thymus is essential to the immune response in young animals, as discussed at the end of this chapter.

Lungs. The lungs are the respiratory exchange organs in the cat. The right lung consists of four lobes (anterior, middle, posterior, and accessory); the left lung consists of only three lobes (anterior, middle, and posterior). As noted above, the lungs are covered by a lining called the ***visceral pleura***, while the walls of the pleural cavity are covered by the ***parietal pleura***. The trachea, its bronchial branchings, and the alveoli consitute the "respiratory tree."

Bronchi. The bronchi are posterior continuations of the trachea in the lobes of the lungs. The bronchi subdivide into succeedingly finer branches until they terminate in pockets called the alveoli.

Alveoli. The alveoli, the terminal subdivisions of the bronchi, are not visible on gross inspection. These tiny sac-like structures are the actual sites of the exchange of carbon dioxide and oxygen in the lungs.

Diaphragm. The diaphragm is a muscular sheet which separates the thoracic from the abdominal cavities. It attaches to the xiphoid process of the sternum, the cartilages of the lower ribs, and the lumbar vertebrae.

ABDOMINAL STRUCTURES (Fig. 25-5)

Locate the point on the midventral line at which the diaphragm joins the abdominal wall. From this point, cut along the midventral line caudally to the level of the pelvis. This superficial incision should pierce the abdominal wall without damaging underlying structures. Without further disturbing the diaphragm, cut laterally, separating it from the body wall to facilitate reflection of lateral wall flaps.

Caudally, separate the abdominal muscle sheets from the medial thigh muscles to allow for full reflection of the body wall flaps. Students with male specimens must take care not to disturb the spermatic cord and its associated genital blood vessels that enter the abdominal wall in this area.

Observe the superficial structure of the abdominal cavity, noting the **greater omental mesentery**. *This convoluted, lobular structure, frequently filled with fat, covers and interconnects various structures of the digestive system. After observing the undisturbed arrangement, remove the greater omentum. Lift it manually to observe its connections and free it from the coiled small intestine. Cut its connections to the greater curvature of the stomach and the dorsomedial surface of the spleen without disturbing the hepatic portal system veins of this area. Then remove and discard it.*

Identify the structures of the digestive tract as it proceeds to the colon.

Liver. The liver is a large, dark digestive gland located immediately posterior to the diaphragm. It is attached to the diaphragm and divided into right and left portions by the falciform ligament. The right and left lobes are further subdivided into lateral and medial lobes. There is also a caudate lobe close to the kidney.

Falciform ligament. The falciform ligament is a developmental remnant of the ventral mesentery which attached the liver and diaphragm to the ventral body wall.

Gall bladder. The gall bladder is located in the right median lobe of the liver. It is small, green, and pear-shaped. It empties into the cystic duct, which joins the hepatic ducts of the liver to form the ***common bile duct***. These ducts may be difficult to locate.

Common bile duct. The common bile duct is the continuation of the fused ducts from the gall bladder and the liver; it extends to the small intestine. The duct terminates caudal to the pyloric sphincter.

Spleen. The spleen is the other large, dark organ in the abdominal cavity. It lies in the greater omentum on the left side of the body cavity, near the greater curvature of the stomach. The spleen is not actually part of the digestive system; it is an organ of the vascular system and is discussed in Chapter 27.

Gastrosplenic ligament. This ligament is the part of the greater omentum extending from the spleen to the stomach.

Lesser omentum. This is a mesentery attachment between the liver and the stomach and duodenum. It is made up of the hepatogastric and hepatoduodenal ligaments.

Stomach. The large organ on the left side of the peritoneal cavity is the stomach. It is located just caudal to the esophagus after the latter pierces the diaphragm. The stomach is divided into three parts: the ***cardiac***, ***fundic***, and ***pyloric*** regions. Note the longitudinal ridges of tissue found on the internal surface of the stomach.

Cardiac portion. The cardiac is the most cranial portion of the stomach, where the relatively narrow esophagus expands into the stomach chamber.

Fundus. This portion of the stomach is caudal to the cardiac portion. It is characterized by a large bulge on the left side of the stomach.

Pyloric portion. The most caudal part of the stomach, the pyloric portion, is immediately cranial to the small intestine.

Pylorus. This is the opening between the stomach and the duodenum. It is guarded by the ***pyloric sphincter***, a thickened muscular region at the caudal end of the stomach.

Greater and lesser curvatures. The large, convex surface on the left side of the stomach is called the ***greater curvature***, while the shorter, concave right lateral side is called the ***lesser curvature***.

Small intestine. The small intestine is the continuation of the digestive tract after the pyloric sphincter. Moving caudally from this point, the small intestine

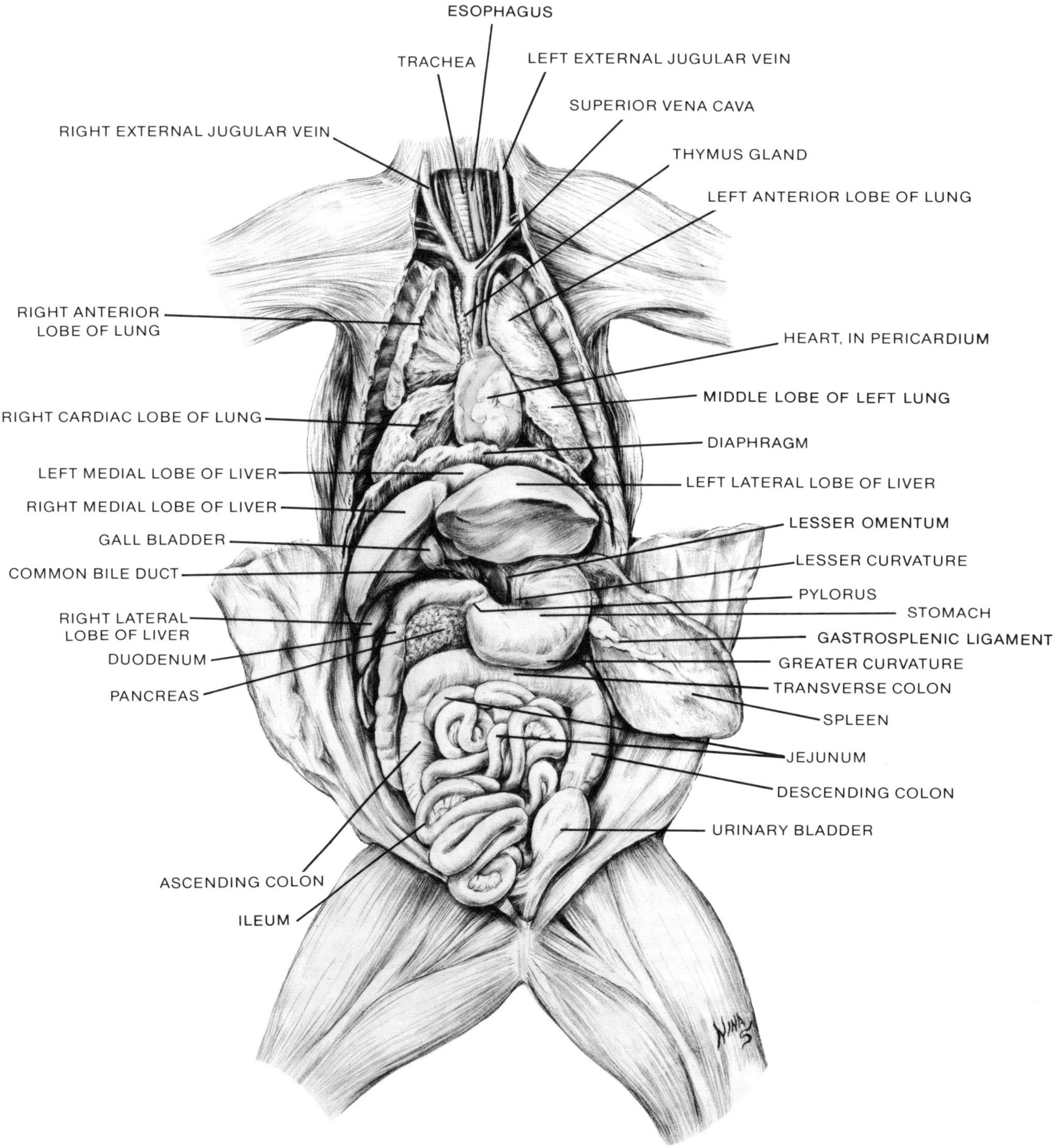

Figure 25-5. Ventral view of the abdomen of a cat, with emphasis on the abdominal and thoracic viscera after removal of the greater omentum.

consists of the ***duodenum***, the ***jejunum***, and the ***ileum***. These three segments are not differentiable by gross examination and are determined histologically.

Duodenum. The duodenum is the continuation of the digestive tract caudal to the pyloric sphincter. It is the portion between the pyloric sphincter and the jejunum. The pancreatic duct joins the common bile duct just before the latter enters the duodenum.

Jejunum. The second segment of the small intestine, the jejunum, comprises approximately two-fifths of the remaining small intestine.

Ileum. This is the tightly coiled, posterior segment of the small intestine. The small intestine terminates in the ileocecal junction.

Mesoduodenum. This is the part of the dorsal mesentery which supports the duodenum.

Pancreas. The pancreas is an important digestive gland. It is elongate and lobular in texture. The pancreas bends abruptly, forming two parts: the head, which lies in the curve formed by the pylorus and the duodenum; and the body, which lies dorsal to the pyloric stomach.

Pancreatic ducts. The principal pancreatic duct proceeds from the pancreas to the duodenum. The junction of the pancreatic duct and the common bile duct, prior to their entry into the duodenum, is called the ***ampulla of Vater***. This junction can be recognized as a slight swelling in the ducts. There is an accessory pancreatic duct which enters the duodenum approximately an inch caudal of the principal duct, but this additional duct may be absent in some animals.

Large intestine. This portion of the digestive tract follows the ileum of the small intestine and terminates in the caudal opening of the digestive tract, the ***anus***. The large intestine consists of the ***cecum***, the ***ascending colon***, the ***transverse colon***, the ***descending colon***, and the ***rectum***.

Cecum. (Not shown in Fig. 25-5) The cecum is a short, dead-end sac which projects posteriorly from the ileocecal junction, where the ileum joins the ascending colon. Here a vermiform appendix is found in humans, but it is lacking in cats.

Ileocecal junction. This is the junction between the ileum of the small intestine and the cecum of the large intestine.

Ileocecal valve. A longitudinal incision at the ileocecal junction will reveal the ileocecal valve. This muscular sphincter prevents the return of digested materials from the large intestine into the small intestine.

Lymph nodules. Lymphatic tissue, usually embedded in fat, is dispersed throughout the walls of the intestines and the surrounding mesenteries.

Ascending colon. Subsequent to the ileocecal juncture, the large intestine runs cranially along the right side of the body. This portion is termed the ascending colon.

Transverse colon. This is the segment of the large intestine which passes transversely across the body cavity.

Descending colon. This is the continuation of the large intestine after the transverse colon. It starts from the left side of the peritoneal cavity and moves toward the middorsal line, where it becomes the ***rectum***.

Anus. The anus is the caudal opening of the digestive tract, leading from the rectum to the outside of the body.

FUNCTION

Digestive Organs

The storage and processing of food ingested by the animal is the principal function of the digestive system. In mammals, chemical processing begins with the salivary glands. In many mammals, these glands secrete enzymes, ***glucosidases***, which convert starch into maltose and ***glucose***. These enzymes start the digestive process prior to the action of the main digestive enzymes in the stomach and intestines. The primary constituent of the saliva, however, is ***mucin***, a glycoprotein responsible for lubricating the food. The saliva also mixes with the food during mastication, so it will enter the digestive tract as a semi-solid mass which is easier to digest. Saliva also aids in keeping the mouth and teeth clean, dissolves the molecules which stimulate the taste buds, lubricates the mouth, aids in swallowing, and buffers the acidity of the mouth to maintain the integrity of the tooth calcium.

The ***teeth*** act to break down the food to a size which is more easily swallowed and digested. The cat is called a ***thecodont*** because each root of each tooth is inserted in its own deep socket. It is also a ***heterodont***, whose teeth are not shaped identically, and a ***diphyodont***, because it has temporary or deciduous dentition that precedes permanent dentition.

The centrally located filiform papillae on the tongue have a horny layer or ***corneum*** and in some cases, spines, which adapt them for scraping. They are used for grooming and cleaning as well as for scraping food from bones. Taste buds are abundant on the sides and at the bases of all the papillae except the filiform ones. In cats and some other vertebrates, the tongue is used as an aid in handling food; in humans it is also instrumental in speech.

The ***palate*** separates the anterior portion of the respiratory pathway from the route by which food moves down the digestive tract. This evolutionary adaptation allows breathing and eating to occur simultaneously. The epiglottis deflects food from the glottis, the respiratory opening in the larynx. It performs this function automatically as a part of the swallow reflex. Thus, it prevents food and saliva from entering the respiratory pathway. Subsequent to swallowing, food enters the esophagus, which conducts it to the stomach using peristaltic activity.

The stomach serves to store food while preliminary digestion occurs prior to entry into the duodenum. The stomach lining is arranged in folds, or rugae, which increases its secretory surface area and expandability.

Gastrin, a hormone secreted by the pyloric region, stimulates secretion of hydrochloric acid by cells of the stomach. The acid aids digestion by making the environment optimally acidic for the action of the enzyme ***pepsin***.

The epithelium of the fundus of the stomach consists of two cell types: the chief cells, which produce ***pepsinogen***, the inactive form of pepsin; and the parietal cells, which secrete hydrochloric acid. The acid activates the pepsinogen, enabling pepsin to hydrolyze food proteins. The combination of hydrochloric acid and pepsin is called ***gastric juice***. Food mixes with gastric juice and becomes acid ***chyme***, which passes from the stomach into the small intestine in a partially digested state.

The ***small intestine*** functions in the digestion of food and the absorption of nutrients. Numerous macroscopic and microscopic folds (***villi*** and ***microvilli***) and the extreme coiling of the intestine in the peritoneal cavity greatly increase the surface area of the organ available to absorb the nutrients in the food. Food in contact with the duodenum elicits the secretion of ***enterogastrone***, a hormone which inhibits gastric secretions. Distension of the duodenal wall or contact with the acidity of the acid chyme also stimulates the duodenal mucosa to produce ***secretin***, a hormone which stimulates the secretion of pancreatic juice. Other enzymes secreted by the duodenal mucosa are ***pancreozymin*** and ***cholecystokinin***. Pancreozymin, like secretin, affects the pancreas. It elicits the release of carbohydrases, lipases, and proteolytic enzymes in their inactive forms (trypsinogen and chymotrypsinogen). These inactive enzymes are transformed to their active states after they enter the small intestine. On the other hand, cholecystokinin secretion results in contraction of the muscles of the gall bladder and release of bile into the small intestine.

The liver is a major regulation site for intermediate metabolism of nutrients absorbed by the digestive tract. Blood from the intestinal capillaries, rich in the nutrients broken down during the digestive processes, proceeds immediately to the liver via the hepatic portal system. Thus the liver can begin to process and store essential nutrients in proper amounts. In particular, carbohydrates may be stored in the liver. The sugar glucose is stored in the form of glycogen, to be released as body needs dictate. In addition to regulating intermediate carbohydrate metabolism, the liver also functions in steroid metabolism, detoxification of drugs and toxins, production of plasma proteins, formation of urea from nitrogenous wastes, fat metabolism, inactivation of polypeptide hormones, and the production of bile.

The gall bladder stores bile after it is produced in the liver. The bile is then conveyed via the common bile duct to the small intestine. Its role in the digestive tract is the emulsification of fats. In addition, bile contains many substances destined for excretion; among these are the breakdown products of proteins and hemoglobin.

The pancreas consists of endocrine cells and exocrine cells. The glandular ***exocrine*** cells are responsible for producing the pancreatic juice secreted into the duodenum. The pancreatic juice contains enzymes which act on proteins, carbohydrates, fats and nucleic acids.

The endocrine cells are found in small scattered nests called ***islets of Langerhans***. Several types of endocrine cells occur, the predominant ones being A (alpha granule containing) and B (beta granule containing) cells. ***Insulin***, an important hormone regulating carbohydrate metabolism, is produced by B cells. Insulin is released into the blood stream in response to elevated blood glucose levels. It lowers the glucose level in the blood plasma by stimulating the conversion of glucose to glycogen and fatty acids, and by increasing the permeability of cell membranes to glucose. The importance of insulin is reflected in the disease ***diabetes mellitis***, a human disorder characterized by a deficiency in insulin production.

The A cells of the islets of Langerhans produce ***glucagon***, a hormone antagonistic to insulin. Release of this hormone results in increased blood sugar levels by stimulating breakdown of glycogen to glucose in the liver. Although insulin is present in all vertebrates, glucagon is only found in reptiles, birds, and mammals.

In the final portion of the digestive tract, the ***large intestine***, water and salts are resorbed. The indigestible matter continues on through the posterior part of the intestine and is eliminated from the body as feces.

Endocrine Organs

There are several glands which originate as outgrowths of the gut that are functionally part of the endocrine system. Among these are the thyroid and the parathyroid.

The ***thyroid gland*** originates embryonically from the floor of the pharynx. The epithelial cells of the thyroid are arranged in hollow vesicles called follicles. Because of this arrangement, the thyroid hormones can be both produced and stored in extracellular form in the same gland. It has been hypothesized that this arrangement of cells and the capacity to store iodine is an important evolutionary adaptation, inasmuch as iodine can become scarce in certain environments. Iodine is an essential structural component of two of the three main thyroid hormones, thyroxine and triiodothyronine. These two hormones are both stored in the thyroid gland, bound to a glycoprotein called thyroglobulin. The hormones are activated when their bonds to the thyroglobulin are broken.

The proper functioning of the thyroid gland is essential to numerous important bodily functions. Thyroxine has dramatic effects on the body's metabolism. Thyroid dysfunctioning results in changes in oxygen consumption and heat production in the tissues, nervous system malfunctions, reproductive irregularities, unusual secretions of glands, and disruption of growth and development patterns. Thyroxine also affects cholesterol metabolism and the rate of absorption of carbohydrates from the digestive tract. In many vertebrates, specific thyroid levels are necessary for normal skin development and pigmentation. In birds and mammals, abnormal thyroid functioning results in inhibition of hair or feather growth and melanin deposition.

The third main thyroid hormone is calcitonin. Calcitonin acts to lower blood calcium, when it is too high, by promoting bone formation. Calcitonin is involved in calcium and phosphate homeostasis. This activity is generally antagonistic to the effects of the parathyroid hormone.

The ***parathyroid gland*** secrets parathyroid hormone, a substance that acts to raise blood calcium levels. Demineralization of bone occurs because the hormone increases the number and activity of the natural bone-degrading cells, the osteoclasts. In addition, parathyroid hormone causes increased calcium absorption from the intestine, while inhibiting calcium excretion from the kidney. All these results act in concert to produce an elevated blood calcium level. This is important because any drastic fall in blood calcium eventually results in paralysis and death.

While increasing the calcium levels in the blood, parathyroid hormone causes an increase in phosphate excretion from the kidneys; thus it maintains the normal phosphate titer in the blood.

Respiratory Organs

Air enters the nose and mouth and passes via the pharynx into the trachea. The larynx contains the vocal cords and so is the main site of sound production in mammals. To produce sound, the vocal cords vibrate under appropriate conditions when air is passed over them. The type and degree of sound produced is dictated by the shape of the mouth, the tautness of the vocal cords themselves, and by actual changes in the shape of the larynx. It is likely that, in human beings, the shape of the adjacent pharynx, posterior to the tongue, is also crucial in production of certain vowel sounds that are important components of our language.

The trachea and bronchi convey the air to the lungs, which are the principal or sole sites of respiratory exchange in primarily terrestrial vertebrates. The pulmonary tree—the progressively smaller branching of air passages—varies widely among land animals. In birds, for instance, the points of gas exchange are minute air capillary tubes through which the gases flow in one direction only. In mammals, on the other hand, gas exchange occurs in terminal sacs, the ***alveoli***. Stagnation in the mammalian alveoli is a potentially severe problem that is counteracted by smooth muscle cells in the walls of the lung. When these muscles contract, the lung surface is moved about so that adjacent air masses are kept in a turbulent state.

An important feature of the mammalian body is the diaphragm. When relaxed, this sheet of muscular tissue forms a crown-shaped end to the abdominal cavity. When the diaphragm contracts, the floor of the chest cavity is lowered, and the volume of the pleural cavity increases; inspiration of air takes place. Relaxation of the diaphragm results in decreased volume of the pleural cavity and aids in expelling air from the lungs.

Immune System

The ***thymus*** is a lymphatic organ of the immune system. Young animals in which the thymus has been removed appear to be unable to recognize alien tissue grafts and will not react against them. The thymus produces lymphocytes and probably controls the ability to carry out certain immunological reactions. Though the thymus is necessary to establish certain classes of immune reaction in a young animal, that job is a transient one; the gland normally degenerates during the lifespan of an organism and may be virtually absent in old animals.

The lymph nodes, masses of lymphoid tissue interposed along the lymphatic vessels throughout the body, act as filters. As a part of the circulatory system, they help prevent certain antigens from entering the blood stream and contain aggregations of nongranular white corpuscles. (These white corpuscles are also called agranular, or lymphoid leukocytes.) Certain kinds of lymphoid leukocytes, called small and large lymphocytes, are produced from stem cells in the bone marrow. They are then "processed" in an unknown way (in the Bursa of Fabricus in birds; the equivalent mammalian site is unknown) so they can give rise to antibody-producing clones. Frequently these lymphocytes reside in the lymph nodes themselves. Interestingly, the thymectomized animals referred to above usually develop a normal capacity to produce such circulating antibodies. This suggests that there are two distinctive origins for the cellular (thymus) and the antibody (Bursa) immune responses.

NATHAN SCHAFER

26

The Cat: Urogenital System

The mammalian urogenital system is a product of two series of changes, one involving the kidney and the other involving structures for internal fertilization and fetal development.

The most posterior portion of the embryonic nephric tissue, the metanephros, is the functional kidney of adult mammals. The evolutionary trend of posterior movement of the functional kidney observed earlier continues in the establishment of the metanephric kidney. In mammals, the anterior portions of the embryonic nephrogenic tissue give rise to the archinephric (or pronephric, or Wolffian) ducts. During embryonic development, these ducts either become part of the male genital system or degenerate. The metanephros is drained by the ureter, a new duct derived as an outgrowth from the caudal portion of the archinephric duct, near the point where it enters the cloaca.

Another important alteration in the mammalian kidney from that of reptiles is an increase in the number of kidney nephrons per unit area. Mammals also have a distinctive nephron structure, particularly with respect to the loops of Henle, which function to allow mammalian urine to be hypertonic to blood and body fluids. This property reflects the necessity of conserving water in the dry terrestrial habitat. It is the opposite of the condition seen in some fresh-water fishes, amphibians, and even aquatic reptiles that excrete large volumes of hypotonic urine to void excess water from the body. In mammals, the kidney retains its osmoregulatory and excretory functions, but the kidney tissue that carries out these processes is located at a new site.

Another system which has undergone significant modification in vertebrates is the urinary duct system. In sharks, a small mesodermal bladder may develop at the posterior end of the archinephric ducts. Amphibian and reptilian endodermal bladders develop from the cloaca. In the adult, urine passes from this bladder into the cloaca, which also serves as a channel for intestinal wastes and gametes. In amniotes—reptiles, birds, and mammals—with the advent of the ureter, urine flows from kidney to ureter to bladder to urethra. This route and the separate rectum-anus route for intestinal wastes are associated with disappearance of the cloaca. Digestive tract waste products and gametes are effectively separated in mammals.

Another major area of change is the manner in which the fetus develops. In sharks and their relatives, young develop either within the female body or in special egg cases that permit advanced development prior to hatching. Most amphibians employ external, aquatic fertilization and development. Reptiles are variable, but most reproduce oviparously, by internal fertilization and a cleidoic egg that is passed from the body of the female. (A cleidoic egg is one covered with protective membranes and a leathery or calcareous shell.) Mammals, such as the cat, reproduce by internal fertilization and development. This method of procreation has many advantages: (1) it protects the young at a time when they are most vulnerable; (2) it prevents desiccation and eliminates the need for reproducing in an aquatic environment; (3) it maintains proper body temperature until the young organism is better able to regulate its temperature by itself; (4) it enables the female to be mobile during pregnancy; (5) it guarantees a continuous nutritional supply to the young and thus eliminates dependence on the limited quantities of food present in the yolk of a cleidoic egg; and (6) largely because of better nutrition, it permits long gestation and, thus, more advanced development prior to birth. To effect internal fertilization and development, structural changes evolved in both male and female organisms.

The gonads of both sexes develop embryonically from the two genital ridges, located medial to the embryonic kidneys. In primitive vertebrates such as cyclostomes, the gametes are freely expelled into the coelom. They subsequently leave the body through posterior abdominal pores.

Subsequent vertebrates have developed ways to facilitate fertilization. For instance, the eggs pass via the coelom into some type of oviduct in females whereas sperm pass through another closed duct system in males. Many—but not all—land vertebrates have developed a penis to facilitate internal fertilization. The claspers of the shark and some bony fish also serve a similiar function, suggesting that evolution of male intromittent organs is more associated with the advantages of internal fertilization than with life on land.

Some male mammals also exhibit a novelty not found in other vertebrates. Their testes may descend from the body cavity, through the abdominal wall, to a position outside the main body to reside in the scrotum. This external carriage may have reflected the high internal body temperatures found in mammals. These high temperatures apparently disrupt sperm production and so make housing the testes outside the body an advantage. In birds, the same ends are met by locating the testes close to the wall of the cool abdominal air sac.

In female mammals structural changes have also occurred to assure efficient internal fertilization and development. The infundibulum and the oviduct receive the egg after ovulation and guide it to the uterus. As in many other vertebrates, fertilization takes place in the oviduct; in some forms, very early development occurs there also. The horns of the uterus are greatly reduced or nonexistent in the primates, so development occurs in the body of the uterus. In both the cat and human, the zygote implants itself in the uterine walls where an intimate apposition of fetal and maternal blood vessels called the placenta develops. The placenta is derived from both maternal and fetal tissues. The endometrium, or uterine lining, is the maternal contribution to the placenta, while the chorion (of trophoblast derivation) is the fetal contribution to this structure. The placenta is highly vascularized; it brings nutrients and oxygen to the fetus from the mother and carries away fetal metabolic wastes. The cat nourishes its young by means of a zonary placenta, a band-shaped, thickened vascular organ (Fig. 26-5). The placenta in humans is disc-shaped.

Mammals are also characterized by the development of the vagina from the distal portion of the reproductive tract. The vagina serves as the birth canal and as the receptacle for the penis during copulation. The appearance of a vagina is also part of the evolution of the lower urogenital tract which resulted in loss of the cloaca.

THE KIDNEYS (Figs. 26-1, 26-2)

Reflect digestive tract structures laterally to one side as far as the connecting mesentery will allow. A bean-shaped kidney becomes visible; it appears to be submerged in the peritoneum of the dorsal body wall.

Beginning at the cranial surface of the kidney, pierce the thin peritoneal membrane. Begin peeling it and any associated fat deposits away from the muscular body wall. Do not disturb the ureter, a thin white tube coursing caudomedially from the medial portion of the kidney to the urinary bladder.

To observe the urethra and associated structures, remove the medial thigh musculature from the pubis and ischium, 1.5 cm from the midline on each leg. With bone shears, cut through the body region of the pubis and the ramus portion of the ischium of each innominate bone; then cut the entire pelvic symphysis. Remove the cut pieces of bone.

Make a frontal section through one of the kidneys. Identify the following structures.

Kidney. The kidneys are paired organs found in the dorsolateral body walls. They are usually embedded in masses of fat which may be quite copious.

Cortex. In a frontal section of the kidney, the cortex is the outer, lighter-colored layer of tissue. It contains the glomeruli and convoluted tubules.

Medulla. The darker, central tissue of the kidney is the medulla. It contains collecting ducts and loops of Henle.

Renal pyramid. In humans, the medulla contains conical areas of tissue called renal pyramids. The base of each cone is oriented toward the outer cortical layer; the apex of each cone is oriented toward the renal sinus. In the cat, there is only one large renal pyramid in each kidney, the apex of which is the renal papilla.

Renal papilla. Located in the medullary portion of the kidney, the renal papilla is the apex of tissue through which the urine passes into the renal pelvis.

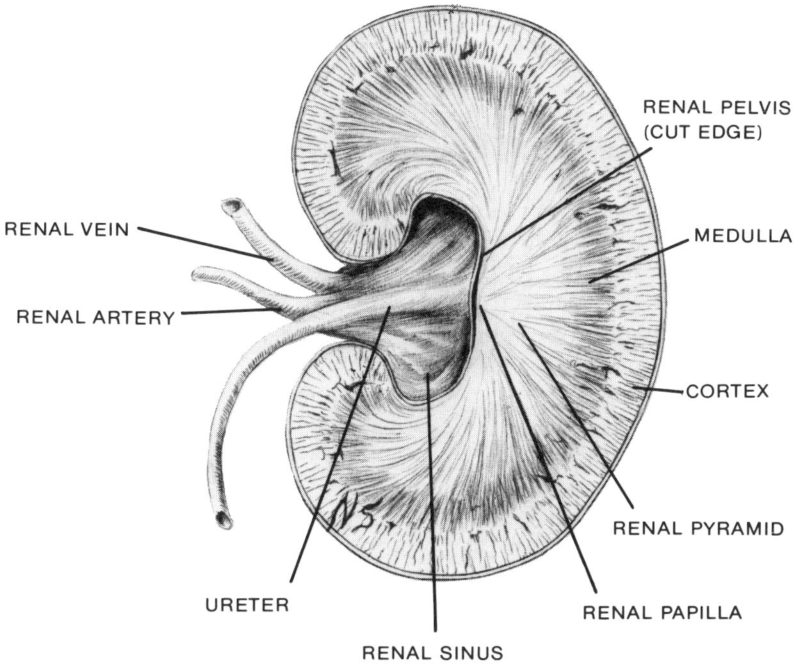

Figure 26-1. Idealized view of a frontal section of a cat kidney.

Renal pelvis. This is the portion of the ureter within the renal sinus; it appears as a funnel with a long, curved stem. Urine passes from the renal papilla into the pelvis, then down the ureter.

Hilus. The hilus is an indentation on the medial side of each kidney. It is the point at which the ureter, the renal vein, and the renal artery all pass into the substance of the kidney.

Renal sinus. The renal sinus is a cavity in the hilus which houses the renal pelvis. It is usually filled with fat.

Adrenal glands. Adrenal glands can be found at the cranial end of each kidney. They are slightly detached from the kidneys, whitish in color, and usually embedded in fat.

Ureters. A ureter connects each kidney with the urinary bladder. Each ureter begins as the renal pelvis at the renal sinus of the kidney. The thin, tubular ureter leaves the hilus and terminates in the neck of the urinary bladder.

Urinary bladder. The bladder is a sac-like, muscular structure of variable volume. The walls of the bladder can appear to be thick when it is empty and thin when it is distended. Smooth muscle cells are present in the bladder wall. The bladder consists of two parts, the fundus and the neck. The ***fundus*** is the large expandable body of the bladder; the ***neck*** is the narrow caudal portion which leads into the urethra.

The bladder is suspended by median and lateral ligaments. The median ligament extends from the midventral line, while the lateral ligaments extend from each side of the bladder to the dorsal body wall.

Urethra. This duct is the caudal continuation of the neck of the bladder.

MALE SYSTEM (Fig. 26-2)

Remove the remaining skin and connective tissue of the scrotum to expose both testes. Locate the small epididymal tube on the surface of each testis and trace it cranially as the ductus deferens within the spermatic cord. Reflect the urinary bladder ventrally and observe paired ductus deferentes looping over the ureters and converging dorsal to the cranial portion of the urinary bladder. The tubules of the ductus deferens converge and then course in tandem within the mesentery just dorsal to the urethral surface until they enter the urethra at the prostate gland. Follow the urethra caudally from the prostate gland and locate the bulbourethral glands on each side of the root of the penis.

Identify the following structures.

Testes. The testes are the paired gonads of the male. They are found outside the body wall in the scrotum, located posterior to the penis.

Scrotum. The scrotum consists of paired sacs located posterior to the penis and ventral to the pelvic region; these contain the testes.

Tunic vaginalis. This is the shiny covering of the testis, epididymis, and ductus deferens. This layer of tissue is derived from the peritoneum; it encapsulates the testes when they descend through the body wall into the scrotum. Thus, the lumen of the sac—the region within the tunic vaginalis—still communicates with the peritoneal cavity.

Epididymis. This structure extends from the anterior end of each testis posteriorly along the lateral testicular surfaces and ends at the posterior end of the testis. Here it is continuous with the ductus deferens. The epididymis is subdivided into three portions: the head, at the cranial end of each testis; body; and tail, at the caudal end of each testis. Associated with the

Figure 26-2. Ventral view of the abdomen of a cat, with emphasis on the structures of the male urogenital system.

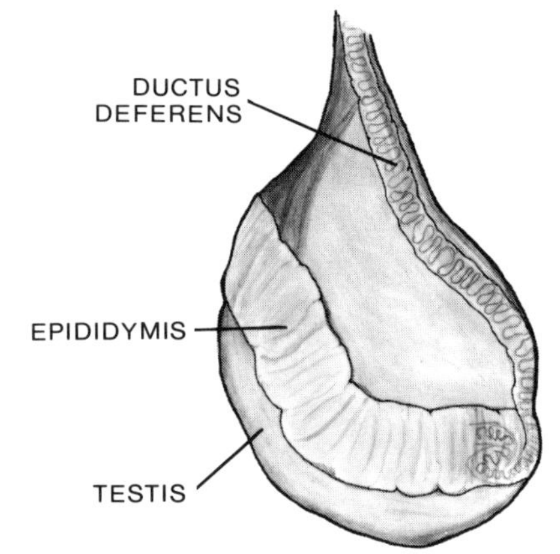

Figure 26-2a. Ventral view of left testis with tunic vaginalis partially removed.

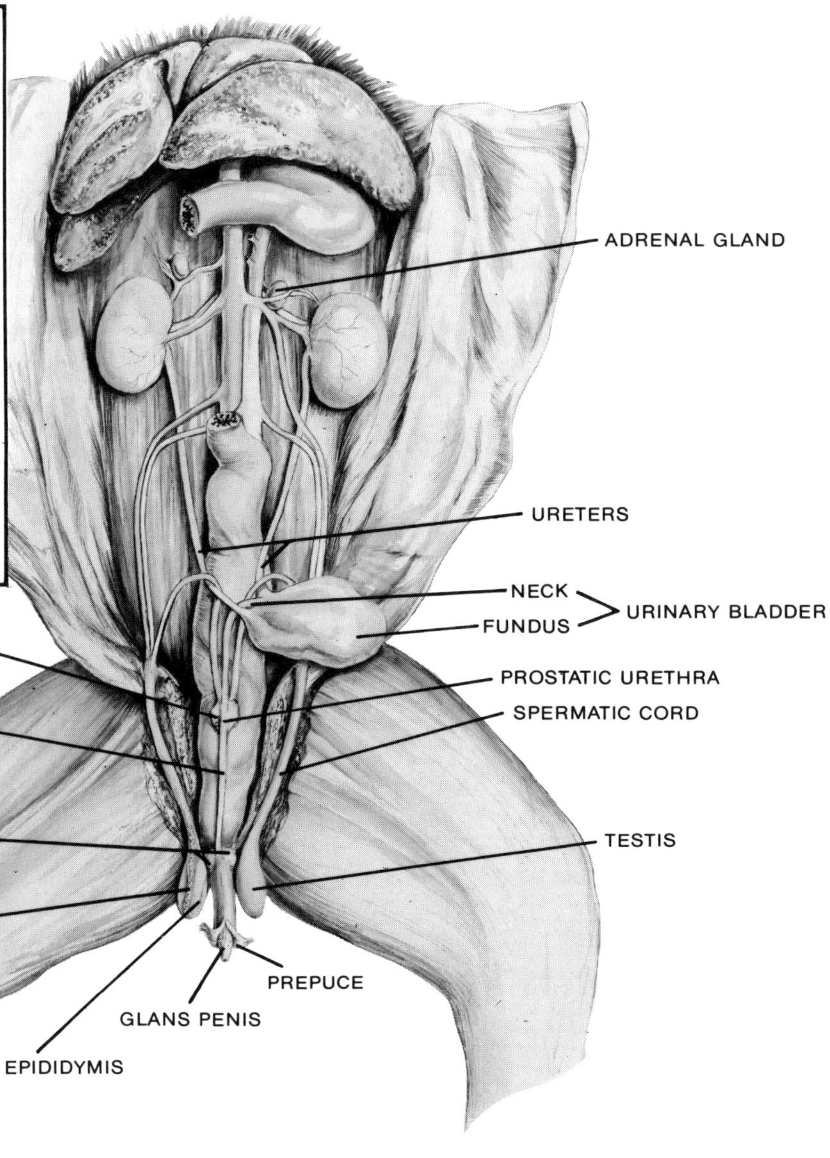

epididymis, but too small to be observed grossly, are the efferent ductules that connect each testis with the head of each epididymis. The tail of the epididymis is continuous with the highly convoluted portion of the ductus deferens.

Ductus deferens. The ductus deferentes are contained in the spermatic cords; each courses cranially from the epididymis, loops over the ureter, and then runs caudally until it terminates in the prostatic urethra.

Spermatic cord. The spermatic cords are paired cords extending from the two testes to the prostatic urethra. Each cord consists of nerves, blood vessels, lymphatic vessels, and ductus deferens, with an epithelial covering.

Inguinal canal. This is the small opening in the abdominal musculature and body wall on each side of the body, through which the spermatic cord passes into the peritoneal cavity.

The Male Urethra and Associated Structures (Fig. 26-2)

Urethra. The male urethra is the narrow passage from the neck of the bladder to the tip of the penis. It is divided into prostatic, membranous, and spongy portions.

Prostatic urethra. The prostatic urethra is that portion of the urethra which passes through the prostate gland. It conveys both urine and prostate gland secretions to the next portion of the urethra. The ductus deferens enters the prostatic urethra.

Prostate gland. This is a bilobed gland located dorsal to the junction of the neck of the bladder and the ductus deferens. The ductus deferens and the neck of the bladder become a common duct within the substance of the prostate gland.

Membranous urethra. This is the segment of urethra between the prostate and the bulbourethral glands.

Bulbourethral glands. (Cowper's glands) These paired glands are connected to the membranous urethra by a pair of almost microscopic ducts.

Spongy urethra. Once the urethra passes into the penis, it is known as the spongy urethra or penile urethra.

The Penis (Figs. 26-2, 26-3)

Penis. Cylindrical in shape, the penis is the external copulatory organ of male mammals. The penis encloses the spongy urethra, the terminal portion of the urethra. The core of the penis is subdivided into three cylindrical bodies, two ***corpora cavernosa*** and the ***corpus spongiosum***. These bodies are termed erectile tissues because they contain blood spaces which become engorged with blood during erection of the penis.

Crura. (Not visible grossly) The crura of the penis are diverging proximal extensions of the corpora cavernosa. They extend laterally and attach to the ischium. They serve to support the erect penis during copulation.

In order to see the following two structures it is necessary to make a cross section through the penis.

Corpora cavernosa. These are the two cylindrical bodies of erectile tissue on the dorsal side of the penis.

Corpus spongiosum. This is the cylindrical body of spongy tissue on the ventral side of the penis which encompasses the spongy urethra.

Glans penis. This is the enlargement at the end of the penis; it is actually a part of the corpus spongiosum. It enfolds and caps the distal ends of the corpora cavernosa. The glans penis is covered by the prepuce.

Prepuce. This is a loose layer of skin which covers most of the glans penis.

Os penis. This is a small bone in the glans penis. With careful dissection, it can be found to one side of the urethra. It helps the penis remain stiff during erection.

Urogenital aperture. The urogenital aperture is the urethral opening at the tip of the penis.

Figure 26-3. Cross section through the body of the penis of a cat.

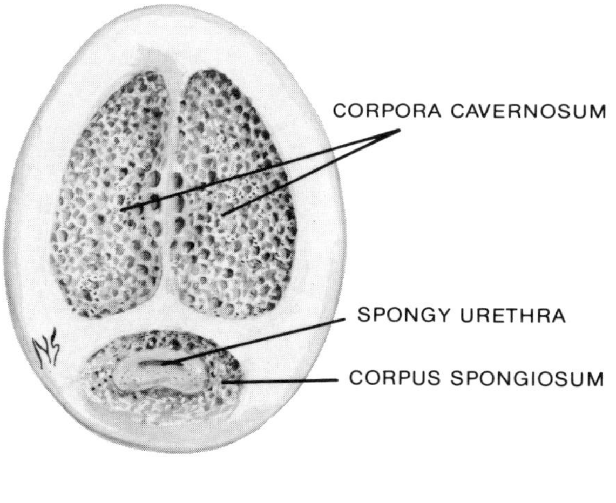

Anal glands. An anal gland can be found near the anus on either side of the rectum. These glands might be regarded to be part of the digestive system; their ducts drain into the anus. However, their odoriferous secretions may serve a reproductive function as sex attractants or stimulants.

FEMALE SYSTEM (Fig. 26-4)

Locate the paired uterine horns, tracing their connections to the mesometrial mesentery sheath along the dorsolateral abdominal walls. Following the uterine horns cranially, locate the oviducts and the cup-like ostia draped over each ovary just caudolateral to each kidney.

Follow each uterine horn caudally. Notice their convergence to form the body of the uterus, which opens into the vagina. The vagina in turn joins the vestibule.

Students with pregnant females may notice a series of enlarged swellings within each uterine horn. These enlargements may each contain an embryo. Isolate each embryo by cutting through the constricted portions of the uterine wall between the embryos. Open each enlarged pouch by a superficial longitudinal incision and examine the contents. Notice not only the

Figure 26-4. Ventral view of the abdomen of a cat, with emphasis on the structures of the female urogenital system. A gravid uterine horn is shown on the right.

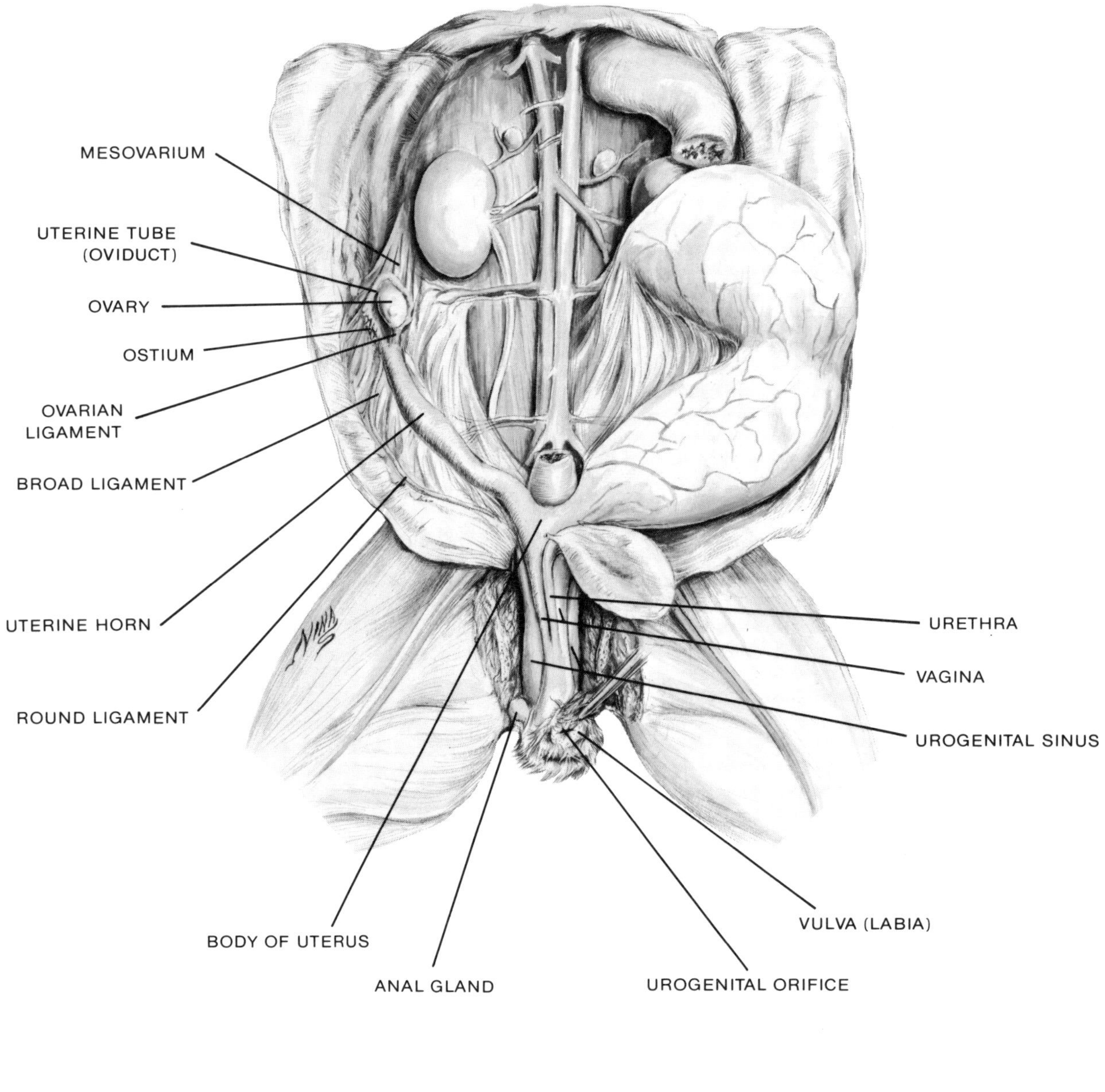

embryo but also the amnionic and chorionic membranes and the placenta. If desired, these embryos can be preserved in 10 percent formalin solution.

Identify the following structures.

Ovaries. The ovaries are the paired female gonads. They are located caudal and slightly lateral to the kidneys.

Corpus luteum. Corpora lutea may be seen, with some difficulty, on the surface of each ovary as small swellings. These structures represent areas from which mature eggs were ovulated and which have participated in sex hormone production.

Mesovarium. The mesovarium is the mesentery sheet which attaches each ovary to the broad ligament of the uterus. It also extends anteriorly to the caudal end of the kidney.

Ovarian ligament. This short, stout ligament connects each ovary to its uterine horn.

The Uterus and Associated Structures

Ostium. (Infundibulum) The ostium is the funnel-shaped, free end of the oviduct or uterine tube. Its opening, situated close to the ovary, is identified by its fringed edge.

Oviduct. (Uterine tube) This is the convoluted tube between the ovary and the horn of the uterus on each side of the body. It lies slightly lateral to the ovary and receives the ova through its free end, the ostium. This free end curves up and around the ovary.

Uterus. The uterus lies dorsal to the urethra and the urinary bladder. It is made up of right and left ***uterine horns***, which fuse medially to form the body.

The uterine horns are the sites of implantation of the developing embryos (Fig. 26-5). The ***placenta***, made up of uterine and embryonic tissues, is of the zonary type. The ***umbilical cord***, containing the umbilical blood vessels, connects the fetus to the placenta. The allantois, an outpocketing of the embryonic hindgut, and the chorion are fused into a ***chorioallantois*** which makes contact with the uterus through microscopic processes of the chorion.

Broad ligament. The broad ligament is the mesentery which connects both the uterine horns and the body of the uterus to the body wall. It supports the entire uterus, a substantial job when heavy embryos are present.

Mesometrium. This is another name for the portion of the broad ligament caudal to the mesovarium.

Round ligament. The round ligament is a fibrous cord extending from each anterior uterine horn to the dorsal body wall. It lies approximately perpendicular to the broad ligament.

Vagina. The vagina is a chamber between the uterus and the vestibule (urogenital sinus). The opening of the urethra into the urogenital sinus marks the boundary between the vagina and the urogenital sinus.

Cervix. (Not shown in Fig. 26-4) The cervix is that portion of the uterus which projects into the vagina. It is best seen by making an incision into the vagina at the point where the vagina and the body of the uterus meet.

Os uteri. (External uterine orifice) (Not in Fig. 26-4) The os uteri is the opening in the cervix which permits communication between the lumen in the body of the uterus and the lumen of the vagina.

Urogenital sinus. (Vestibule) The urogenital sinus is the portion of the reproductive tract posterior to the entry of the urethra. The urethra enters on the ventral side of the sinus. The urogenital sinus is, therefore, a passageway of both the excretory and reproductive systems.

Urogenital orifice. This is the external opening of the urogenital sinus. It is located at the caudal end of the urogenital sinus, ventral to the anus.

Vulva. This is a general term referring to the external genitalia ventral to the anus. This region includes the labia majora and the clitoris.

Labia majora. The labia majora are the longitudinal folds of skin bordering the urogenital orifice.

Clitoris. The clitoris is a small mass of erectile tissue on the ventral wall of the urogenital sinus near the urogenital orifice. The clitoris is homologous to the glans penis of male mammals.

Anal glands. As in the male, anal glands can be found near the anus, on either side of the rectum. Presumably they function as accessory reproductive organs.

FUNCTION

The urogenital system serves both excretory and reproductive functions. The excretory system removes nitrogenous wastes from the body, maintains water balance, and controls the flow of some essential molecules and ions from the body.

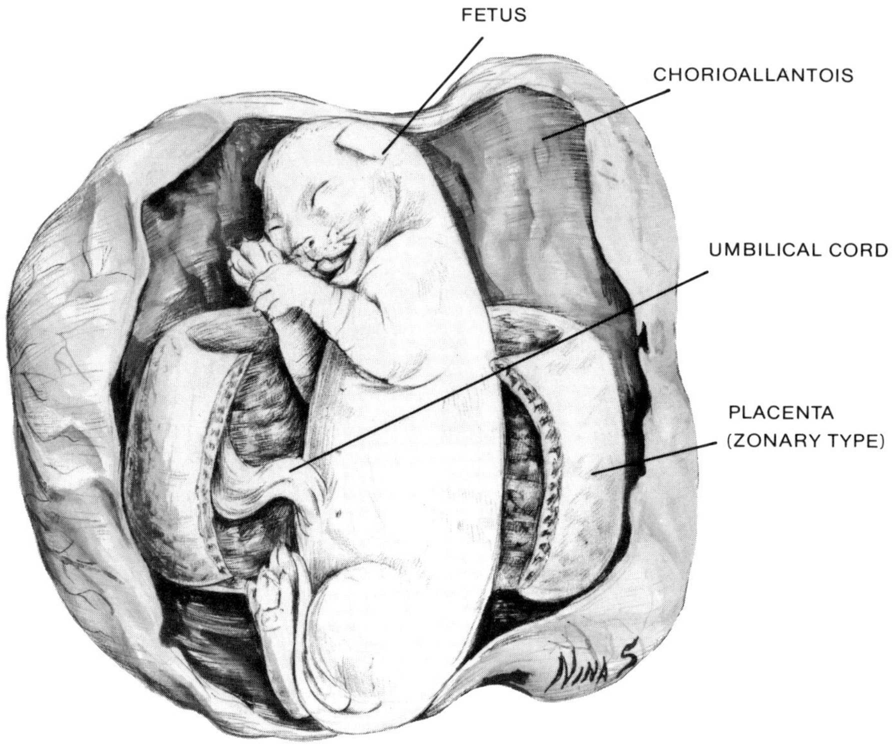

Figure 26-5. Fetus of a cat after removal from the uterus. Fetal membranes and placenta are partially removed in this view.

In mammals, the kidney forms a urine which is usually produced in a form hypertonic to the body fluids; thus it helps to conserve water. Water conservation is an important adaptative problem which terrestrial vertebrates must solve in order to survive on land.

The adrenal glands develop in close proximity to the kidneys. The adrenals are endocrine glands which consist of two types of tissue. The adrenal cortex, the outer tissue layer, manufactures and secretes steroid hormones which control a wide range of body maintenance functions. Among the hormones secreted are aldosterone, which controls salt and water balance, and cortisol and corticosterone, which modify metabolism. There is evidence that small quantities of the sex hormones progesterone and testosterone may also be secreted by the adrenal.

The other portion of the adrenal glands is the adrenal medulla, the inner tissue surrounded by the cortex. In reality it is a portion of the nervous system which has been modified to serve an endocrine function. The medulla develops from neural crest cells, the source of the neurons of spinal and sympathetic ganglia. When stimulated by the autonomic nerve fibers which innervate it, the adrenal medulla secretes epinephrine (adrenalin) and norepinephrine (noradrenalin). These hormones are involved in rapid mobilization of body functions in stressful emergency situations.

The route of urine flow and storage shows widespread evolutionary variation in vertebrates. In mammals, the ureters carry urine directly to the bladder, thereby eliminating passage of the urine through the cloaca. The urinary bladder originally developed in vertebrates as a temporary storage site for urine prior to excretion from the body through the cloaca. In animals such as frogs, water can be resorbed from the bladder to combat desiccation of the body. The urethra also shows considerable variation between the sexes. In male mammals it not only transports urine but also receives sperm from each ductus deferens and carries the seminal fluid to the exterior.

The Male System

The testes manufacture the male gametes, the sperm. Embryonically, the testes develop inside the abdominal cavity; but in many mammals, they subsequently descend through a passage in the abdominal wall called the inguinal canal. The testes then rest in the scrotal sacs outside the body wall, although there is still a cavity of communication—the lumen of the processus vaginalis—between the scrotum and the abdomen.

Table 26-1 lists the sequence of major structures through which the urine and male gametes pass.

The Female System

The ovaries produce the female gametes, the eggs or ova. Mammalian ova are barely macroscopic; they contain much less yolk than the eggs of vertebrates like sharks or reptiles. In mammals, the ovaries are relatively smaller than in lower forms. In the cat, small vesicles called Graafian follicles may be visible on the surface of the ovaries. Each follicle contains an egg, several of which are released into the uterine horns during each estrus cycle. This differs from the human condition where, generally, only one egg is released per estrus cycle.

The corpus luteum is a temporary endocrine organ which forms after ovulation from the ruptured vesicle of the Graafian follicle. It secretes progesterone, a hormone important in maturation and maintenance of the uterine lining. If fertilization does not occur, the corpus luteum degenerates to a nonfunctional state; it is then called the corpus albicans.

The ostium, oviduct, uterus, and vagina are the route of egg and embryo transport. Ovulated eggs enter the infundibulum presumably because of the beating of the cilia which line the structure. Fertilization usually occurs in the oviducal portion of the system; in fact, early developmental events such as cleavage may go on as the tiny embryo is transported posteriorly toward the uterus.

Once in the uterus, the embryo implants in the richly vascularized wall. In the cat this occurs in the uterine horns; gestation occurs there also. Human embryonic development takes place in the body of the uterus.

Besides serving as the birth canal, the vagina plays a crucial role in mammalian copulation. It is the

Table 26-1. Sequence of major structures through which urine and gametes pass in the male cat.

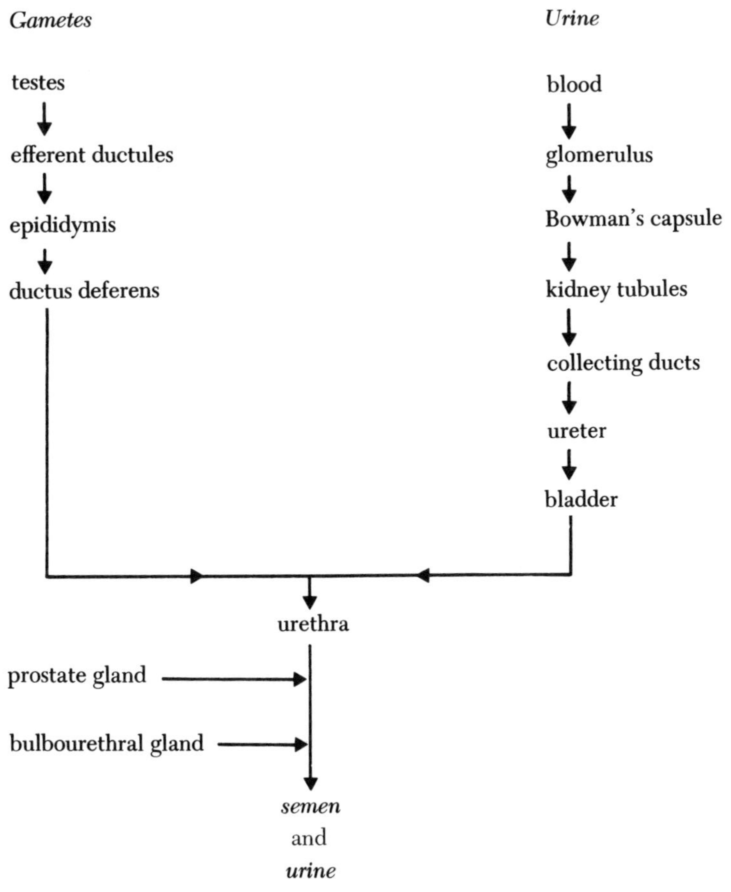

Table 26-2. Sequence of major structures through which urine and gametes pass in the female cat.

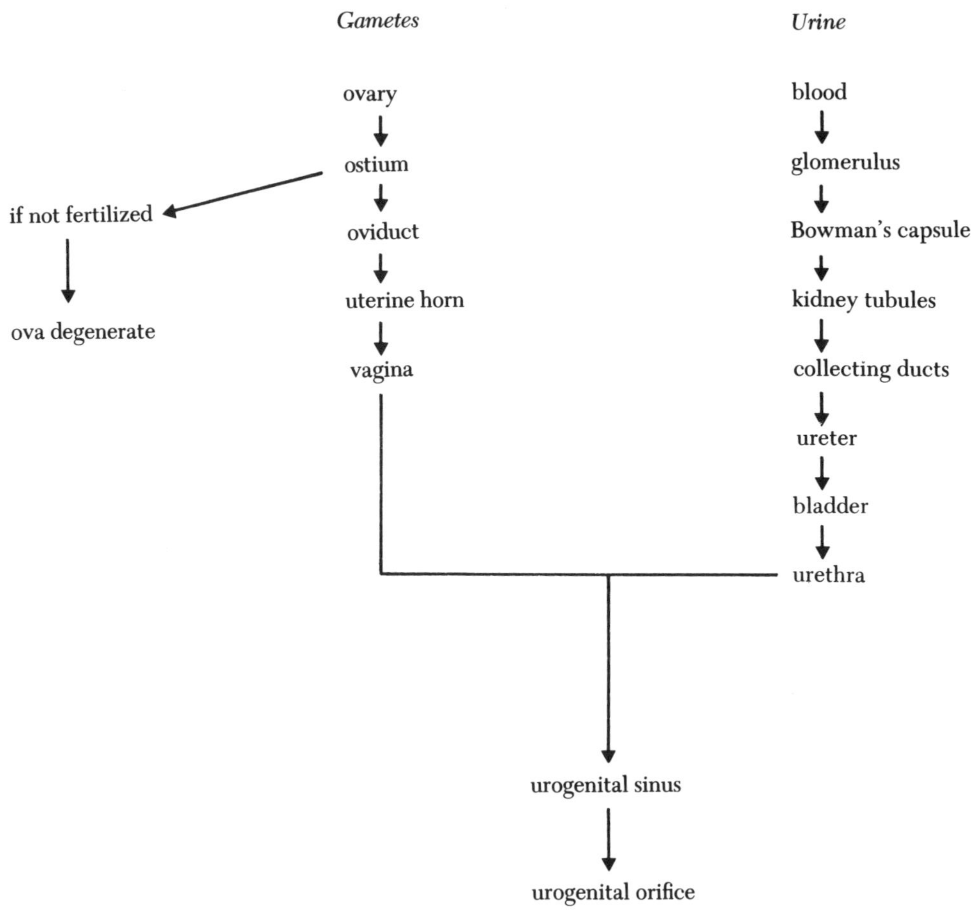

receptacle for the penis of the male and the site for deposition of the sperm. The vagina forms as a result of fusion of the posterior ends of the embryonic Müllerian ducts. This fusion is associated with elimination of the cloaca of earlier vertebrate forms; it results in formation of the single genital sinus of the vagina and of the body of the uterus. The more anterior uterine horns represent unfused derivations of the Müllerian ducts.

The routes of gametes and urine in a female cat are indicated in Table 26-2. In contrast to the male, the female system has no true common duct portion that transports gametes and urine for any significant distance.

MARY M. COOKE

27

The Cat: Circulatory System

Most vertebrates have blood with pigment-containing cells that circulates in a closed vascular system. The vessels leading away from the heart, the arteries, are very often very muscular; they are under both nervous and hormonal control. Veins, returning blood to the heart, tend to be larger in diameter and more thin-walled than arteries. Capillaries are tiny vessels with walls only one cell thick that connect arteries to veins in the body tissues. The capillaries are sites of nutrient and waste product exchange between blood and tissue fluids. In a mammal or bird, the pumps which drive the blood to the respiratory organs and to the general body are the right and left ventricles of the heart, respectively. The atria of the heart serve as primer pumps, adding 30 percent to the ventricular blood volume; this makes the ventricles function more efficiently.

COMPONENTS OF THE MAMMALIAN CIRCULATORY SYSTEM

The original function of the circulatory system was transport of nutrients, such as oxygen, to the tissues, and removal of waste products, such as carbon dioxide. In higher vertebrates, the system has several newer functions, which are of great importance. The circulation of lymphocytes in the bloodstream facilitates the prompt recognition of invading microbes and toxins. Differential peripheral circulation assists the body in controlling the core temperature. And the system serves as the vehicle for hormonal communication and regulation.

The Lymphoid System

A second vascular system in the bodies of higher vertebrates contains the lymph, a fluid similar to blood plasma but lacking the cellular components found in whole blood. The lymphatic vascular system is not a complete "circulatory" system, for the lymph flows in one direction only, from the periphery toward the heart. The lymph vessels, or lymphatics, have thin walls like the veins. The development of the lymphatics is correlated with increasing blood pressure in the evolution of vertebrates, for the lymph system serves as a route for returning to the venous system fluid that is expressed out of the capillaries by hydrostatic pressure. Lymphatics are seen first in bony fishes; they are present in all higher vertebrates. Teleosts, amphibians, and reptiles have "lymph hearts," muscular chambers that propel lymph into the veins from the lymph vessels. Birds and mammals use a combination of valves and skeletal muscle compression to force the lymph toward the heart.

Organs Associated with the Circulatory System

Several visceral organs are associated with maintaining the homeostasis of blood. Red blood cells are subject to severe mechanical stress and destruction; a continual process of replacement is necessary. The site of blood formation, in adults, as well as embryos, is remarkably variable. In vertebrates other than mammals, the capsule of the kidney is a major embryonic source of blood cells; mammalian embryos utilize the liver and spleen. Various adult vertebrates may use the spleen or the bone marrow to produce a continuing supply of new red blood cells. In adult mammals, the spleen, though no longer the site of red blood cell formation, continues to be associated with the circulatory system, destroying old and defective blood cells.

THE EVOLUTION OF THE MAMMALIAN CIRCULATORY SYSTEM

One of the most important steps in the evolution of the vertebrate circulatory system was the change in locus of blood oxygenation from gills to lungs. This shift coincides with the appearance of a double circulation through the heart and a separation of arterial and venous blood flow. Septa developed dividing the cardiac chambers; the ventral aortic vessel from the heart to the gills was lost; and complete separation occurred between systemic and pulmonary blood flow.

Other important trends include a general increase in blood pressure, reduction of the number of portal systems, and modifications in the pattern of venous flow to the heart. The increased pressure and rates of flow are essential to the rapid delivery of oxygen required by birds and mammals living with body temperatures of 37–40°C.

In order to clarify the relationship between the heart of the dogfish shark and that of the cat, the major evolutionary changes will now be described. A detailed description is beyond the scope of this manual; it can be found in texts of comparative anatomy and vertebrate evolution.

Jawed Fishes

Jawed fishes have a circulatory system exemplified by that of the elasmobranchs (sharks). It is characterized by a heart without atrial or ventricular septa; hepatic and renal portal systems; and a myogenic heartbeat which can operate under parasympathetic inhibition. Paired pectoral girdle vessels and a simplified visceral circulatory pattern are present. Advances not seen in elasmobranchs are apparent in bony jawed fishes, however.

In vertebrate embryos, six pairs of aortic arches are usually found connecting the ventral and dorsal aortae. The fate of these arches in the adults is highly variable but of interest because, in a sense, it mirrors what occurred during vertebrate evolution. Bony fishes always lack arch I (the efferent spiracular arch in the shark) and arch II (the afferent spiracular and hyoidean in a shark) is often missing. Arches III through VI supply the gills, as in sharks. Portions of the dorsal aortae supply the head as the internal carotid arteries.

Lungfishes (Dipnoi)

Approximately 300 million years ago, placoderms developed lung-like structures. These were oxygen-acquiring organs auxiliary to the gills, which allowed survival under conditions of low partial pressure of oxygen in the water. These lungs were probably supplied with venous blood from arch VI. Oxygenated blood probably returned to the heart through the hepatic vein into the sinus venosus; later, a new vessel from the lungs, the pulmonary vein, entered the atrium directly. Partial atrial and ventricular septa developed. In modern lungfish, the conus arteriosus, from which blood leaves the heart, is divided so that lung-oxygenated blood flows mostly from the left side of gill arch III and thence to the head and neck, while deoxygenated blood flows in the right half to the gills and through aortic arch VI to the lungs. Accompanying these major changes in the pattern of blood flow are an increase in the sigmoid shape of the heart and modifications of the posterior venous flow.

Amphibians

Some amphibians continue the trend toward a double circulation. Separation of oxygenated from deoxygenated blood in the atria is ensured by a complete atrial septum. Venous blood empties into the right atrium via the sinus venosus; oxygenated blood from the lungs is carried via the pulmonary veins to the left atrium.

The pattern of blood flow through the conus seen in the lungfish is preserved in amphibians, with oxygenated blood and deoxygenated blood in the ventral and dorsal halves, respectively. The aortic arches become much more specialized in the amphibians. Aortic arch III, supplied with oxygenated blood from the lungs, is substantially modified to become the carotid artery, supplying the head. Aortic arch V is lost, and the entire body caudal to the neck is supplied with blood flowing over aortic arch IV. In the anurans—frogs and toads—the cardinal veins have been lost, and venous return occurs through the inferior vena cava. Sympathetic innervation of the heart is also found in the amphibians, allowing increased cardiac output as a response to stress.

Reptiles

The major change in the circulatory system found in reptiles concerns aortic arch IV and its relationship with the heart. The conus arteriosus is divided and reduced so that the systemic arches arise directly from the ventricles. Except for the left systemic arch, double circulation is functionally complete in reptiles. The ventricles are much more highly muscled than the atria, and the muscular continuity between the sets of chambers is lost. This necessitates the development of special tissue, in an area called the atrio-ventricular (A-V) node, to conduct the impulses from the atria to the ventricles.

Birds and Mammals

Birds and mammals have an anatomically double circulation with a complete septum separating the ventricles. Apparently, two paired systemic arches carrying oxygenated blood to the body are redundant; the right systemic arch is lost in mammals, while the left systemic arch is lost in birds. The remaining half of the systemic arch is represented by the aorta. The ventricles of birds and mammals, particularly the left ventricle, are very muscular, and consequently less distensible. Whereas lower vertebrates can increase cardiac output by increasing ventricular volume, birds and mammals must depend in part on increased heart rate.

The renal portal system shows many changes among land vertebrates. Reduced in reptiles, it can be shut off by appropriate muscular valving in birds. In mammals it is absent; arterial blood flows to the glomeruli of the nephrons and thence directly to the peritubular capillaries of those nephrons, where it participates in resorbtive processes. Flow from the "tail" region of the body to the kidney capillaries is seen in embryos and then disappears.

The Evolution of the Venous System

The evolution of the mammalian veins is also somewhat complicated. The principal change in the anterior venous system from the dogfish shark to the cat is the altered flow from the paired anterior cardinal veins into the single median superior vena cava and the paired jugular veins. The mammalian posterior venous system is the evolutionary result of a fusion of the hepatic vein, the posterior cardinal veins, and the new subcardinals. With these changes, venous systemic blood returns to the heart via a median vessel, the inferior vena cava. The inferior vena cava ultimately replaced the posterior cardinal veins, and in mammals only a vestige—the azygous vein—remains.

THE HEART (Fig. 27-1)

The dissection of any mammalian circulatory system requires patience and care, as the vessels are often heavily invested with fat which must be carefully picked off with blunt forceps. Although this work is tedious, it is essential that each important vessel be thoroughly cleaned so that it can be positively identified. The main blood vessels must remain intact; nerves, however, may be cut. A second complication may be incomplete injection of the vessels, particularly veins. In spite of this, it is still possible to find and identify the vessels; they can be easily differentiated from nerves as they contain clotted blood and thus appear as continuous or splotchy brown stripes. Note: the organization of the circulatory system, particularly the venous component, is highly variable. The diagrams in this section represent a common or generalized condition. If your specimen appears to vary, use the brief description of the vessel rather than the diagram to make an identification.

The thoracic cavity was opened during the dissection of the respiratory system. If any tissue is still present, use blunt forceps to remove adherent thymus tissue. Make a longitudinal incision in the pericardium along the midventral line and reflect the pericardium to expose the heart. Locate and identify the vessels of the heart and familiarize yourself thoroughly with the appearance of both the ventral and dorsal surfaces of the heart. The dorsal surface is best exposed by reflecting the heart cranially. Do not remove it from the chest.

Anterior vena cava. This large vein, carrying venous blood to the heart from all areas anterior to the diaphragm except the lungs, can be seen coursing caudad toward the heart and entering the right atrium dorsal to a flap of spongy tissue, the ***right atrial appendage*** (auricle).

Posterior vena cava. This large vein, carrying blood pooled from all parts of the body posterior to the diaphragm, can be seen running craniad toward the heart. It enters the right atrium though its dorsal face, slightly posteromedial to the entrance of the superior vena cava and is best seen by gently reflecting the heart cranially.

Right atrium. This chamber lies in the upper left quadrant of the heart *when seen in a ventral view.* The bulk of the chamber lies slightly caudal to the right atrial appendage.

Right ventricle. This chamber receives venous blood from the right atrium. It lies cranial to the interventricular sulcus, an oblique groove bearing a prominent coronary artery, visible in the ventral exposure.

Pulmonary artery. This large blue-injected vessel carries venous blood from the right ventrical to the lungs. Note that the pulmonary artery is ventral to the aorta. Immediately prior to passing under the arch of the aorta, the pulmonary artery splits, sending branches to the left and right lungs. Trace the branches of the pulmonary artery to both lungs, removing adherent pericardium where necessary.

Pulmonary veins. These red-injected vessels carry oxygenated blood from the vessels to the left atrium. Several of these vessels enter the heart through the dorsal face of the left atrium.

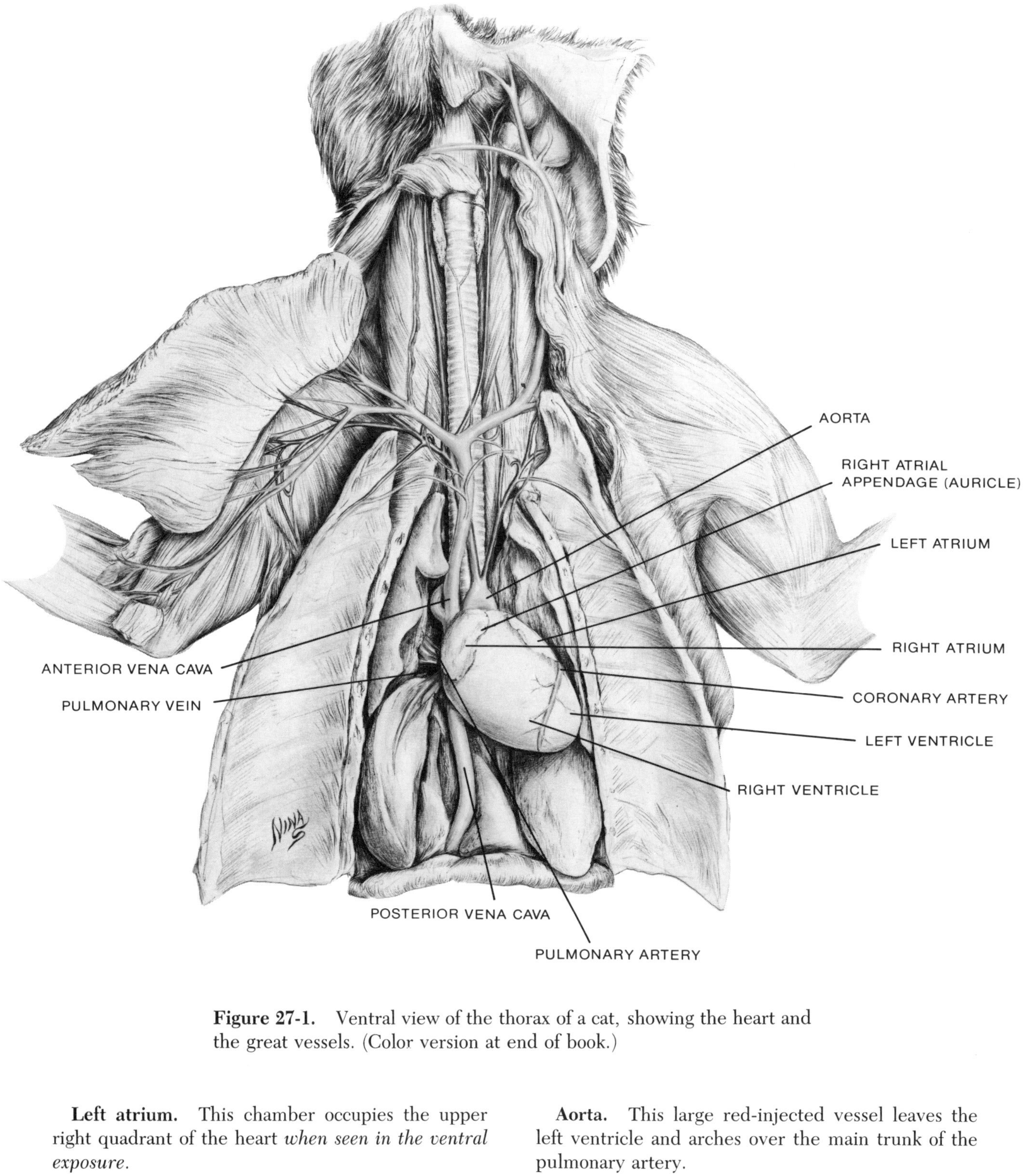

Figure 27-1. Ventral view of the thorax of a cat, showing the heart and the great vessels. (Color version at end of book.)

Left atrium. This chamber occupies the upper right quadrant of the heart *when seen in the ventral exposure.*

Left ventricle. The left ventricle is caudal to the interventricular sulcus. This ventricle has the thickest, most muscular walls, and it is responsible for pumping the systemic blood.

Aorta. This large red-injected vessel leaves the left ventricle and arches over the main trunk of the pulmonary artery.

Coronary arteries. Several arteries can be seen running over the surface of the heart. These vessels supply the heart muscle with oxygenated blood.

DISSECTION OF THE SHEEP HEART (Figs. 27-2 to 27-6)

At this point, the anatomy of the sheep heart will be examined, as an example of a typical mammalian heart. The purpose of this dissection is to identify the chambers of the heart and to establish the relationship of the vessels carrying blood to and away from these chambers. The dissection will follow the path of blood flow through the heart; you should note which chambers and vessels hold deoxygenated blood and which carry oxygenated blood. Identify all the chambers and vessels noted in the previous section as well as the features of internal anatomy described below.

Remove all fat from the surface of the heart, particularly from around the vessels. Familiarize yourself thoroughly with the external anatomy; learn to identify the vessels of the heart without probing to see where they go. Look at several hearts to become familiar with the typical condition.

Ventral surface of the heart. (**Fig. 27-2**) This is the surface of the heart that was exposed in the cat when the chest was opened. It can be identified in the isolated heart by the prominent ***interventricular sulcus*** and by the two large vessels springing from the anterior end, or ***base***, of the heart. No vessels enter or leave the center of the ventral face. The arteries arising from the heart are the ***aorta*** and the ***pulmonary artery***. The pulmonary artery leaves the heart slightly to the left of and ventral to the aorta. A band of tissue is frequently found connecting the pulmonary artery and the aorta. This is the ***ligmentum arteriosum***, a remnant of the embryonic ductus arteriosus. This ductus is the portion of aortic arch VI extending between the pulmonary artery and the aorta on the left side of the body. The closure of these ducts near the time of birth effectively separates pulmonary (arch VI) from systemic (arch IV) circulation. The superior vena cava, located in a more dorsal position, should not be confused with the great arteries, as it is a typical thin-walled vein.

Dorsal surface of the heart. (**Fig. 27-3**) The dorsal face of the heart receives the ***superior vena cava***; the ***inferior vena cava***, at a position slightly anterior to the center of the face; and the ***pulmonary veins*** in the center of the left quadrant. Thus, the

Figure 27-2. Ventral surface of a sheep heart.

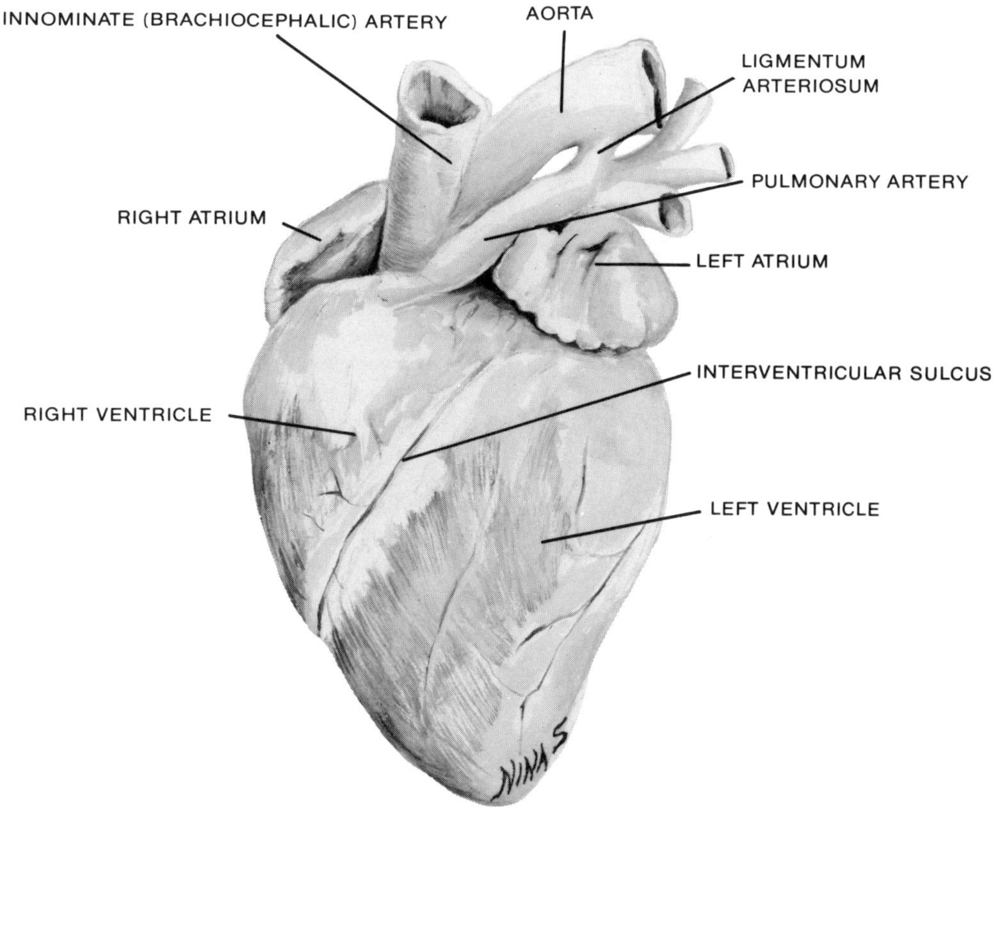

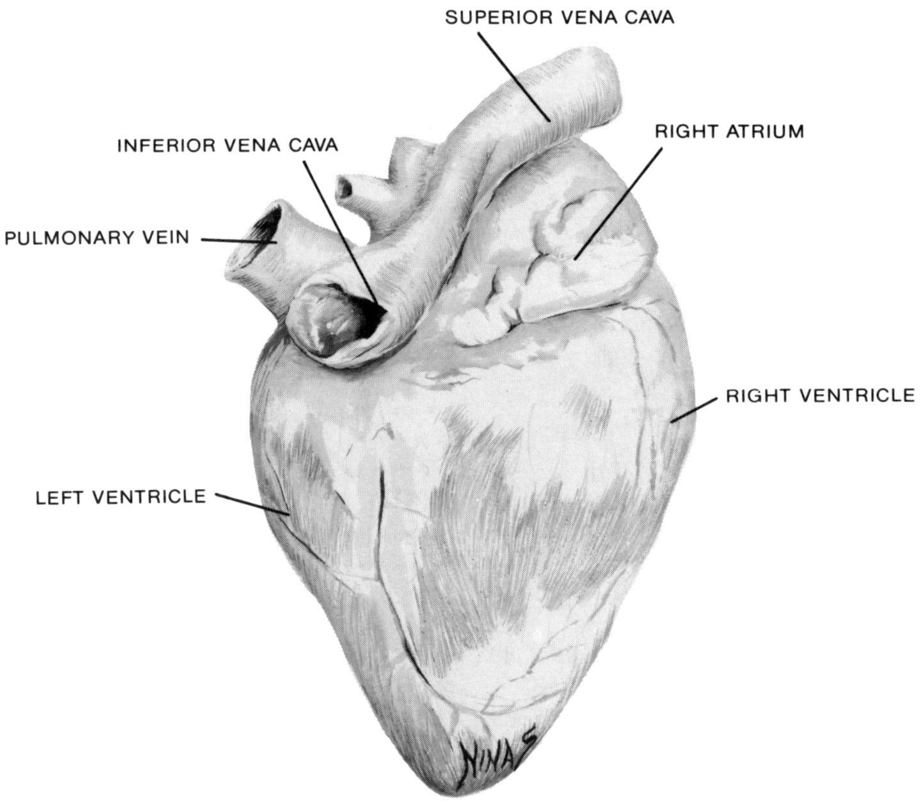

Figure 27-3. Dorsal surface of a sheep heart.

faces of the heart may be easily distinguished on the basis of the vessels they support and where these vessels originate or terminate. The faces may also be distinguished by the fact that the free edges of both auricular appendages can be seen in the ventral view.

Right Atrium and Ventricle (Figs. 27-4, 27-5)

Looking at the dorsal surface of the heart, insert the point of your scissors into the superior vena cava. Cut through the dorsal wall of this vessel and of the inferior vena cava. Cut through the right atrium in a medial-lateral direction, extending the cut laterally to the base of the pulmonary artery. Then cut through the right ventricle toward the apex of the heart, cutting about one-half the length of the ventricle. Separate the chambers and examine the internal structure (Fig. 27-4).

Endocardium. This smooth endothelial layer lines all the chambers of the heart.

Tricuspid valve. (Right atrio-ventricular valve) This valve, made up of three flaps of endocardium, prevents backflow of blood from the ventricle to the atrium during systole (ventricular contraction). It can be seen at the caudal end of the right atrium, between the atrium and the ventricle.

Chordae tendinae. These strands of connective tissue can be seen running from the right ventricle to the leaflets of the atrio-ventricular valve. They serve to prevent eversion of the valve into the atrium during systole.

Fossa ovalis. This aspirin-sized darkened depression on the atrial septum is the vestige of an opening between the atria in the embryo. This opening allowed oxygenated blood to pass from the right to the left atrium, and thereby to bypass the nonfunctional fetal lungs.

Papillary muscle. The chordae tendinae are attached to these bundles of ventricular muscle, which contract with the ventricle and prevent eversion of the valve at systole.

Trabeculae carneae. These are muscular cords found on the internal walls of the ventricles.

Moderator bands. These are bands of tissue which stretch across the ventricle. They are thought to aid in preventing overdistension.

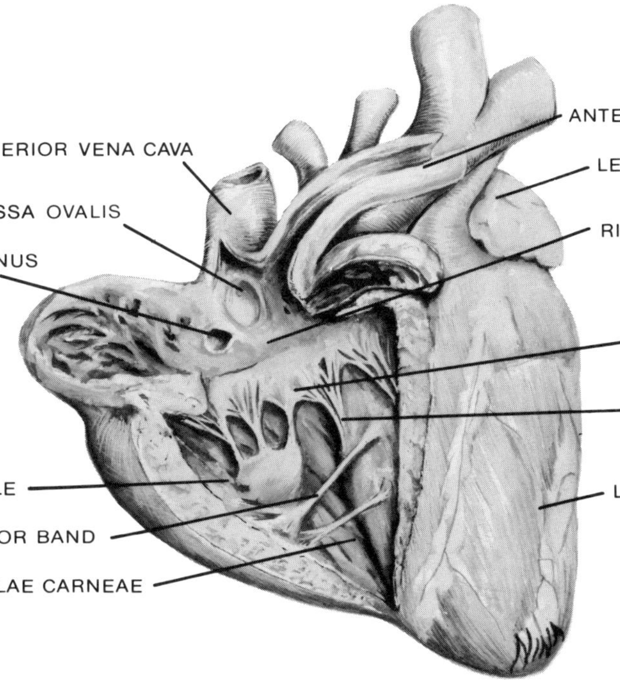

Figure 27-4. Interior of the right atrium and ventricle of a sheep heart.

Figure 27-5. Interior of the right atrium and ventricle of a sheep heart, showing the aortic semilunar valve and the pulmonic semilunar valve.

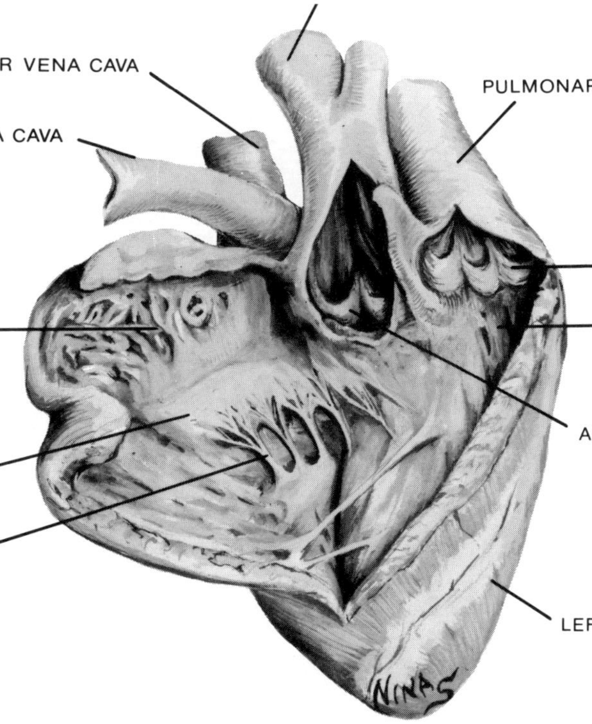

Orifice of the coronary sinus. The cardiac veins, returning blood from the heart tissue itself, empty into this sinus.

To expose the semilunar valves of the major arteries, insert your scissors into the base of the pulmonary artery (and subsequently into the aorta) and cut the vessel wall anteriorly 1 to 2 cm (see Fig. 27-5).

Pulmonic semilunar valve. This valve, composed of three crescent-shaped leaflets of endocardium, can be seen at the base of the pulmonary artery.

With your probe, trace the path of blood flow from the vena cavae through the right side of the heart to the pulmonary artery.

Left Atrium and Ventricle (Fig. 27-6)

Looking at the ventral surface, open the left side of the heart. Beginning about 1 or 2 cm from the medial edge of the left atrium, cut laterally though the atrium. Then cut posteriorly through the left atrium and left ventricle to the apex of the heart. Spread the chambers apart. Cut through the bicuspid (mitral) valve in the region where the aorta leaves the left ventricle to expose the aortic valves on this face.

Except for the increased musculature of the left ventricle, the interior of the left side of the heart is similar to the right.

Bicuspid or mitral valve. (Left atrio-ventricular valve) The left atrio-ventricular valve is two-leaved or bicuspid. like the right atrio-ventricular valve, it is supported by chordae tendinae and papillary muscle.

Aortic valve. This semilunar valve, which prevents backflow of aortic blood into the left ventricle, can be located at the base of the aorta. Study the structural differences between the atrio-ventricular valve and the two semilunar valves. Since there is little danger of valve eversion, the semilunar valves are not supported by chordae tendinae and papillary muscle.

Pulmonary veins. These thin-walled vessels can be seen opening into the left atrium.

With your probe, trace the path of blood flow from its return to the heart from the lungs via the pulmonary veins, through the left heart, to the aorta.

Figure 27-6. Interior of the left atrium and ventricle of a sheep heart.

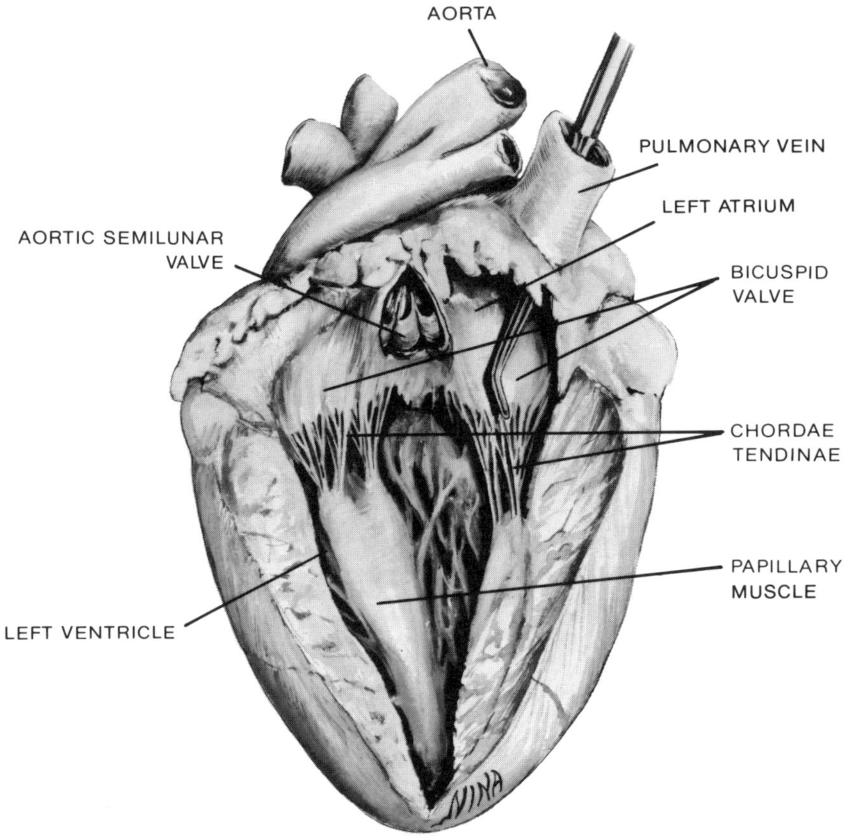

VESSELS ANTERIOR TO THE DIAPHRAGM IN THE CAT

Return to the thoracic cavity of the cat and dissect and identify the vessels of the chest, neck and forelimbs, clearing them of fat and clipping nerves where necessary.

Arteries (Figs. 27-7, 27-8)

Aorta. This vessel leaves the left ventricle running craniad and immediately arches dorsally, giving off two branches at the top of the arch:

Left subclavian artery. The more lateral of the two aortic branches, this large vessel supplies the left chest wall and the left forelimb.

Brachiocephalic artery. This large unpaired artery supplies the head, neck, right chest, and right forelimb. It runs craniad in a medial position, then branches into the right common carotid artery, the left common carotid artery, and the right subclavian artery at the level of the first ribs.

Trace the common carotid arteries cranially from the fork in the brachiocephalic artery.

Common carotid arteries. These are paired vessels, arising from the brachiocephalic, which supply each side of the head and neck. The branches of the common carotid include small branches to the muscles and the following:

Posterior thyroid artery. (Fig. 27-8) This is a small artery running parallel to the trachea. It gives off branches to the esophagus, trachea, and thyroid.

Anterior thyroid artery. This small artery runs medially to the top of the thyroid gland.

Laryngeal artery. This small artery runs medially to the larynx.

Bisect and reflect the digastric muscle. This will reveal the terminal branches of the common carotid artery.

Occipital artery. This is a small branch given off in the anterior region of the neck, immediately cranial to the point at which the common carotid passes deep to a large vein running across the neck (the transverse vein). The occipital artery courses dorsad toward the base of the skull.

Internal carotid artery. Given off just anterior to the occipital artery, this small artery runs mediodorsad to supply the brain. In the cat, the internal carotid is usually ligamentous. However, it is an important vessel in many other mammals.

External carotid artery. After giving off the internal carotid slightly caudal to the mandible, the common carotid becomes the external carotid artery. Curving laterally, it supplies the face. Its branches are the ***lingual artery***, a large artery coursing mediad on the ventral surface of the jaw to the tongue, and the ***external maxillary artery***, a prominent branch running dorsad along the side of the face, supplying the lips, teeth, salivary glands, and ear.

Return to the fork of the brachiocephalic artery and locate the right subclavian artery. As you trace it into the arm, cut off the many nerves of the brachial plexus, after verifying that none of them are actually uninjected blood vessels.

Right subclavian artery. The most lateral of the three branches of the brachiocephalic artery, this large artery feeds the thorax and axillary region. Its branches are the following:

Internal mammary artery. This vessel may have been severed when the chest was opened. It courses ventrad to supply the sternum and ventral chest. Locate the proximal portion, leaving the right subclavian artery, and the distal portion, running into the ventral chest wall.

Vertebral artery. This artery is given off shortly after the bifurcation of the brachiocephalic into the right common carotid and the right subclavian arteries, as the right subclavian passes over the ribs. It runs craniad for a short distance and then dorsad to supply the brain. (It passes through the transverse foramen seen earlier and ultimately feeds the basilar artery.)

Costocervical artery. This small artery is given off close to or immediately dorsal to the vertebral artery and arches dorsocaudad, giving off branches to the ribs and deep neck.

Thyrocervical artery. This large artery runs toward the shoulder as the right subclavian artery passes into the axilla. The branch of the thyrocervical supplying the shoulder is the ***transverse scapular artery***.

Axillary artery. After giving off the thyrocervical artery, the subclavian artery is directed toward the arm and becomes the axillary artery. Its branches are:

Anterior (ventral) thoracic artery. This vessel may have been destroyed in the dissection of the muscles. It is the first branch of the axillary artery in the axilla; it supplies the anterior regions of the pectoralis muscles.

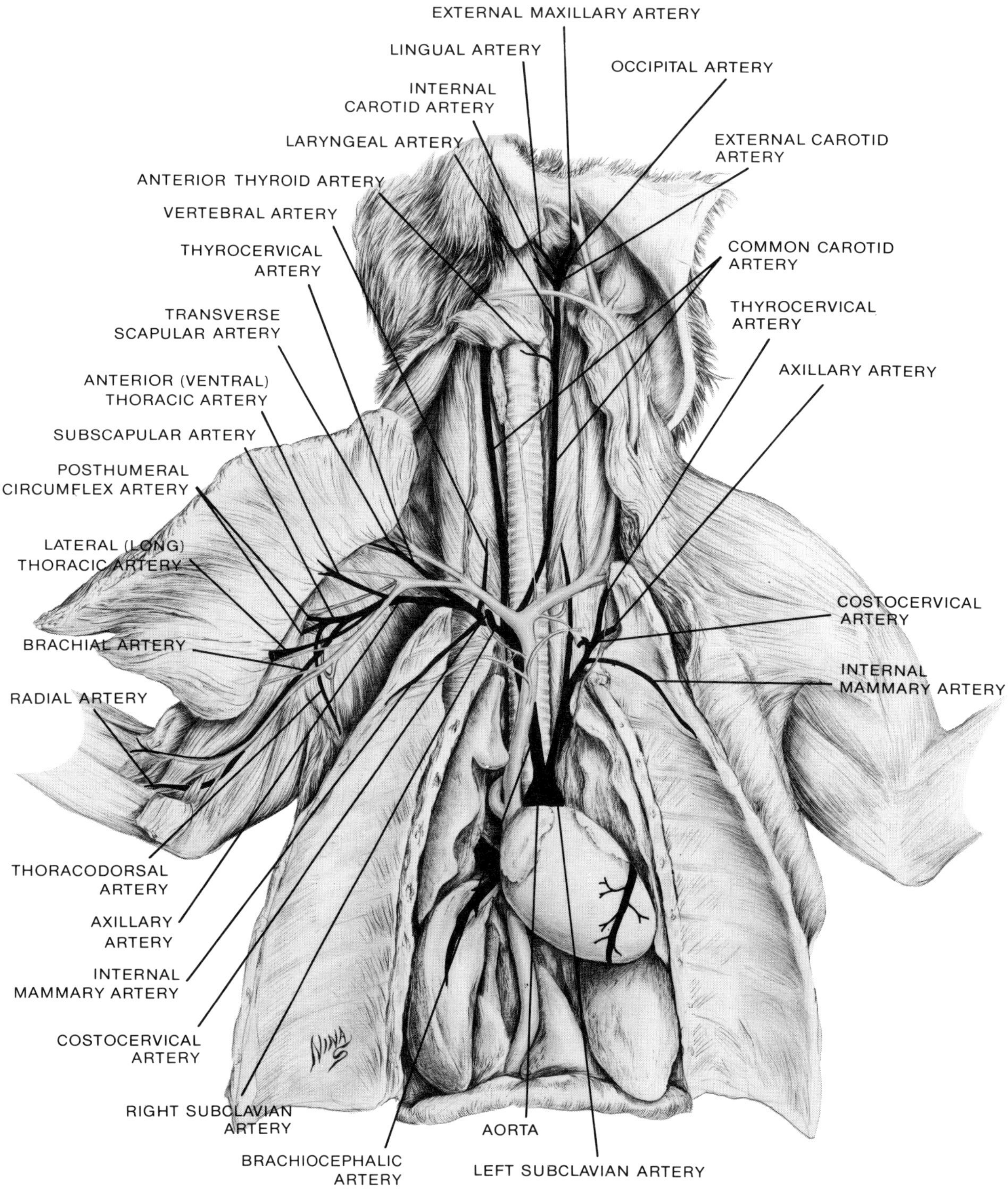

Figure 27-7. Ventral view of the thorax of a cat, showing arteries anterior to the diaphragm. (Color version at end of book.)

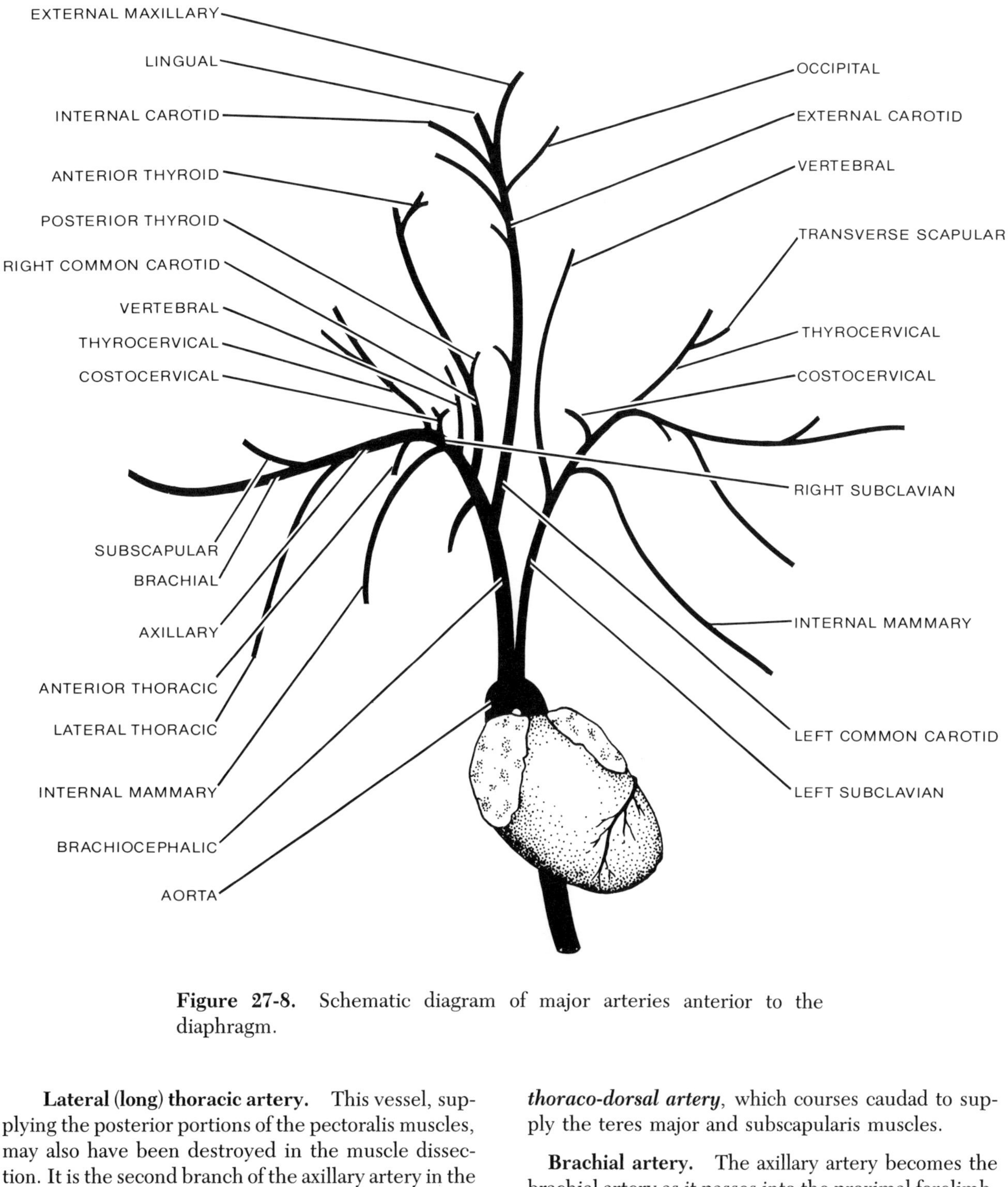

Figure 27-8. Schematic diagram of major arteries anterior to the diaphragm.

Lateral (long) thoracic artery. This vessel, supplying the posterior portions of the pectoralis muscles, may also have been destroyed in the muscle dissection. It is the second branch of the axillary artery in the axilla.

Subscapular artery. As the axillary artery enters the forelimb proper, it forks, forming the subscapular and brachial arteries. The subscapular artery, the cranial of the two branches, itself has two offshoots. These are the ***posthumeral circumflex artery***, a branched vessel running deep into the arm, and the ***thoraco-dorsal artery***, which courses caudad to supply the teres major and subscapularis muscles.

Brachial artery. The axillary artery becomes the brachial artery as it passes into the proximal forelimb. It gives off branches to the muscles and the humerus, passes through the supracondyloid foramen at the distal end of the humerus, and continues into the lower forelimb as the ***radial artery***.

Note: The branches of the left subclavian artery are the same as the right, but the origin of the left subclavian artery from the arch of the aorta rather than from the brachiocephalic artery produces some asymmetry. Therefore the thoracic and axillary branches of the left subclavian artery should also be dissected and identified. On the left (in some cats), the subscapular artery is a continuation of the axillary, between the axillary and the brachial.

Veins (Figs. 27-9, 27-10)

These veins, like the arteries, will be traced from the heart craniad. It is important to remember that this is opposite the direction of normal blood flow; the smaller veins are tributaries, not branches, of the larger ones. Keep in mind that some veins may not be injected and that often a vein courses beside the artery of the same name. All of the following veins are paired except the vena cava, azygous, sternal, thyroid, and transverse veins.

Anterior vena cava. This is the large median vessel which drains the anterior regions of the body. Tracing this vessel craniad, the tributaries encountered are:

Azygous vein. This vessel runs craniad lateral to the spine on the right side, arches ventrally, and enters the anterior vena cava from the right, immediately anterior to the heart.

Sternal vein. This veins drains the sternum and ventral chest; it enters the anterior vena cava about the level of the third rib. Its tributaries are the right and left ***internal mammary veins***, which should be located and dissected from the ventral chest wall.

Right costovertebral vein. This vessel enters the anterior vena cava on the right, near the fork of the brachiocephalic artery; in some specimens it may join the brachiocephalic. Its tributaries, found almost immediately craniolateral to its entrance into the anterior vena cava, are the ***right costocervical vein***, draining the deep chest, and the ***right vertebral vein***, draining the deep back.

Brachiocephalic veins. The left and right brachiocephalic veins unite to form the anterior vena cava. Tracing the ***left brachiocephalic vein*** laterally, identify its tributaries:

Thyroid vein. This is a very small median vessel formed by the union of two veins draining the thyroid gland. The thyroid vein may empty into either the right or left brachiocephalic vein.

Left costovertebral vein. This vessel, the functional counterpart of the right costovertebral vein already dissected, empties into the left brachiocephalic vein rather than directly into the anterior vena cava. It receives as tributaries the appropriate costocervical vein and vertebral vein.

Left subclavian vein. This vein drains blood from the forelimb region. Its union with the ***left external jugular vein*** forms the left brachiocephalic vein.

Left external jugular vein. This paired vein, the more medial of the two large tributaries of the left brachiocephalic vein, courses caudad from the head and neck. The tributaries of the left external jugular, tracing cranially, are:

Thoracic duct. This uninjected vessel, a ***lymphatic*** rather than a vein, is large, thin-walled, and somewhat lumpy. It can be located lateral to the esophagus, at the level of the aortic arch, and traced cranially to its entrance into the external jugular vein, slightly distal to the brachiocephalic vein. The thoracic duct is an unpaired vessel which returns lymph to the blood vascular system from the posterior regions of the body and the left side of the anterior regions.

Left lateral lymphatic(s). This is a vessel or vessels returning lymph from the left side of the head and neck regions. These lymphatics join the external jugular vein immediately lateral to its union with the subclavian vein. The right external jugular also receives uninjected lymphatic vessels draining the right anterior regions. The most important of these vessels is the ***right lymphatic duct***.

Left internal jugular vein. This very small vein, possibly uninjected, joins the left external jugular vein with the lateral lymphatics. Both jugulars run parallel and lateral to the common carotid arteries.

Transverse scapular vein. This large vessel can be seen entering the external jugular ventromedial to the scapula. An important source of venous blood in the transverse scapular vein is the ***cephalic vein***, a large superficial vein which is apparent near the anterior margin of the lateral surface of the forelimb.

Immediately anterior to the entrance of the transverse scapular vein, the external jugular passes through the musculature of the neck to lie superficial to the sternomastoid muscle. Continuing to trace the external jugular cranially, locate the union of the following:

Posterior facial vein. This prominent vein drains the side of the head. It is the most lateral of the three main tributaries of the external jugular vein.

Anterior facial vein. A large vein, draining lips and teeth, this is the median vein at the point of union of three veins to form the external jugular vein.

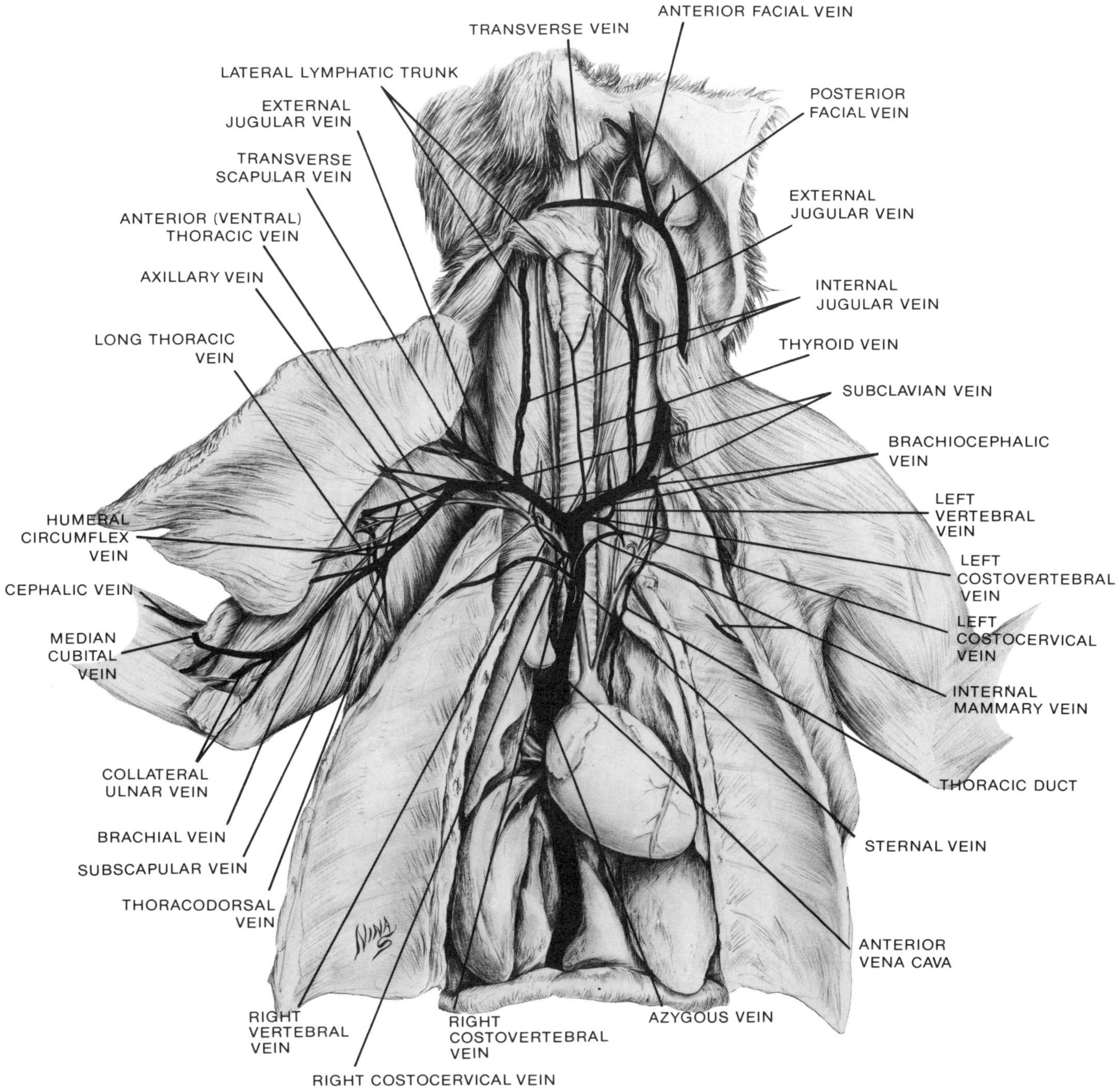

Figure 27-9. Ventral view of the thorax of a cat, showing veins anterior to the diaphragm. (Color version at end of book.)

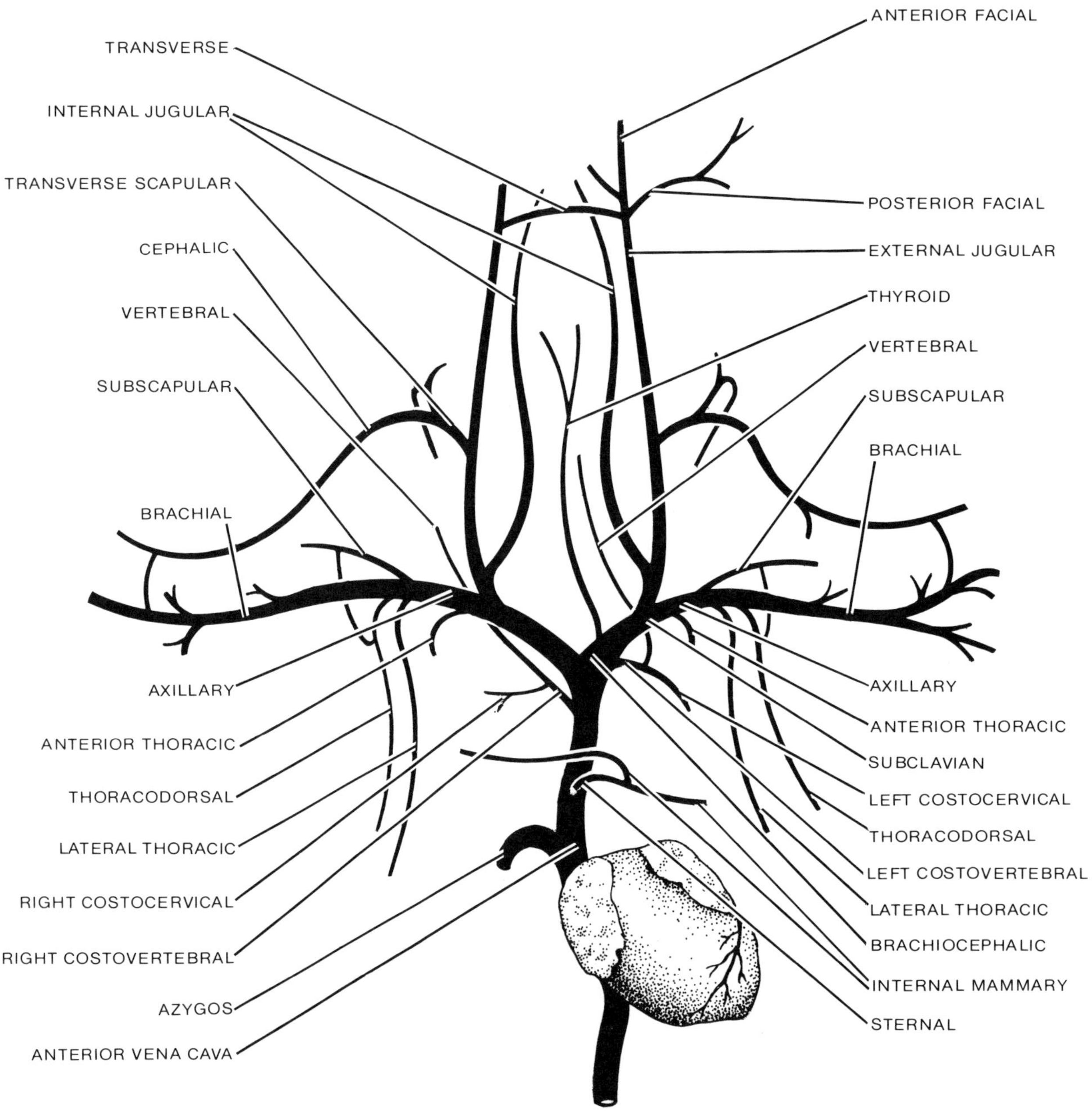

Figure 27-10. Schematic diagram of major veins anterior to the diaphragm.

Transverse vein. This large vein runs across the neck to form an anastomosis between the two external jugular veins. The union of this vein with the two facial veins forms the external jugular vein.

Left subclavian vein. Return to the point of union of the left subclavian vein and the left external jugular to form the left brachiocephalic vein. Trace the left subclavian vein laterally out of the chest.

Axillary vein. This is the distal portion of the subclavian vein, which courses in the axilla toward the thoracic cavity. It tributaries are the following:

Anterior (ventral) thoracic vein. Running close to the artery of the same name, this vessel drains the cranial regions of the pectoral muscles.

Subscapular vein. This vein, also coursing close to the artery of the same name, drains the muscles of the medial surface of the scapula. Its tributaries are the ***thoracodorsal vein***, from the teres major and subscapularis muscles, and the ***humeral circumflex veins***, draining the deep muscles of the upper forelimb.

Lateral (long) thoracic vein. This vein, draining the more posterior regions of the pectoralis muscles runs parallel to the artery of the same name.

Tracing the axillary vein distally into the forelimb, where it becomes the ***brachial vein***, observe the tributaries to the muscles. Proximal to the elbow, the brachial vein is joined by the ***ulnar vein***. In the crook of the elbow (the cubital fossa), the brachial vein is closer to the anterior border of the forelimb than the artery. At that point, locate the vein which connects the brachial and cephalic veins, the ***median cubital vein***.

VESSELS POSTERIOR TO THE DIAPHRAGM

Arteries (Figs. 27-11, 27-12)

Return to the arch of the aorta. Follow the aorta caudally, observing the intercostal arteries springing from it. Dissect away the connective tissue adhering to the aorta as it passes though the diaphragm. Free the ventral margin of the diaphragm from the chest wall and cut the diaphragm from the aorta ventrolaterally to the free ventral border, so the diaphragm can be reflected to expose the aorta. Clear away the fat adhering to the aorta and the bases of its abdominal branches. Reflecting the viscera to the animal's right, trace the abdominal aorta caudally. The first branch encountered, immediately caudal to the diaphragm, is the celiac trunk. Dissect away the mesentery adhering to this vessel and to its three branches. Then, allowing the stomach to fall back into its normal position, but retaining the liver reflected ventrally, expose the terminations of the first two branches: the hepatic to the liver, and the left gastric to the stomach. Reflect the stomach to the right again to observe the course of the third celiac branch, the splenic artery. In all dissections of the mesentery, take care not to harm large, uninjected vessels filled with brown clotted blood. These are the portal veins.

The branches of the abdominal aorta are:

Celiac trunk. (**Coeliac trunk**) The first branch of the aorta after it passes through the diaphragm, this short unpaired vessel supplies the anterior viscera via its three branches:

Hepatic artery. This is an unpaired vessel which courses ventrad past the lesser curve of the stomach in the mesentery. At the level of the pylorus, it terminates in several branches: the ***cystic artery***, an unpaired artery running craniad to the gall bladder; several branches to the liver; and the ***gastroduodenal artery***, a short, branched vessel coursing caudad to the pylorus and terminating as the ***anterior pancreaticoduodenal artery***. All these branches of the hepatic artery should be dissected and identified.

Left gastric artery. This second branch of the celiac trunk leaves the trunk immediately posterior to the hepatic artery. It courses ventrad to the lesser curve of the stomach.

Splenic artery. The largest and most distal branch of the celiac trunk, this artery courses ventrocaudad along the dorsomedial surface of the spleen. It gives off branches to the spleen, the pancreas, and the greater curvature of the stomach.

Anterior mesenteric artery. Tracing caudally, the next branch of the abdominal aorta encountered is this large unpaired vessel. Given off at the level of the anterior end of the kidney, it courses ventrad in the mesentery, giving rise to the following branches to the posterior viscera:

Posterior pancreaticoduodenal artery. This branch arises from the anterior surface of the anterior mesenteric artery at the level of the termination of the duodenum. It curves laterally and ends in the wall of the duodenum, supplying it and the posterior portions of the pancreas.

Colic arteries. Three distinct large branches—the right, middle, and ileocolic—leave the posterior face of the anterior mesenteric artery proximal to, slightly distal to, and at the level of the posterior pancreaticoduodenal artery. These arteries supply the ascending and transverse colon and the cecum. The right colic artery is sometimes found as a branch of the ileocolic.

Intestinal arteries. The anterior mesenteric artery terminates in many branches, the intestinal arteries, which course in the mesentery supplying the jejunum and ileum.

Note: See Fig. 27-12 for the following branches.

Adrenolumbar arteries. These paired vessels leave the aorta at the level of the cranial end of the kidney, course laterad on the dorsal abdominal wall, supplying blood to the adrenal gland and dorsal abdominal wall.

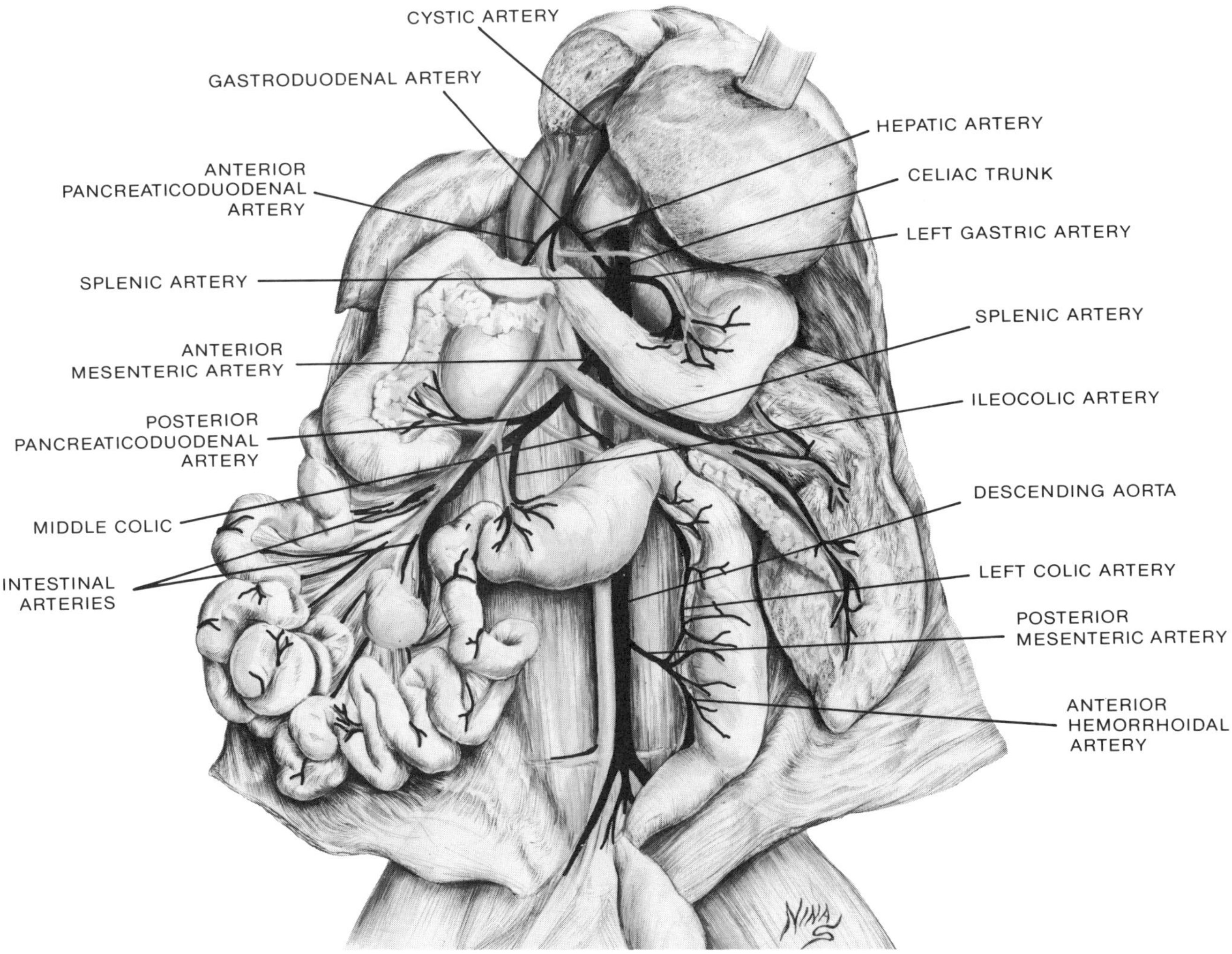

Figure 27-11. Ventral view of the abdomen of a cat, showing arteries supplying the viscera. (Color version at end of book.)

Renal arteries. These are paired vessels leaving the aorta at the level of the hilus of the kidneys. These arteries course directly laterad to the hilus.

Lumbar arteries. These are several pairs of arteries which arise from the dorsal surface of the aorta. They course dorsad toward the spine. The first pair may be seen at the level of the adrenolumbar veins; several more pairs follow caudally, and the last are at the level of the pelvic ilium.

Internal genital arteries. These are small paired arteries which course laterad and then caudad along the dorsal abdominal wall toward the gonads. These vessels arise from the aorta approximately at the level of the top of the bladder.

Posterior mesenteric artery. A large unpaired artery, the posterior mesenteric artery leaves the aorta at the level of the middle of the descending colon. It courses ventrad in the mesentery, giving off the following branches (see Fig. 27-11):

Left colic artery. This artery, the anterior branch of the posterior mesenteric artery, courses craniad in the mesentery, giving off branches to the descending colon. It forms an anastomosis with a colic branch of the anterior mesenteric artery.

Anterior hemorrhoidal artery. This artery, which supplies the rectum, can be seen running caudad from the posterior mesenteric artery along the dorsal surface of the colon.

Iliolumbar arteries. These paired vessels leave the aorta and course laterad along the dorsal abdominal wall to supply posterior portions of the body wall.

Immediately after giving off the iliolumbar arteries, the aorta divides into three branches—two external iliac arteries and a medial common trunk of the internal iliacs. With blunt forceps, trace one of the external iliac arteries caudally. Isolate it from adherent connective tissue. Cutting through overlying muscles when necessary, reflect the abdominal wall to expose the external iliac from the aorta to the thigh (Fig. 27-12). Be sure you dissect the vessels in one of the two hindlimbs.

Common trunk of the internal iliac arteries. The medial of the three terminal branches of the abdominal aorta, this vessel divides promptly into three branches, dorsal to the entrance of the urethra into the bladder. The branches are the two paired ***internal iliac*** (hypogastric) arteries and the ***caudal artery***.

Caudal artery. This is an unpaired artery in a medial position supplying the tail.

Right and left internal iliac arteries. These vessels, medial and deep to the external iliac arteries, course dorsolaterad, giving off several branches: the paired ***umbilical arteries***, one arising from each internal iliac and running ventrad then craniad to the bladder; the ***middle hemorrhoidal arteries***, paired vessels, one arising from each internal iliac and running mediocaudad to supply the rectum; and the ***gluteal branches***, given off at the level of the middle hemorrhoidal arteries and running laterocaudad to supply the dorsal muscles of the flank.

External iliac arteries. Return to the aorta and, on the dissected side, trace the external iliac artery laterocaudad. ***Deep femoral arteries*** leave the external iliac arteries as they pass through the abdominal musculature; the deep femorals give off several branches immediately after leaving the external iliacs. Branches are: the ***posterior epigastric artery***, a small artery running craniad on the ventrolateral body wall; a branch to the bladder; and branches to the muscles of the thigh.

Continue your dissection by finding the vessels which supply one hindlimb.

Femoral artery. The external iliac artery continues into the hindlimb as the femoral artery. The following branches arise from it:

Femoral circumflex artery. Arising shortly after the deep femoral artery branches off, this artery and its branches course laterodistad to supply the muscles of the upper hindlimb. It passes between the vastus medialis and the rectus femoris muscles.

Saphenous artery. This very superficial artery is given off just proximal to the separation between the adductor femoris and the semimembranosus muscles. It runs distad on the medial surface of the leg toward the posterior margin of the limb.

Popliteal artery. After giving off the saphenous artery, the femoral artery curves dorsal to the semimembranosus muscle and continues as the popliteal artery.

Bisect the sartorius and gracilis muscles, if this has not already been done, and reflect them. Separate the adductor femoris and semimembranosus muscles. Distal to the entrance of the femoral, bisect the semimembranosus. Reflect the distal half, exposing the popliteal artery as it runs behind the knee, in the popliteal fossa.

Veins (Fig. 27-13)

Return to the heart and locate the posterior vena cava as it enters the right atrium. Trace it caudally, freeing the posterior vena cava from adherent connective tissue as it passes through the diaphragm. Using blunt forceps and scalpel, carefully pick away liver tissue until the ventral half of the vessel is exposed from the level of the diaphragm to its emergence from the posterior margin of the liver. It may be easier to work from the posterior margin of the liver cranially to the diaphragm, by inserting blunt forceps or a probe into the liver just ventral to the vena cava.

Note: The unpaired arteries supplying the viscera do not have corresponding veins draining directly into the posterior vena cava. As you dissect and identify the tributaries of the posterior vena cava, notice which organs are drained by paired tributaries and which vessels do not drain directly into the posterior vena cava.

Posterior vena cava. This large vessel is the route of return to the heart of all venous blood caudal to the diaphragm. Tracing the vena cava caudally, its tributaries are:

Hepatic veins. These vessels—there may be two or three—are the last tributaries into the posterior vena cava before it passes through the diaphragm and empties into the heart. They feed into it as it passes through the liver.

Adrenolumbar veins. These paired vessels empty into the posterior vena cava at the level of the

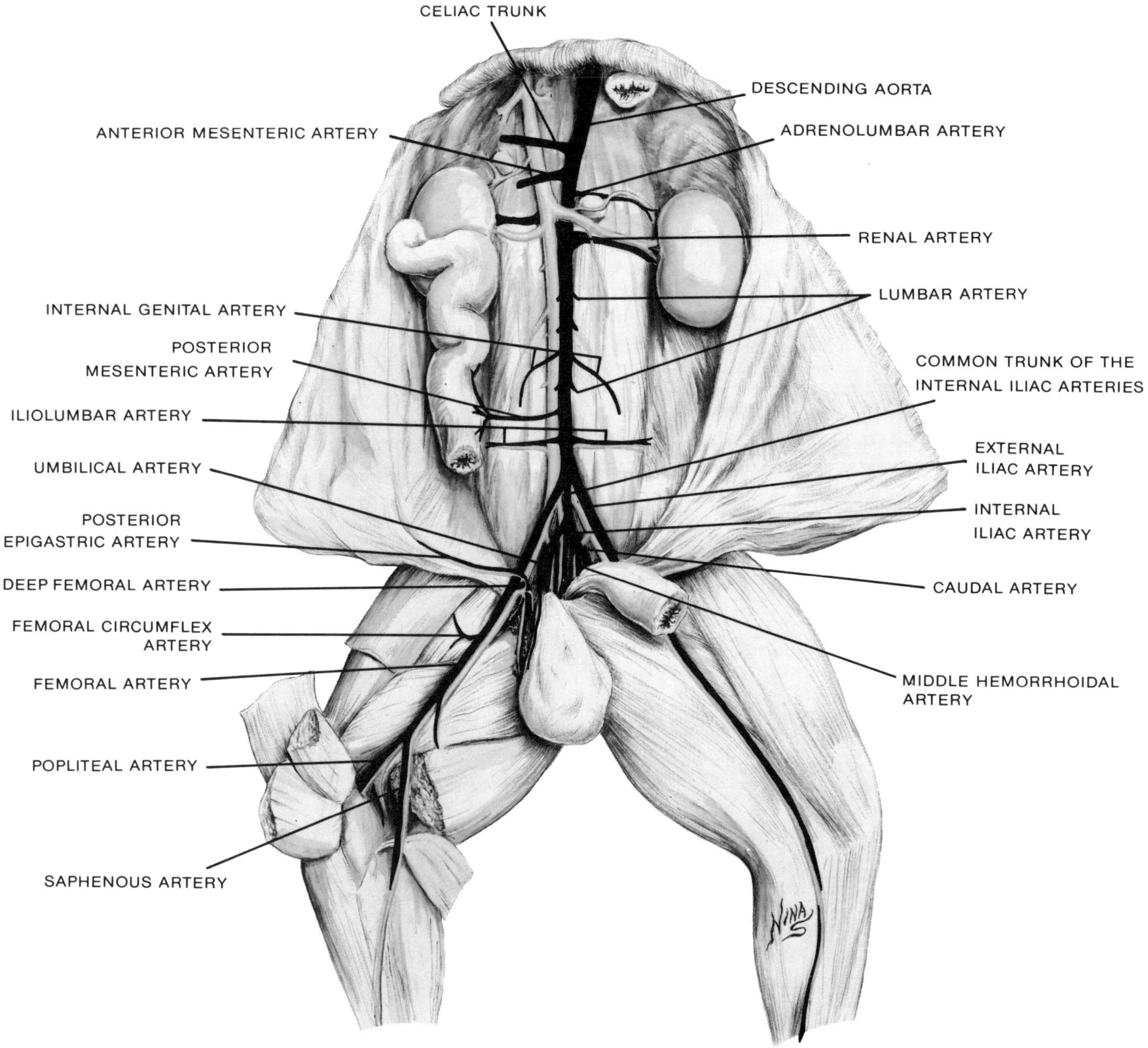

Figure 27-12. Ventral view of the abdomen and hindlimbs of a cat, showing arteries posterior to the diaphragm. (Color version at end of book.)

adrenal glands. One or both may empty into the appropriate renal vein rather than directly into the vena cava. They drain the adrenal glands and dorsal regions of the abdominal cavity.

Renal veins. These large paired vessels course with the renal arteries. They drain the kidneys, emerging at the hilus.

Lumbar veins. Several small unpaired veins, possibly uninjected, run ventrad from the middorsal line to empty into the posterior vena cava. They drain the regions supplied by the lumbar arteries.

Internal genital veins. This small vein or pair of veins on each side run with the artery of the same name in the lower abdominal region. The left internal

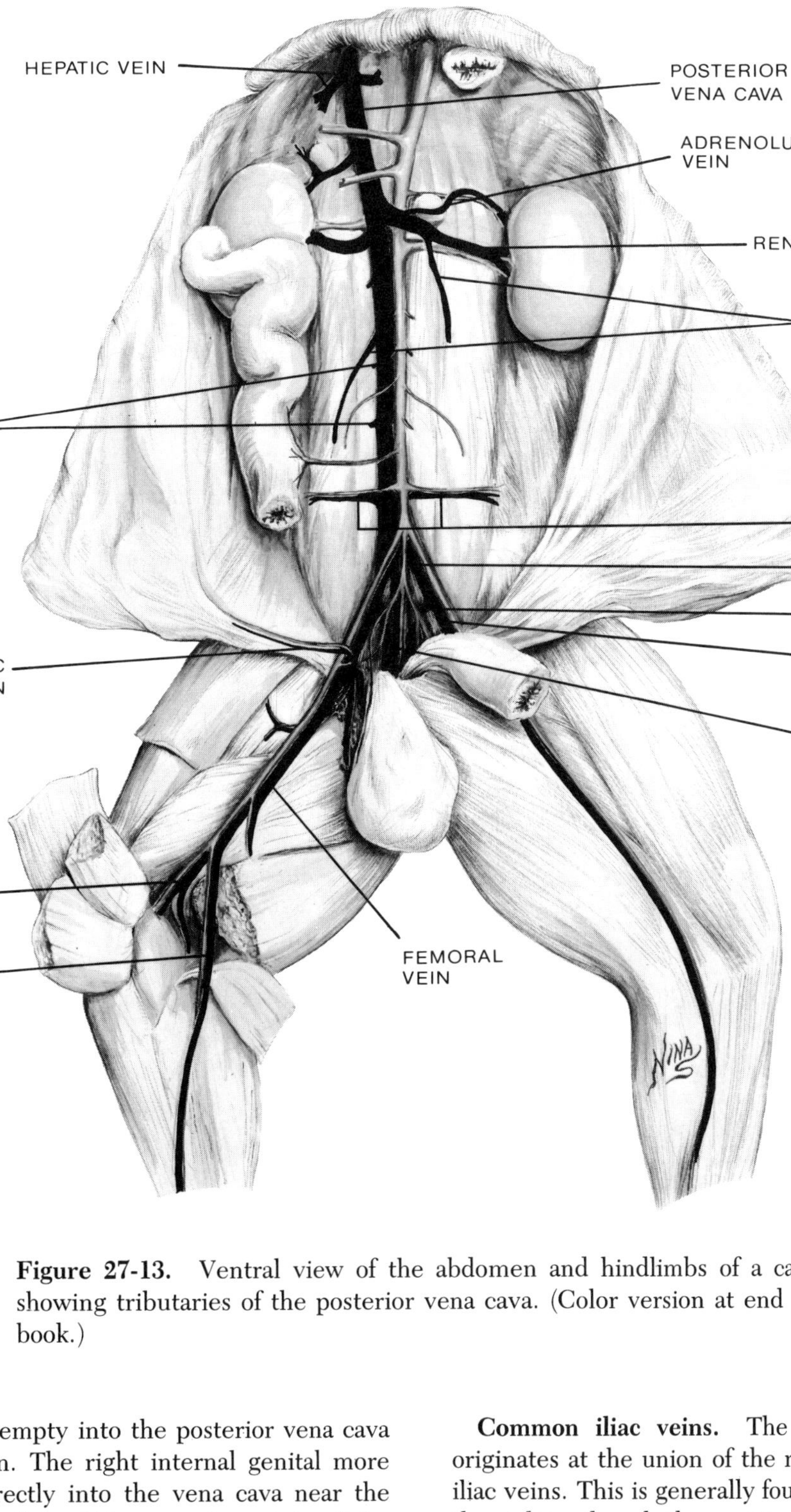

Figure 27-13. Ventral view of the abdomen and hindlimbs of a cat, showing tributaries of the posterior vena cava. (Color version at end of book.)

genital vein may empty into the posterior vena cava via the renal vein. The right internal genital more often empties directly into the vena cava near the posterior margin of the kidney or the origin of the internal genital artery from the aorta.

Iliolumbar veins. These large paired vessels, draining the posterior abdominal wall, enter the posterior vena cava at the level of the iliolumbar arteries.

Common iliac veins. The posterior vena cava originates at the union of the right and left common iliac veins. This is generally found slightly anterior to the pelvis, though the position of the union is variable. The tributaries of the common iliac veins are:

External iliac veins. These paired veins course from the hindlimbs; they are the source of most of the blood in the common iliac veins.

Internal iliac veins. One of these paired vessels enters the dorsomedial side of each common iliac vein from the deep pelvis, slightly posterior to the level at which the urethra leaves the bladder. Each internal iliac vein runs closely beside the appropriate internal iliac artery.

Caudal vein. The small, unpaired caudal vein drains the tail. It enters the right common iliac vein from the midline, proximal to the entrance of the internal iliac vein. It may be uninjected.

Return to the common iliac vein. Trace the external iliac vein laterocaudally into the leg. Work on the same side as your dissection of the external iliac artery.

Femoral vein. Distal to the abdominal cavity, the external iliac on each side of the body is called the ***femoral vein***. Dissect and identify the following tributaries of the femoral vein:

Deep femoral vein. As the femoral passes though the abdominal wall, it receives this vessel from a medial position. The deep femoral vein carries blood received from the ***posterior epigastric vein***, which drains the abdominal wall coursing parallel to the ***posterior*** epigastric artery. The deep femoral also receives a tributary from the bladder and a tributary from the external genitalia.

Saphenous vein. This is a large superficial vein on the medial surface of the leg. It originates at the foot and empties into the femoral vein superficial to the separation of the adductors longus and femoris.

Popliteal vein. This large vein runs behind the knee, with the popliteal artery. It can be seen by reflecting the bisected semimembranosus muscle. As the popliteal vein courses mediocraniad to join the saphenous vein, it becomes the femoral vein.

The Hepatic Portal System (Fig. 27-14)

All the veins of the unpaired visceral organs drain into the hepatic portal system rather than directly into the posterior vena cava. Portal veins carry nutrient-rich blood directly from the capillary beds of the digestive organs to the capillary beds of the liver. Thus the liver, the principal organ of metabolism, has immediate access to the products of digestion. The fact that the portal veins lie between two capillary beds means however, that they are not injected unless the animal has been especially prepared to display the portal system. In that case the latex in the portal veins is, by convention, yellow. In animals which have not been prepared this way, the veins of the portal system are filled with clotted blood; they appear as wide, light brown stripes in the mesentery.

To locate the portal system, reflect the ileum and jejunum caudally so that the head and neck of the pancreas are exposed, the mesentery supporting the small intestine is fully fanned out, and the anterior mesenteric artery runs caudad in a medial position. Now, locate in the mesentery a large, thin-walled vessel filled with brown clotted blood running directly craniad. It is most apparent in the mesentery at the level of the duodenum or anterior jejunum. Uninjected veins are more fragile than injected vessels, so it is important to dissect the portal system with care.

Anterior mesenteric vein. This is the large vein just located in the mesentery, medial to the reflected duodenum. It drains the posterior visceral organs via the following tributaries, which can be located by tracing the vein caudally:

Posterior mesenteric vein. This is a branch formed by the union of the left colic vein and the anterior hemorrhoidal vein. It joins the anterior mesenteric.

Colic veins. Like the colic arteries, there are several colic veins. These tributaries drain venous blood from the ascending, transverse, and descending colon into the anterior mesenteric vein. One colic vein is a short vessel running from the cecum to the anterior mesenteric vein. A much longer colic vein runs craniad along the dorsal surface of the descending colon, close to the left colic and superior hemorrhoidal arteries, and through the mesentery to join the anterior mesenteric vein at the level of the hilus of the kidney.

Reflect the spleen so its medial surface is up. Carefully dissecting away any fat, find the uninjected vein running close to the splenic artery.

Gastrosplenic vein. This large vein runs along the medial surface of the spleen and then, through the mesentery dorsal to the stomach, to the anterior mesenteric vein, which it joins dorsal to the pylorus, forming the hepatic portal vein.

Hepatic portal vein. Formed by the union of the gastrosplenic vein and the anterior mesenteric vein, this large, short vein runs craniad into the sinusoids of the liver. Blood then enters the hepatic veins, which in turn drain into the posterior vena cava.

Gently reflect the pyloric portion of the stomach and the anterior duodenum cranially and dissect the hepatic portal vein away from the mesentery as far cranially as possible. Replace the stomach and duo-

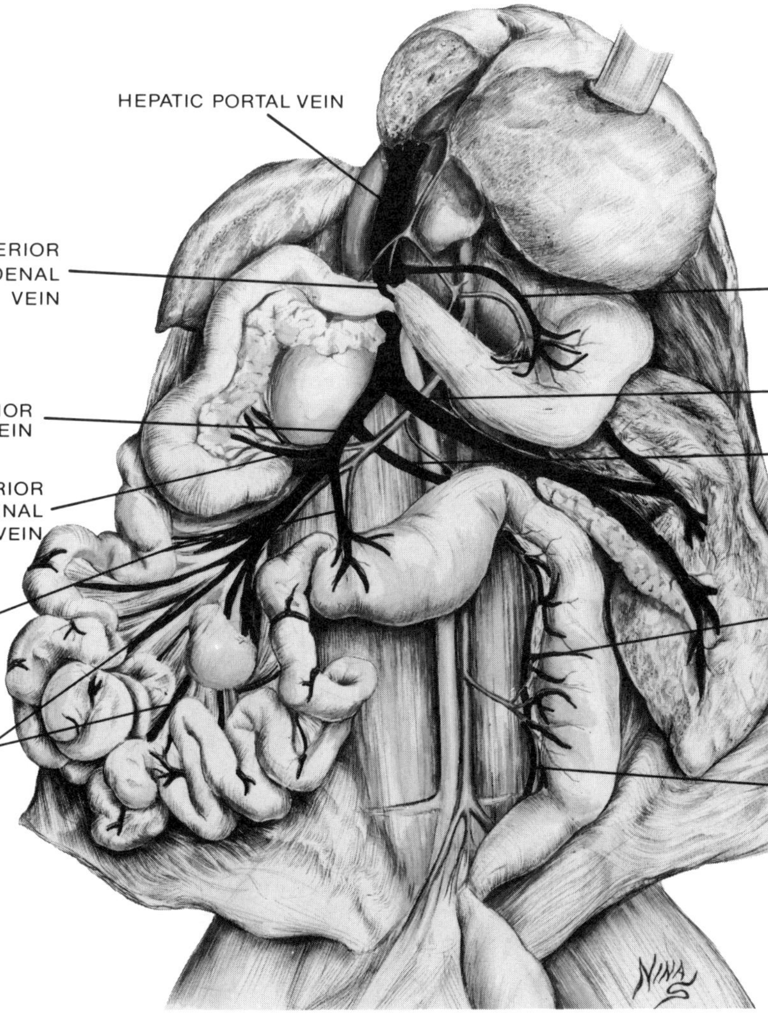

Figure 27-14. Ventral view of the abdomen of a cat, showing the hepatic portal system of veins. (Color version at end of book.)

denum and continue the dissection by spreading the duodenum and pyloric stomach slightly and locating the hepatic portal vein as it runs craniad, dorsal to the pylorus.

The following small tributaries may be seen:

Coronary veins. These veins, draining the lesser curve of the stomach, run mediad in the lesser omentum and empty into the hepatic portal vein just anterior to the pylorus.

Anterior pancreaticoduodenal vein. Draining the most anterior pancreas and duodenum, the short anterior pancreaticoduodenal vein courses mediad and enters the hepatic portal vein at the level of entrance of the coronary veins.

Trace the ***hepatic vein*** into the liver, coursing medial to the bile duct.

THE PHYSIOLOGY OF THE MAMMALIAN CARDIOVASCULAR SYSTEM

The heart is both the pump which moves the blood and its own signal generator. Histologically, the heart is composed of a form of striated muscle called cardiac muscle. The individual muscle cells branch and line up end to end, and the intervening cell membranes are modified to vastly reduce their electrical impedance. Thus, the cells of the heart are connected so that electrical impulses can be conducted very rapidly. In fact, in higher vertebrates, there are two low-impedance systems, one atrial and the other ventricular; they are separated by a thin bank of high-impedance connective tissue. Atrial impulses are transmitted to the ventricles by the atrio-ventricular (A-V) node, a collection of conductive cells with small diameter and slow transmission. This system ensures that the atria are stimulated at the same time and then, after a short pause because the signal is delayed in the A-V node, the ventricles contract.

The S-A Node: The Pacemaker

The signal which causes the sequential contraction first of the atria and then of the ventricles is myogenic; it originates within the heart muscle itself. In the normal heart, the "pacemaker" is the sino-atrial (S-A) node, located in the wall of the right atrium near the entrance of the anterior vena cava. The cells of the S-A node, located in the wall of the right atrium near the rest of the heart, are more permeable to sodium. Sodium ions leak into the cells, causing the resting potential to decay. At a set threshold value of depolarization, self-excitation occurs. There are other foci of rhythmicity in the heart, notably the A-V node and the conductive cells of the ventricle. Normally, however, the S-A node, which has the highest intrinsic rate, controls these secondary pacemakers by depolarizing them before their resting potentials have decayed to the self-excitation level.

Heart Output—Intrinsic Regulation

Different physiological states demand variation both 4,6,11in the total rate of blood oxygenation and delivery and
in differential partitioning of blood flow to various organs. Exercise, for example, requires a great increase in the rate of blood flow and the delivery of a higher proportion of the blood to active muscles. Modulation of total heart output is achieved by variation of heart rate and ventricular contractility. One of the simplest factors regulating contractility is the total amount of venous blood returning to the heart. Cardiac muscle fibers, like those of skeletal muscle, contract more vigorously from a longer relaxed length. The increased amount of venous return causes a greater distension of the ventricles each time they are filled. A more vigorous contraction results. This relationship is called the Frank-Starling Law. A second form of intrinsic regulation results from the presence of atrial stretch receptors which increase heart rate and atrial contractility when highly stimulated.

Heart Output—Neural and Hormonal Regulation

Heart output can also be regulated by extrinsic mechanisms. The vasomotor center in the medulla, which receives input from the vascular system and the higher brain centers, regulates the heart through the autonomic nerves. Heart function in higher vertebrates is regulated by both types of autonomic nerves, parasympathetic and sympathetic. The parasympathetic nerves, represented primarily by postganglionic fibers in the vagus, liberate acetylcholine at their ends; they serve a depressor function, reducing the heart rate. The rate change is apparently achieved by hyperpolarization of the S-A and A-V nodes, which lengthens the period of decay of polarization before self-excitation is achieved. The parasympathetic nerves to the heart are tonically active, so that under normal circumstances the heart is beating at a rate slower than its own myogenic rate.

The sympathetic nerves to the heart innervate both the atria and the ventricles. They liberate norepinephrine (noradrenalin) and have a strong pressor effect on the heart, increasing heart rate and contractility. The increase in heart rate is apparently due to an increase in sodium ion permeability, which in turn shortens the time required for decay of polarization to the self-excitation level. At the same time the sympathetic nervous system is affecting the heart directly, it also stimulates the adrenal medulla to release epinephrine and norepinephrine into the general circulation, causing systemic vasoconstriction, emptying of venous reservoirs, and increase in blood pressure. Thus sympathetic excitation of the heart is accomplished by appropriate changes throughout the vascular system.

Partitioning of Blood in the Systemic Circulation

Differential partitioning of blood among organs is controlled, like heart output, both by intrinsic mechanisms and by neural and hormonal stimulation. Local autoregulation is achieved by reflex modulation in the tone of the walls of the meta-arterioles, tiny blood vessels immediately preceding the capillaries, and by the sphincter muscles around the beginning of each capillary. High local oxygen concentration causes the sphincter muscles to constrict, preventing the influx of fresh blood. Sympathetic stimulation, both by direct innervation and by circulating adrenal hormones, causes vasoconstriction in the arterioles of the digestive tract and skin and emptying of venous reservoirs, thus raising blood pressure and heart output. Active tissues such as muscles involved in exercise, do not appear to react to sympathetic stimulation with vasoconstriction. In these tissues, the smooth muscle cells on meta-arterioles and sphincter muscles relax, ignoring the presence of epinephrine; they thereby allow increased blood flow to sites at which oxygen is being consumed. As a result of localized relaxation, a differential in peripheral resistance arises that shunts blood from quiet to active tissues. A particularly important neural control mechanism regulates the amount of blood flowing to and from the skin, and thus controls the extent to which the blood and body core are cooled or heated.

Control of Blood Flow in the Lungs

The rate of blood flow to the lungs is directly dependent on the rate of systemic venous return. The local distribution of deoxygenated blood within the lungs is controlled by an oxygen-sensitive autoregulatory mechanism very similar to that in the systemic capillaries. In the pulmonary circulation, however, blood is shunted to regions of high local oxygen concentration, ensuring that it will be oxygenated in the most highly ventilated areas of the lung.

Control of Blood Flow in the Coronary Arteries

Continuous adequate delivery of oxygenated blood to the heart muscle itself is essential throughout all variations in heart activity. The capillaries in cardiac muscle demonstrate autoregulation identical to that in the systemic capillaries. More important, however,

Table 27-1. Major systemic vessels of the mammalian circulatory system.

VEINS

1. *Anterior*

axillary → *subclavian* and *external jugular*

internal jugular → *subclavian* and *external jugular*, or → union of *two brachiocephalics*

subclavian and *external jugular* → union of *two brachiocephalics*

right atrium ← *anterior vena cava* ← union of *two brachiocephalics*

sternal → *anterior vena cava*

azygous (unpaired) → *anterior vena cava*

internal mammary (paired) → sternal

costovertebral (may join brachiocephalic or anterior vena cava) receives costocervical and vertebral → union of *two brachiocephalics*

2. *Posterior*

a) Abdominal veins *draining into the posterior vena cava*. (These are usually paired and course with abdominal arteries.)

right atrium ← posterior vena cava ←
- hepatic (two or three)
- adrenolumbar
- renal
- genital
- lumbar (unpaired)
- iliolumbar
- common iliac

external and internal iliacs drain into the common → common iliac

femoral → external and internal iliacs

saphenous → femoral

deep femoral → femoral

popliteal → femoral

b) The hepatic portal vein, draining the digestive organs and spleen, empties into the liver sinusoids.

Its main tributaries are:
- anterior mesenteric
- posterior mesenteric
- gastrosplenic

ARTERIES

1. *Anterior*

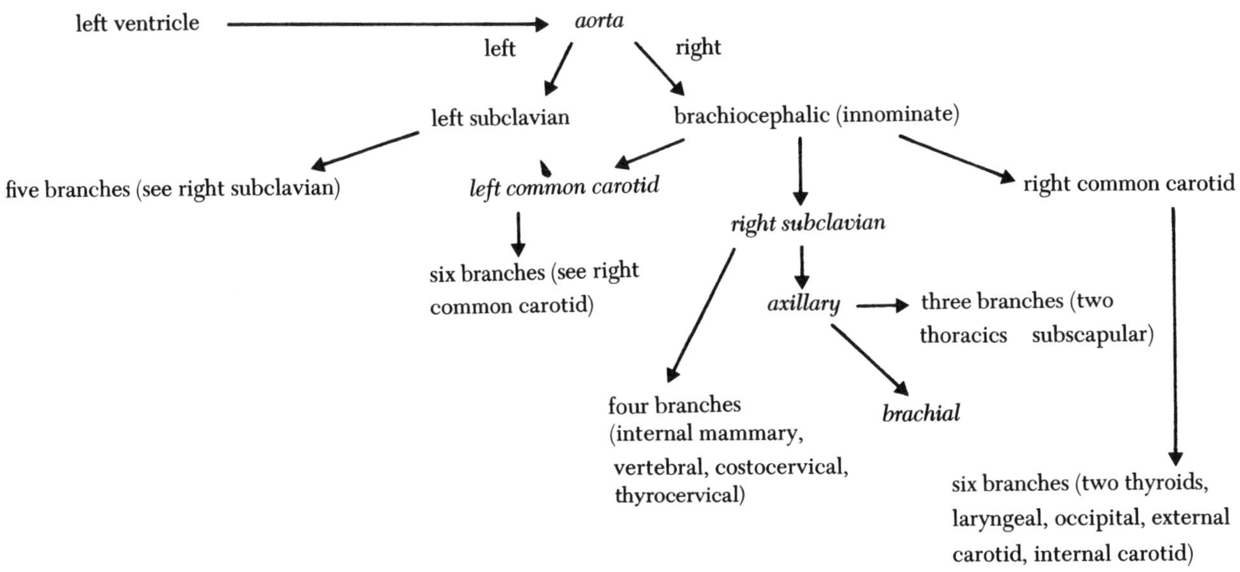

2. *Posterior*

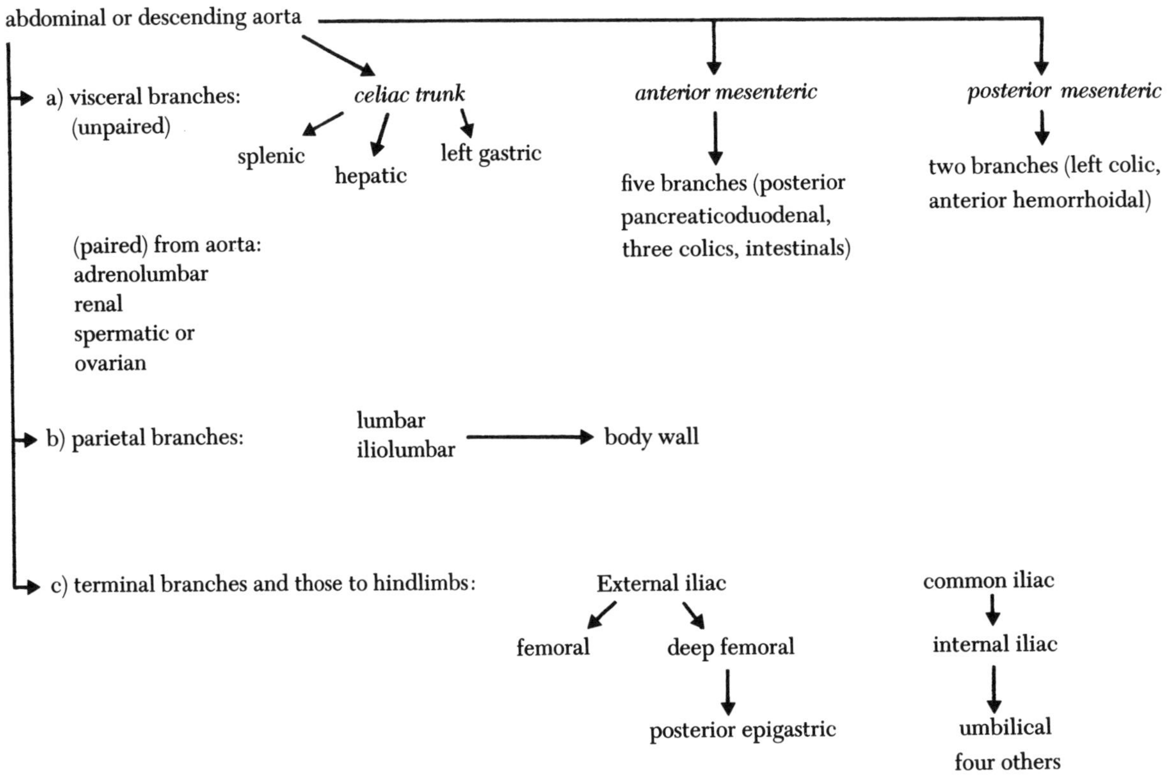

is the effect of sympathetic stimulation on coronary arteries and arterioles. In these vessels of the heart wall, epinephrine paradoxically causes vasodilation; and thus the pressor response, which constricts vessels elsewhere, increases the flow to the heart muscle itself.

The Relationship of Blood Volume to the Extracellular Fluid

Another parameter of the circulatory system which is subject to control is total blood volume and the volume of the extracellular fluid. In normal capillary flow, fluid is expressed into the extracellular spaces by hydrostatic pressure at the arterial end of the capillary. Almost all of the fluid is recovered at the venous end by reflux along an osmolarity gradient produced by retention of serum proteins within the capillary. The recovery is incomplete in higher vertebrates with high blood pressure, because of the high capillary pressure and because some proteins escape at the arterial end of the capillary and do not reenter the venous end. They are returned to the blood, along with the excess fluid, by the lymphatics, which fill with increasing extracellular fluid (ECF) pressure. Lymph flows through the lymphatics to the thoracic duct and the right lymphatic vessel; these empty into the external jugular veins. If the constant correction of the volume of ECF is disturbed by excessive capillary pressure, lymphatic block, or low concentration of serum proteins, edema—swelling due to excessive interstitial fluid—results. The total blood volume and the ECF volume are autoregulated by the kidney. If the blood volume is excessive, then arterial blood pressure rises and increased renal filtration results. The reverse occurs in hypovolemia (low blood volume). This system is somewhat slow to equilibrate; it is augmented in higher vertebrates by powerful hormonal regulators of kidney action.

Glossary

MARY BRYAN McNABB

(L. denotes a Latin origin; Gr. indicates a Greek origin.)

Abductor (L.*abductus*, "to lead away") A muscle that acts to draw an extremity away from the median line of the body, or part of an extremity away from the axis of the extremity.

Adductor (L. *adductus*, "to lead to") A muscle that acts to draw an extremity toward the median line of the body, or part of an extremity toward the axis of the extremity.

Adenohypophysis The anterior portion of the pituitary gland, arising from the embryonic pharyngeal region. It is a principal site of hormone production.

Afferent (L. *ad*, "to;" *fero*, "bear") Conducting inward from a peripheral part toward a center.

Amniote (Gr. *amniox*) Possessing an amnion around the embryo. Reptiles, birds, and mammals are amniotes.

Ampulla (L., "flask") A flask-like bulge.

Annularis (L., "ring") Circular muscles, such as the buccal muscle ring around the oral hood of the lamprey.

Aponeurosis (Gr. *aponeurosis*, "to pass into a tendon") Any tendinous or fascial expansion or thickness, characteristically flat, which forms attachments for origins or insertions of muscles.

Appendicular (L. *appendicul[a]*, "small appendage") Pertaining to an appendage or limb.

Arachnoid membrane (Gr. *arachnoidees*, "like a cobweb") A delicate membrane enclosing the brain, found between the dura mater and the pia mater. Also called the *arachnoidea*.

Archicortex (Gr. *archi*, "ancient") One of the three major portions of the pallial roof and walls of the cerebrum. Also called the *archipallium*.

Arcualia The primitive cartilages and structural supports from which part of a typical vertebra may be formed.

Ascidian (Gr. *askidion*, "little wineskin") Generally used to refer to any tunicate, but specifically, those tunicates whose larval forms show specific chordate characteristics but whose adult forms have no chordate tail.

Atriopore The posterior opening of the atrial cavity of the amphioxus, which protects the gills.

Atrium (L., "court") 1. In cephalochordates and urochordates, a chamber that temporarily holds water that has passed through the gills. Also called the *atrial cavity*. 2. A principal chamber of the vertebrate heart.

Autonomic (Gr. *autos*, "self") Self-controlling or functionally spontaneous, as the autonomic nervous system. See, *parasympathetic* and *sympathetic*.

Axial Pertaining to the axis formed by the head and trunk.

Axon (Gr. *axis*) The cylindrical extension of a nerve cell which conducts impulses away from the cell body.

Bowman's capsule (after Sir William Bowman, an English surgeon) A thin, double-walled capsule surounding the glomerulus of a nephron.

Buccal (L., "cheek") Pertaining to the mouth or cheek region.

Buccal cirri (L. *cirrus*, "curl") Tentacles around the mouth.

Cardinal sinus (L. *sinus*, "bay") In some vertebrates, the thin-walled cavity or space through which blood returns to the heart.

Cardinal veins (L. *cardinalis*, "principal") The major veins of the body in ancient vertebrates and many amniote embryos.

Cartilage (L. *cratis*, "wickerwork") A supportive tissue composed largely of an extracellular matrix in which living cells are embedded; may contain collagenous or elastic fibers and become calcified.

Cecum (L. *caecus*, "blind") An intestinal pouch which marks the transition between the small and the large intestine.

Chitin (Gr. *chiton*, "tunic") An inelastic, chemically inactive polysaccharide similar to cellulose. It is found in many invertebrates.

Choana (L. and Gr. *choane*, "funnel") Generally, any funnel-shaped cavity; specifically, the paired passages between the nasal cavity and the pharynx.

Cloaca (L., "sewer") A sewer at the end of the gut in some vertebrates, which receives the ducts of the reproductive, digestive, and urinary systems.

Coelom (Gr. *koilos*, "hollow") The body cavity in which the visceral organs are suspended. Also called the *perivisceral* cavity.

Condyle (Gr. *kondylos*, "knuckle") A rounded projection of bone found at points of articulation.

Conus arteriosus A prominent conical extension occurring where the ventral aorta leaves the ventrical of the heart. If the expansion is an extension of the blood vessel, it is termed the *conus arteriosus*. If it is an extension of the heart musculature as well, it can also be called the *bulbus cordis*.

Cornea (L. *corneas*, "horny") The transparent tissue which covers the front of the eyeball.

Countercurrent flow Flow of fluids such as blood and sea water in opposite directions. Exchange of oxygen, heat, or waste products is more efficient with this pattern of flow.

Demibranch (L. *dimidus*, "half;" Gr. *branchia*, "gills") A gill arch and the lamellae on one surface of the interbranchial septum. Also called a *hemibranch*.

Dendrite (Gr. *dendron*, "tree") An extension of a nerve cell, usually branched, which conducts impulses toward the cell body.

Dentigerous Having or containing teeth.

Detritus (L., "worn down") Loose, partially disintegrated organic materials.

Diencephalon (Gr. *enkephalos*, "brain") The posterior subdivision of the forebrain. Also called the *thalamencephalon*.

Disjunct Not joined; separated.

Dorsal lamella (L. *lamina*, "thin sheet") A ciliated groove running dorsoventrally on the side of the tunicate gill basket. It is the site for collection of food-laden mucous strands.

Duct of Cuvier (after Georges Cuvier, a French naturalist) Either of two short venous vessels which, in some vertebrates, conduct blood from the cardinal veins to the sinus venosus. The right one becomes part of the anterior vena cava in humans.

Dura mater (L., "hard mother") The outermost and most fibrous of the three membranes covering the brain.

Efferent (L., "to bear away from") Conducting away from a center.

Endocardium (Gr. *endon*, "within;" *kardium*, "heart") The thin endothelial layer lining the chambers of the heart.

Endocrine Producing secretions that are distributed by way of the blood stream.

Endolymph (Gr. *endo*, "within;" L. *lympha*, "water") The fluid contained in the membranous labyrinth of the inner ear.

Endostyle (Gr. *endo*, "within;" *stylos*, "column") A longitudinal fold in the pharyngeal region of lower chordates. The ciliated cells of this fold produce mucus.

Epithelium The cellular covering of any free surface in an organism, such as the outer surface of the skin and the innermost cellular surface of the gut.

Esophagus (Gr. *oisophagos*, "gullet") A tube which connects the pharynx and the stomach.

Exocrine Producing secretions that are distributed into the lumen of organs or of the gut, or to the outside of the body.

Extensor (L. *extensus*, "extension") A muscle which extends a joint or any other body part by increasing the angle between the bones involved.

Facet (Fr. *facette*, "small face") A smooth, flat, or rounded area of articulation on a hard surface, usually bone.

Fascia (L., "band") A sheet or band of fibrous tissue enveloping, supporting, or connecting structures of the body.

Flexor (L. *flexus*, "to bend") A muscle which moves a joint or extremity toward a bent position.

Foramen (L., "an opening") A natural opening or passage, generally though bone.

Fossa (L., "ditch") Any hollow or depressed area.

Ganglion (Gr., "tumor") Any group of nerve cells, mainly cell bodies, found outside the central nervous system.

Gill arches A set of curved skeletal bars which support gill structures in some vertebrates. Also called *branchial arches*.

Gill raker One of the processes on branchial arches of fish which projects into the pharynx and helps to prevent food from clogging up the gill slits.

Gill slit A perforation of the body wall into the pharynx. Also called a *branchial cleft*.

Glomerulus (L. *glomus*, "a ball of yarn") The capillary tuft within Bowman's capsule of a nephron.

Gular (L. *gula*, "throat, gullet") Pertaining to, or situated in, the upper throat.

Holobranch (Gr. *holos*, *"entire;"* branchia, *"gills"*) A gill arch and the lamellae on both surfaces of the interbranchial septum.

Hydrosinus (L. *hydra*, "water;" *sinus*, "bay") An anterior cavity off the pharynx of a lamprey.

Insertion (L. *insertio*, "to join into") The site of attachment of a muscle to the bone, organ, or part which it moves.

Internuncial neuron (L. *internuntius*, "messenger") A neuron which connects two or more neurons.

Intestine (L. *intestinus*, "inside") The tubular portion of the alimentary canal lying posterior to the stomach. Digestion and the absorption of the end products of digestion take place in the intestine.

Joint (L. *juncto*, "joining" or "connection") The site of articulation between two or more bones.

Lagena (L., "flask") The posteriormost of the three sacs in the vertebrate inner ear. It is associated with hearing in some vertebrates.

Lamellae (L. *lamina*, "small plate" or "disc") Minute transverse plates located in the gills, containing tiny capillaries. Also called *gill filaments*.

Meninx (Gr., "membrane") A membrane, especially one of the three membranes (meninges) enveloping the brain and spinal cord.

Mesencephalon The middle division of the brain.

Mesenchyme (Gr. *mesos*, "middle;" *enkyma*, "infusion") An embryological term referring to an aggregation of cells in the mesoderm from which such structures as connective tissue, blood, blood vessels, and the heart are derived.

Mesentery A membranous fold of the peritoneum supporting various organs and attaching them to the body wall.

Mesonephros (Gr. *mesos*, "middle;" *nephros*, "kidney") An excretory organ generally found in embryos of higher vertebrates. It develops posterior to the pronephros and is therefore the "second" kidney.

Metanephros (Gr. *meta*, "after") The most posterior embryonic excretory organ, persisting in higher vertebrates as the definitive kidney.

Metapleural folds (Gr. *meta*, "after;" *pleura*, "side") Ventral longitudinal folds occurring in amphioxus, once thought to be the forerunners of paired appendages in more advanced vertebrates.

Motor (L., "mover") Producing or assisting in motion.

Müllerian ducts (after J. Peter Müller, a German physiologist) Two embryonic ducts that give rise in some female vertebrates to oviducts, uteri, and vaginae.

Musculi pectinati (L. *pectinatus*, "to comb") Small cones of muscle fiber projecting from the inner walls of the heart.

Myogenic (Gr. *mys*, "muscle;" *gennan*, "to produce") A regular, orderly contraction of heart muscle taking place because of inherent properties of the muscle rather than by specific nervous stimulation.

Myomere (Gr. *mys*, "muscle;" *mere*, "limit") A muscle block delineated by connective tissue.

Myoseptum The connective tissue between adjacent myomeres.

Nasohypophysial pouch (L. *nasus*, "nose;" *hypophein*, "to grow up from below," relating to the pituitary) A chamber ventral to the nostril of the lamprey, which appears to pump water over the nasal epithelia. Also called the *nasopharyngeal pouch* or *pituitary pouch*.

Neocortex (Gr. *neos*, "new;" L. *cortex*, "shell") That portion of the cerebrum showing cellular stratification of the most highly developed type. Also called the *neopallium*.

Neoteny The attainment of sexual maturity and reproductive capability while retaining, in general, a larval morphology.

Nephridium A primitive excretory organ, typically consisting of a tube opening to the coelom and discharging via a pore to the exterior of the body.

Nephrocyte (Gr. *kytos*, "hollow vesicle") An excretory cell able to store substances of an excretory nature.

Nephron (Gr. *nephros*, "kidney") The functional unit of the kidney, consisting of a glomerulus, Bowman's capsule, and tubular portions.

Nephrostome (Gr. *stome*, "mouth") The coelomic opening of a nephridium.

Neuron (Gr., "nerve") A nerve cell.

Nostril (L. *naris*, "nostril") The external opening of the nasal passages.

Nucleus (L. *nux*, "nut") A group of nerve cell bodies within the central nervous system that have some function in common.

Ocellus (L. *oculus*, "small eye") A group of light-sensitive cells; a simple eye.

Olfactory bulb (L. *olfacere*, "smell") The anterior bulbous projection of the olfactory lobe of the brain in which the olfactory nerve terminates.

Ophthalmic (Gr. *opthalmos*, "eye") Pertaining to the eye.

Opisthonephric (Gr. *opisth*, "at the back") The kidney of most anamniote adult vertebrates. The primitive and embryonic pronephric kidney is located anterior to the opisthonephros.

Oral hood (L. *os*, "mouth") A prolongation of the metapleural folds surrounding the mouth and bearing cirri.

Orbit (L., "mark of a wheel") The bony cavity that contains the eyeball.

Origin (L. *origo*, "beginning") The fixed end or attachment of a muscle. Also used to refer to the beginning of a nerve in the central nervous system.

Otic (Gr. *ous*, "ear") Pertaining to the ear.

Otolith (Gr. *ous*, "ear;" *lith*, "stone") A hard body in the inner ear, sometims calcareous, and used in the detection of gravity.

Paedogenesis See *neotony*.

Palate (L.) The roof of the mouth. (Note: the secondary palate separates nasal and oral cavities in some vertebrates.)

Pallium (L., "cloak") The gray matter covering the cerebral hemispheres characterized by a distinctive layering of the cellular elements.

Papilla (L., "nipple") A small projection similar in form to a nipple.

Parasympathetic That part of the autonomic nervous system arising in the midbrain and hindbrain and in the sacral region of the spine. These nerves liberate acetylcholine and act antagonistically to those of the sympathetic nervous system.

Parietal pericardium (L. *parietalis*, "pertaining to the walls of a cavity") The membrane lining the outer walls of the pericardial cavity.

Pericardial cavity (L. or Gr. *peri*, "around;" Gr. *kardia*, "heart") A cavity, encompassed by distinct epithelia, containing the heart and separating it from other visceral organs.

Peritoneum (Gr. *peritonos*, "stretched across") the serous membrane lining the abdominal and pelvic walls (parietal peritoneum) and enveloping the visceral organs (visceral peritoneum).

Pharyngeal glands (Gr. *pharynx*, "throat") A pair of glands, lying ventral to the tongue of the lamprey, that secrete anticoagulants through pores in the tongue. Also called *buccal glands*.

Pharynx (Gr., "throat") The part of the alimentary canal extending from the mouth to the beginning of the esophagus. In aquatic vertebrates, it is the site of the gills.

Pia mater (L., "tender mother") The innermost of the three membranes covering the brain.

Pleura (Gr., "rib") The membrane enveloping the lungs and lining the thoracic cavity.

Pleuroperitoneal cavity (Gr. *peritonos*, "stretched across;" *pleuro*, "side") The coelom.

Portal system A blood pathway which begins and ends in capillaries without passing through the heart. (Note: the term is not applied to the pathway between gill and systemic capillaries in fishes and amphibia.)

Posttrematic and pretrematic (L. *post-*, "after;" *pre-*, "before;" Gr. *trema*, "hole") Describing the areas posterior to and anterior to the actual space of a gill slit.

Pronephros (L. or G. *pro-*, "before;" *nephros*, "kidney") A primitive kidney formed of the most anterior embryonic kidney tissue. It is characteristically composed of short, segmentally arranged tubules with openings to the coelom.

Prosencephalon The forebrain.

Pseudobranch (Gr. *pseudo*, "false") An accessory holobranch, usually incomplete, as in the spiracle of sharks and under the operculum of various fishes.

Raphe (Gr. *raphe*, "seam") A line of union of two lateral halves or parts of a structure.

Reflect To bend back.

Retina The light-receiving sensory layer in the eye that is the immediate instrument of vision.

Rhombencephalon (L. *rhombos*, "magic wheel") The hindbrain.

Rostrum (L., "beak") A beak-like appendage or part.

Sacculus (L., "little sac") One of the divisions of the membranous sacs of the inner ear.

Semilunar valve The valve of the aorta, the cusps of which resemble crescents. Sometimes the term *semilunar* is also used with respect to the valves of the pulmonary artery.

Sensory (L. *sensorium*, "to experience") Pertaining to or assisting in sensation.

Sinus venosus (L. *sinus*, "bay") A distinct chamber of the vertebrate heart that receives venous systemic blood and opens into the atrium.

Siphon (L. *sipho*, "pipe" or "tube") A pair of extending tubes in certain molluscs and ascidians; one conducts water to the mouth and gills, the other carries waste water away.

Somatic (Gr. *soma*, "body") Pertaining to or characteristic of the body.

Sphincter (L. and Gr. *sphinkter*, "that which binds tight") An annular band of muscle which constricts or closes off a channel or vessel.

Spiracle (L. *spirare*, "to breathe") A small aperture anterior to the gill region in the shark. As a remnant of a gill slit, it serves as a breathing orifice during some feeding.

Stomochord (Gr. *stoma*, "mouth;" *chord*, "cord") A supporting structure in the proboscis of some hemichordates (the enteropneusts).

Subneural gland (L. *sub*, "under;" Gr. *neuron*, "nerve") A small, oval structure located between the two siphons in ascidians. A small ciliated duct connects it to the pharynx.

Subpharyngeal gland (Endostyle) A mucus-producing gland lying ventral to the pharynx in the ammocoete. It gives rise to thyroid structures in the adult lamprey.

Sympathetic nervous system That part of the autonomic nervous system arising from the thoracolumbar cord. Its neurons secrete epinephrine and act antagonistically to those of the parasympathetic system.

Symphysis (Gr., "growing together") A firm, immobile joint between bony surfaces, constructed of fibrocartilage or bone.

Tela choroidea (L. *tela*, "web") A fold of pia mater which forms a roof to a ventricle of the brain.

Tensor (L. *tensio*) A muscle that stretches a body part.

Thyroid (Gr. *thyros*, "shield-shaped") An endocrine gland in vertebrates arising from the pharynx. Its hormones influence growth and metabolic rate.

Tubercle (L. *tuberculum*, "small hump" or "knob") A rounded prominence, generally on a bone.

Tunic (L. *tunica*) An outer covering of body tissue composed of *tunicin*, a cellulose-like protein. Also called a *test*.

Tunicate A member of the Phylum Urochordata; an ascidian.

Utriculus (L., "small bag") The major organ of the vestibular region of the inner ear which gives information about the position and movement of the head.

Velum (L., "veil," or "curtain") A membrane which closes off the entrance to the respiratory tube in lampreys.

Ventricles (L. *vener*, "a body;" *ventriculus*, "stomach") 1. Narrow cavities in the brain. 2. The most muscular chambers of the heart.

Visceral (L. *visceralis*, "internal organ") Pertaining to internal organs or viscera of the body.

Visceral pericardium The coelomic epithelium covering the heart.

Wheel organ A ring of ciliated tentacles in the oral hood of the amphioxus, used for deterring stagnancy.

Wolffian duct (after Kaspar Wolff, a German anatomist) An embryonic mesonephric duct persisting as the genital duct in the male vertebrate. Also called the *ductus mesonephros*.

Zygapophysis (Gr. *zygon*, "yoke;" *apophysis*, "offshoot") An articular process of a vertebra.

References

Chapter 1

HILDEBRAND, M. 1974. *Analysis of Vertebrate Structure.* New York: John Wiley. *A comprehensive functional and evolutionary approach to the vertebrates.*

MCFARLAND, W. N., *et al.* 1979. *Vertebrate Life.* New York: Macmillan. *A recent comprehensive text on all major aspects of vertebrates.*

PROSSER, C. L. 1973. *Comparative Animal Physiology.* Philadelphia: W. B. Saunders. *A very comprehensive treatment of animal physiology.*

ROMER, A. S., and T. S. PARSONS. 1977. *The Vertebrate Body.* Philadelphia: W. B. Saunders. *An authoritative evolutionary approach.*

WAKE, M. H., ed. 1979. *Hyman's Comparative Vertebrate Anatomy.* Chicago: University of Chicago Press. *This general anatomy textbook contains good, extensive descriptions of structures and organs.*

YOUNG, J. Z. 1962. *The Life of Vertebrates.* New York: Oxford University Press. *An excellent source of information about vertebrates.*

These and other general works have sections which supplement many of the chapters in this book. They are not, however, cited specifically in each instance in which they would be useful. The reader can find useful information on many special topics in these general references.

Chapter 2

BARRINGTON, E. J. W. 1965. *The Biology of Hemichordates and Protochordates.* Edinburgh: Oliver and Boyd. *A detailed treatment of hemichordates, tunicates, and amphioxus with emphasis on structure and minute observations.*

BERRILL, N. J. 1955. *The Origin of Vertebrates.* Oxford: Clarendon Press. *A detailed evolutionary study of the protochordates.*

Chapter 7

ALEXANDER, R. MCNEILL. 1975. *The Chordates.* New York: Cambridge University Press. *Deals with functional aspects.*

BRODAL, A., and R. FANGE. 1963. *The Biology of Myxine.* Oslo: Universitatesforlaget. 1963. *A series of papers, principally on the hagfish, dealing with structures and physiology of the cyclostomes.*

CARTER, G. S. 1967. *Structure and Habit in Vertebrate Evolution.* Seattle: University of Washington Press. *Good treatment of the ammocoete to provide examples of primitive chordate structures.*

HARDISTY, M. W. 1979. *The Biology of the Cyclostomes.* London: Chapman and Hall. *Detailed information on the cyclostomes.*

Chapter 8

GILBERT, S. G. 1973. *Pictorial Anatomy of the Dogfish.* Seattle: University of Washington Press.

KLUGE, A. G., *et al.* 1977. *Chordate Structure and Function.* New York: Macmillan.

Chapter 10

GANS, C., and T. S. PARSONS. 1964. *A Photographic Atlas of Shark Anatomy.* New York: Academic Press.

TURNER, D. C., and J. T. BAGNARA. 1976. *General Endocrinology.* Philadelphia: W. B. Saunders.

Chapters 11 and 12

BUDKER, P. 1971. *Life of Sharks.* New York: Columbia University Press. *A brief discussion of gill structure and respiratory patterns.*

GANS, C., and T. S. PARSONS. 1964. *A Photographic Atlas of Shark Anatomy.* New York: Academic Press.

HOAR, W. S., and D. J. RANDALL. 1970. *Fish Physiology.* Volume IV. New York: Academic Press. *A valuable discussion, but perhaps too technical without preliminary readings.*

KENT, G. C. 1969. *Comparative Anatomy of the Vertebrates.* St. Louis: C. V. Mosby. *A general discussion of hearts, features of the cardiovascular system, gills, and respiratory movements.*

SATCHELL, G. H. 1971. *Circulation in Fishes.* New York: Cambridge University Press. *An excellent discussion of heart action and the cardiovascular system.*

Chapter 14

BUDKER, P. 1971. *Life of Sharks.* New York: Columbia University Press. *A behavioral approach to the sensory systems.*

LINEAWEAVER, T. H., and R. L. BACKUS. 1979. *The Natural History of Sharks.* Philadelphia: Lippincott. *A simple but complete discussion of sensory systems.*

WOLSTENHOLME, G. E., and J. KNIGHT, eds. 1970. *Symposium on Taste and Smell in Vertebrates.* London: Churchill. *A set of definitive papers on current theories of taste and smell.*

________. 1977. "Recent advances in the Biology of Sharks." *Amer. Zool. 17:* 287–515.

Chapter 17

ELLIOTT, H. C. 1969. *Textbook of Neuroanatomy.* Philadelphia: Lippincott.

GORDON, M. S. 1977. *Animal Physiology: Principles and Adaptations.* New York: Macmillan.

GUYTON, A. C. 1969. *Function of the Human Body.* Philadelphia: W. B. Saunders.

NORRIS, H. W., and S. P. HUGHES. 1920. "The Cranial, Occipital, and Anterior Spinal Nerves of the Dogfish." *J. Comp. Neurol. 31:* 293–395.

PICKFORD, G. E., and J. W. ATZ. 1957. *The Physiology of the Pituitary Gland of Fishes.* New York: The New York Zoological Society.

PATT, D. I., and G. R. PATT. 1969. *Comparative Vertebrate Histology.* New York: Harper and Row.

HANSON, S. W., and S. L. CLARK. 1955. *Anatomy of the Nervous System.* Philadelphia: W. B. Saunders.

WEBSTER, D., and M. WEBSTER. 1974. *Comparative Vertebrate Morphology.* New York: Academic Press.

Chapters 18 through 22

BISHOP, S.C. 1926. "Notes on the Habits and Development of the Mudpuppy." *N.Y. State Mus. Bull. 268:* 1–61.

BISHOP, S. C. 1932. "The Spermatophores of *Necturus maculosus* Rafinesque." *Copeia 1932 1:* 1–3.

BISHOP, S. C. 1941. "The Salamanders of New York." *N.Y. State Mus. Bull. 324:* 1–365.

CARROLL, R. L., and R. HOLMES. 1980. "The Skull and Jaw Musculature as Guides to the Ancestry of Salamanders." *Zool. Jour. Linn. Soc. 68(1):* 1–40.

DAWSON, A. B. 1920. "The Integument of *Necturus maculosus.*" *J. Morph. 34(3):* 487–577.

EATON, T. H. 1957. "Motor Areas and Superficial Muscle Action in Salamanders." *J. Elisha Mitchell Soc., 73(1):* 1–10.

GILBERT, S. G. 1973. *Pictorial Anatomy of the* Necturus. Seattle: University of Washington Press.

HARRIS, J. P., Jr. 1959. "The Natural History of *Necturus.* I." *Field and Laboratory 27(1):* 11–20.

HARRIS, J. P., Jr. 1961. "The Natural History of *Necturus* IV." *J. Grad. Res. Center, Southern Methodist Univ. 29(1):* 69–81.

NORTHCUTT, R. G., 1979. "The Comparative Anatomy of the Nervous System and the Sense Organs." *In* M. H. Wake, ed., *Hyman's Comparative Vertebrate Anatomy.* Chicago: University of Chicago Press.

NORTHCUTT, R. G. and E. KICLITER. 1980. "Organization of the Amphibian Telencephalon." *In* S. O. E. Ebbeson, ed. *Comparative Neurology of the Telencephalon.* New York: Plenum.

SMITH, B. G. 1911. "The Nests and Larvae of *Necturus.*" *Biol. Bull. 20(4):* 191–200.

WILDER, H. H. 1903. "The Skeletal System of *Necturus maculatus* Rafinesque." *Mem. Boston Soc. Nat. Hist 5(9):* 387–439.

Chapter 24

ALEXANDER, R. M. 1969. *Animal Mechanics.* Seattle: University of Washington Press.

CROUCH, J. E. 1969. *Text Atlas of Cat Anatomy.* Philadelphia: Lea and Febiger.

FIELD, H. E., and M. E. TAYLOR. 1969. *An Atlas of Cat Anatomy.* Chicago: University of Chicago Press.

GANS, C. 1974 and 1979. *Biomechanics: An Approach to Vertebrate Biology.* Philadelphia: Lippincott; Ann Arbor: The University of Michigan Press.

GRANT, J. C. B. 1958. *A Method of Anatomy*. Baltimore: Williams and Wilkins.

GRAY, J. 1968. *Animal Locomotion*. New York: W. W. Norton.

HILDEBRAND, M. 1974. *Analysis of Vertebrate Structure*. New York: John Wiley and Sons.

WALKER, W. F., JR. 1965. *Vertebrate Dissection*. Philadelphia: W. B. Saunders.

WEICHERT, C. K., and W. PRESCH. 1975. *Elements of Chordate Anatomy*. New York: McGraw-Hill.

Chapters 25 and 26

BLOOM, W., and D. W. FAWCETT. 1969. *A Textbook of Histology*. Philadelphia: W. B. Saunders

CROUCH, J. E. 1969. *Text Atlas of Cat Anatomy*. Philadelphia: Lea and Febiger.

FIELD, H. E., and M. E. TAYLOR. 1969. *An Atlas of Cat Anatomy*. Chicago: University of Chicago Press.

GILBERT, S. G. 1968. *Pictorial Anatomy of the Cat*. Seattle: University of Washington Press.

Chapter 27

AYRES, S. M. 1971. "Cardiac Function." *In* Ayres, S. M., and J. M. Gregory, eds. *Cardiology—A Clinicophysiological Approach*. New York: Appleton-Century-Crofts. *A brief, simple review of human heart physiology.*

CHIDSEY, C. A., III. 1971. "Neural and Hormonal Control of the Circulation." *In* Conn, H. L., Jr., and O. Horowitz, eds. *Cardiac and Vascular Diseases*. Volume 1. Philadelphia: Lea and Febiger. *A brief summary of intrinsic and extrinsic controls of human heart output and blood distribution.*

HOOD, L. E., *et al.* 1978. *Immunology*. Menlo Park, CA: Benjamin/Cummings. *A good introduction to the immune system.*

ROBB, J. S. 1965. *Comparative Basic Cardiology*. New York: Grune and Stratton. *A complete review of the literature of anatomy and physiology of hearts and circulatory systems.*

Index

A

B

D

E

F

G

H

N

O

P

Q

R

S

To the users of this book

This section of *Vertebrates—A Laboratory Text* consists of colored illustrations of blood vascular systems of the dogfish shark and the cat. These color plates duplicate some of the black and white figures that appear earlier in various chapters.

These pages have been perforated so that you can easily remove each color plate and place it alongside the related pages of the text, wherever it will prove most useful to you in your study of the animal being dissected or reviewed.

You may find it helpful to put your name on each color plate you remove from the book. For safe storage and easy retrieval, we suggest that you place each loose color illustration next to its corresponding black and white figure.

The Editors

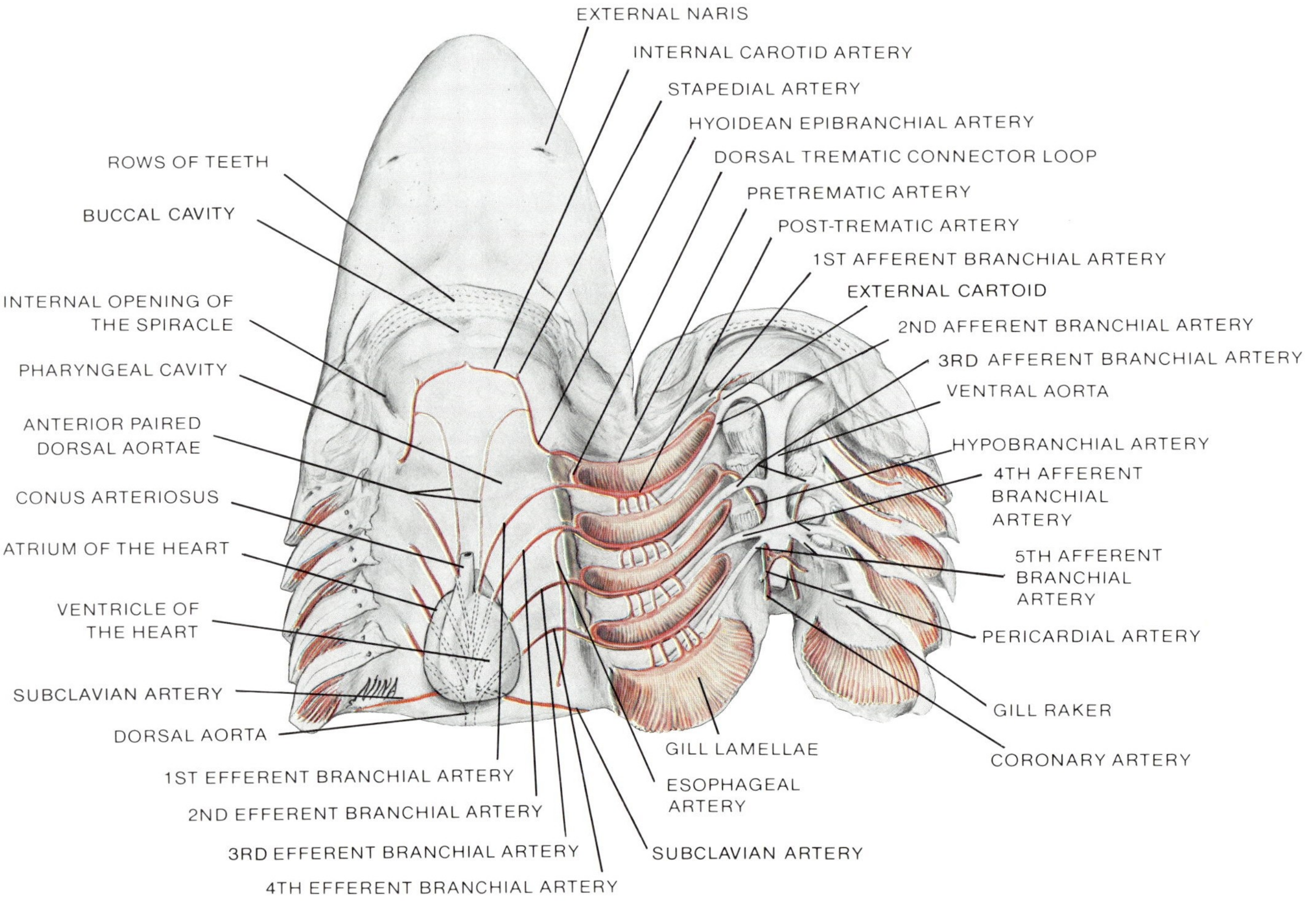
EXTERNAL NARIS
INTERNAL CAROTID ARTERY
STAPEDIAL ARTERY
HYOIDEAN EPIBRANCHIAL ARTERY
DORSAL TREMATIC CONNECTOR LOOP
PRETREMATIC ARTERY
POST-TREMATIC ARTERY
1ST AFFERENT BRANCHIAL ARTERY
EXTERNAL CARTOID
2ND AFFERENT BRANCHIAL ARTERY
3RD AFFERENT BRANCHIAL ARTERY
VENTRAL AORTA
HYPOBRANCHIAL ARTERY
4TH AFFERENT BRANCHIAL ARTERY
5TH AFFERENT BRANCHIAL ARTERY
PERICARDIAL ARTERY
GILL RAKER
CORONARY ARTERY
ROWS OF TEETH
BUCCAL CAVITY
INTERNAL OPENING OF THE SPIRACLE
PHARYNGEAL CAVITY
ANTERIOR PAIRED DORSAL AORTAE
CONUS ARTERIOSUS
ATRIUM OF THE HEART
VENTRICLE OF THE HEART
SUBCLAVIAN ARTERY
DORSAL AORTA
1ST EFFERENT BRANCHIAL ARTERY
2ND EFFERENT BRANCHIAL ARTERY
3RD EFFERENT BRANCHIAL ARTERY
4TH EFFERENT BRANCHIAL ARTERY
GILL LAMELLAE
ESOPHAGEAL ARTERY
SUBCLAVIAN ARTERY

Figure 11-4

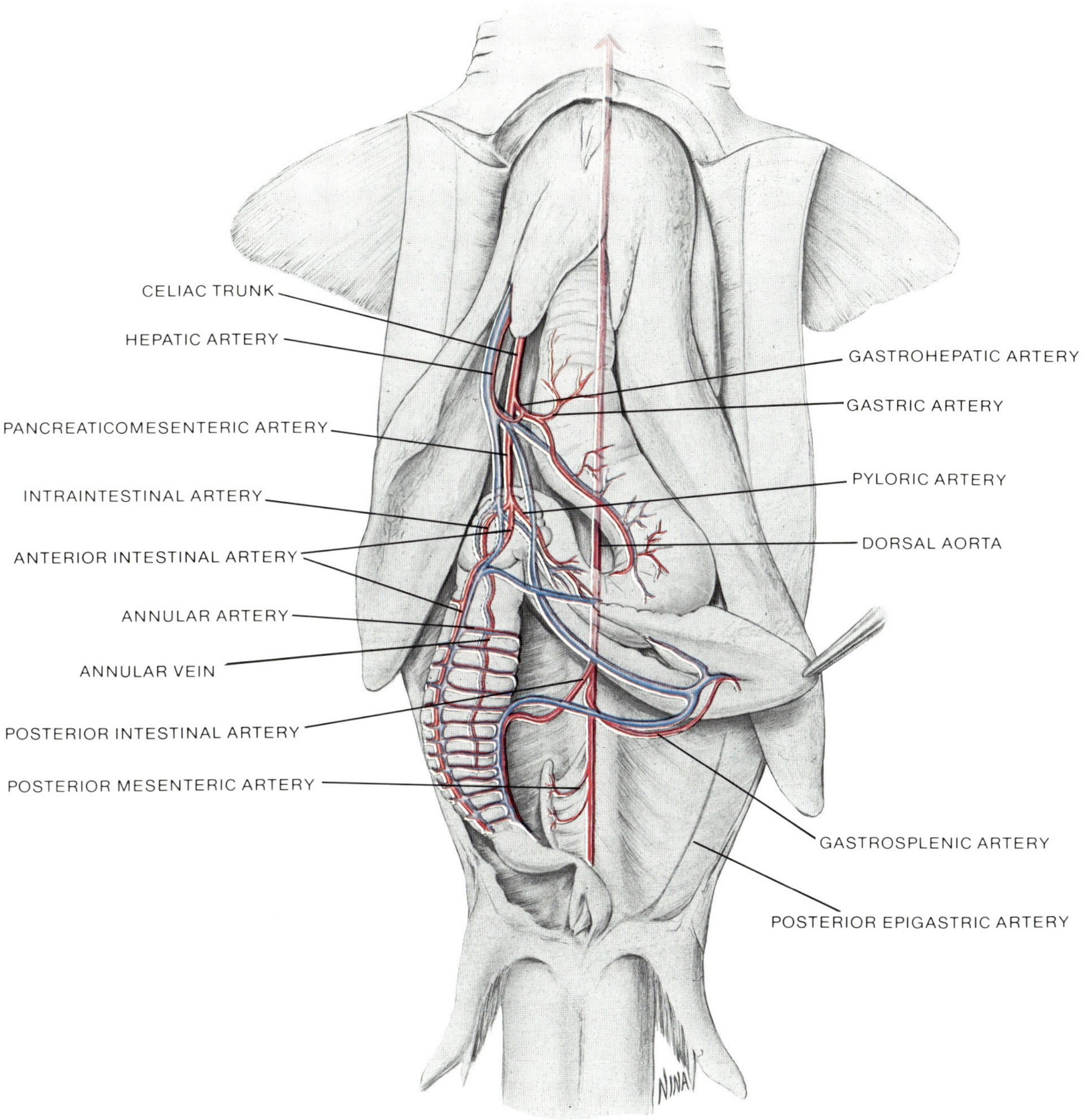

CELIAC TRUNK
HEPATIC ARTERY
PANCREATICOMESENTERIC ARTERY
INTRAINTESTINAL ARTERY
ANTERIOR INTESTINAL ARTERY
ANNULAR ARTERY
ANNULAR VEIN
POSTERIOR INTESTINAL ARTERY
POSTERIOR MESENTERIC ARTERY
GASTROHEPATIC ARTERY
GASTRIC ARTERY
PYLORIC ARTERY
DORSAL AORTA
GASTROSPLENIC ARTERY
POSTERIOR EPIGASTRIC ARTERY
NINA

Figure 11-5

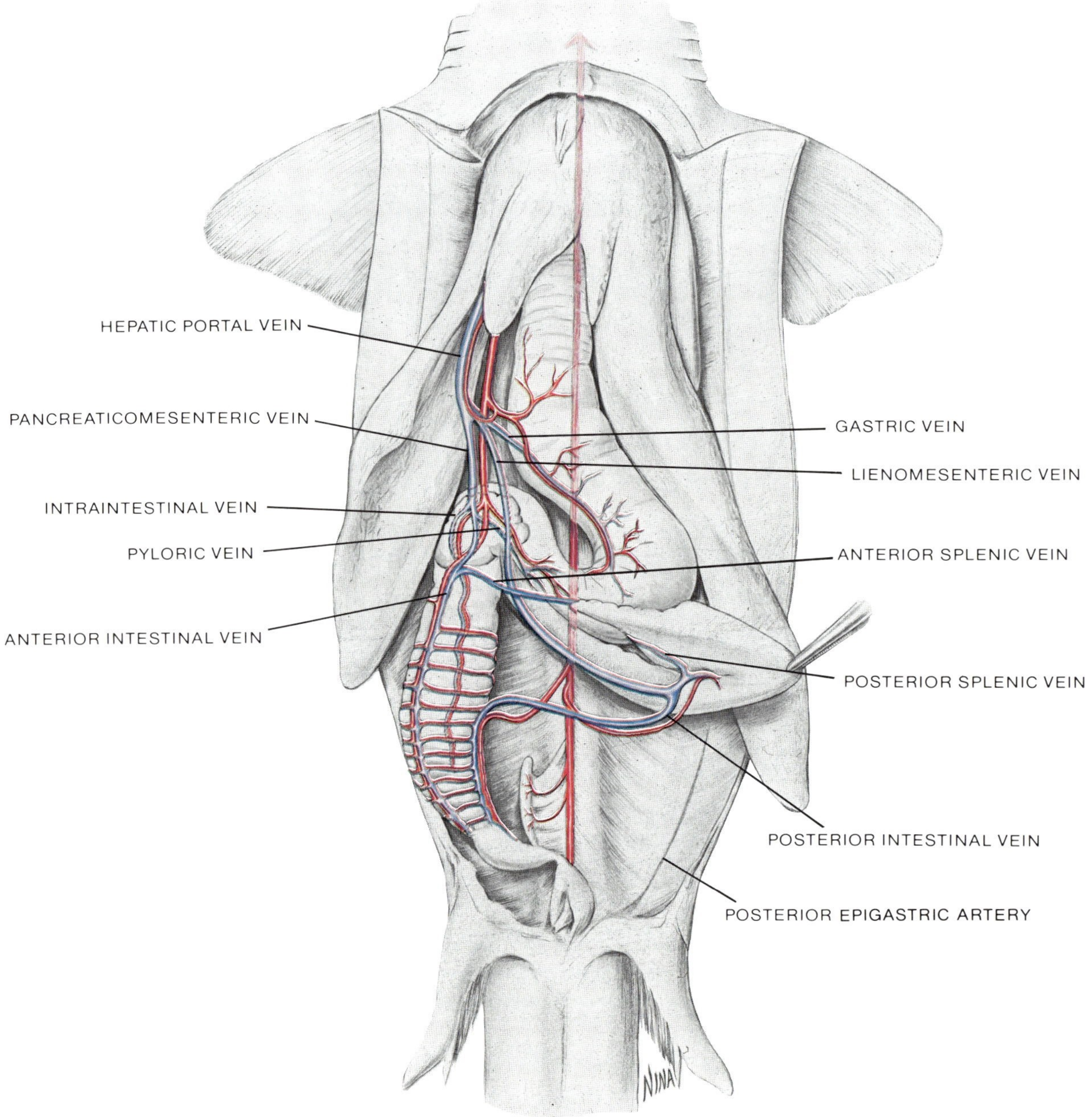
HEPATIC PORTAL VEIN
PANCREATICOMESENTERIC VEIN
INTRAINTESTINAL VEIN
PYLORIC VEIN
ANTERIOR INTESTINAL VEIN
GASTRIC VEIN
LIENOMESENTERIC VEIN
ANTERIOR SPLENIC VEIN
POSTERIOR SPLENIC VEIN
POSTERIOR INTESTINAL VEIN
POSTERIOR EPIGASTRIC ARTERY
NINA

Figure 11-7

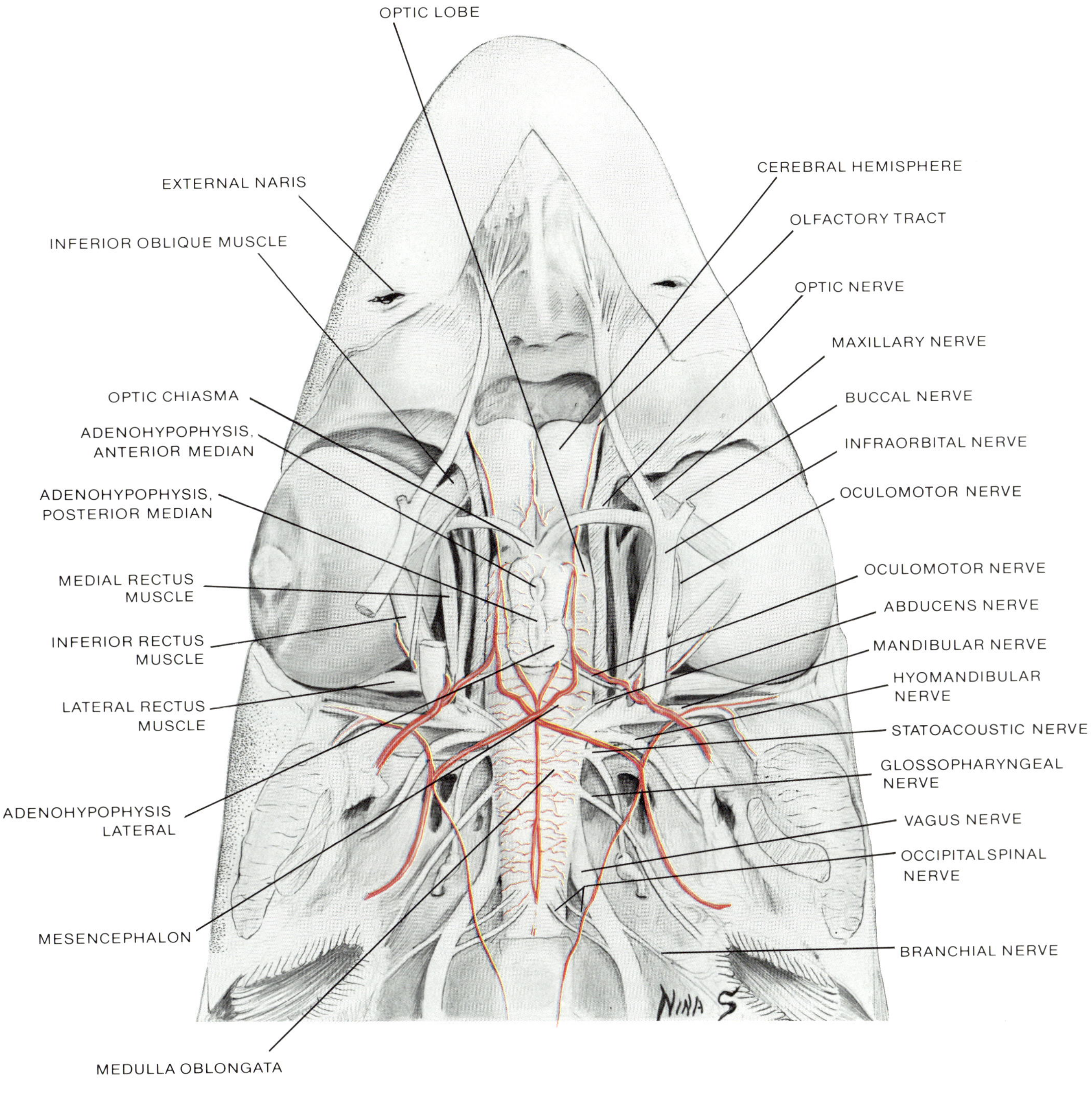
OPTIC LOBE
EXTERNAL NARIS
INFERIOR OBLIQUE MUSCLE
OPTIC CHIASMA
ADENOHYPOPHYSIS, ANTERIOR MEDIAN
ADENOHYPOPHYSIS, POSTERIOR MEDIAN
MEDIAL RECTUS MUSCLE
INFERIOR RECTUS MUSCLE
LATERAL RECTUS MUSCLE
ADENOHYPOPHYSIS LATERAL
MESENCEPHALON
MEDULLA OBLONGATA
CEREBRAL HEMISPHERE
OLFACTORY TRACT
OPTIC NERVE
MAXILLARY NERVE
BUCCAL NERVE
INFRAORBITAL NERVE
OCULOMOTOR NERVE
OCULOMOTOR NERVE
ABDUCENS NERVE
MANDIBULAR NERVE
HYOMANDIBULAR NERVE
STATOACOUSTIC NERVE
GLOSSOPHARYNGEAL NERVE
VAGUS NERVE
OCCIPITALSPINAL NERVE
BRANCHIAL NERVE
NINA S

Figure 15-4

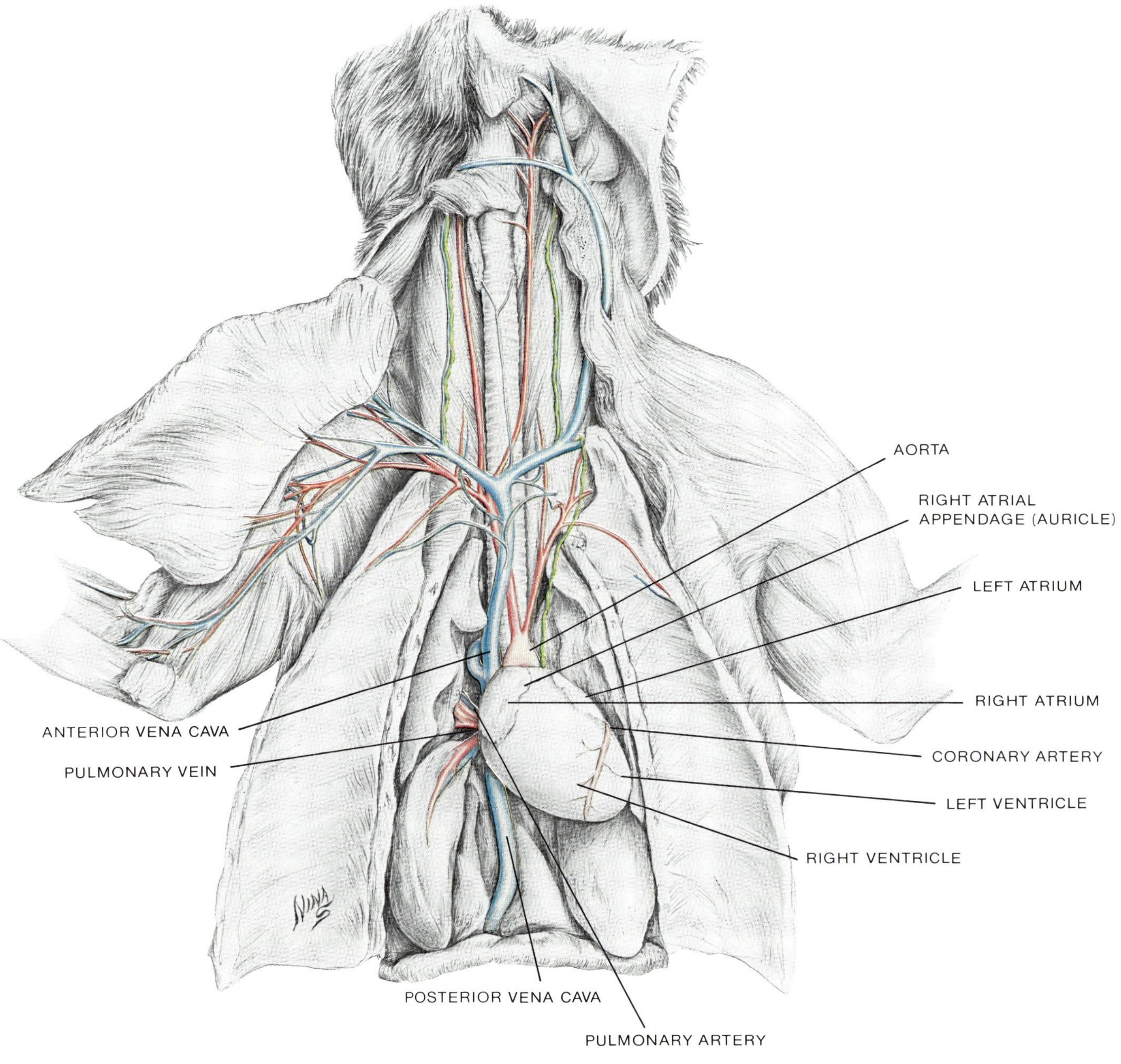
AORTA
RIGHT ATRIAL
APPENDAGE (AURICLE)
LEFT ATRIUM
RIGHT ATRIUM
CORONARY ARTERY
LEFT VENTRICLE
RIGHT VENTRICLE
ANTERIOR VENA CAVA
PULMONARY VEIN
POSTERIOR VENA CAVA
PULMONARY ARTERY

Figure 27–1

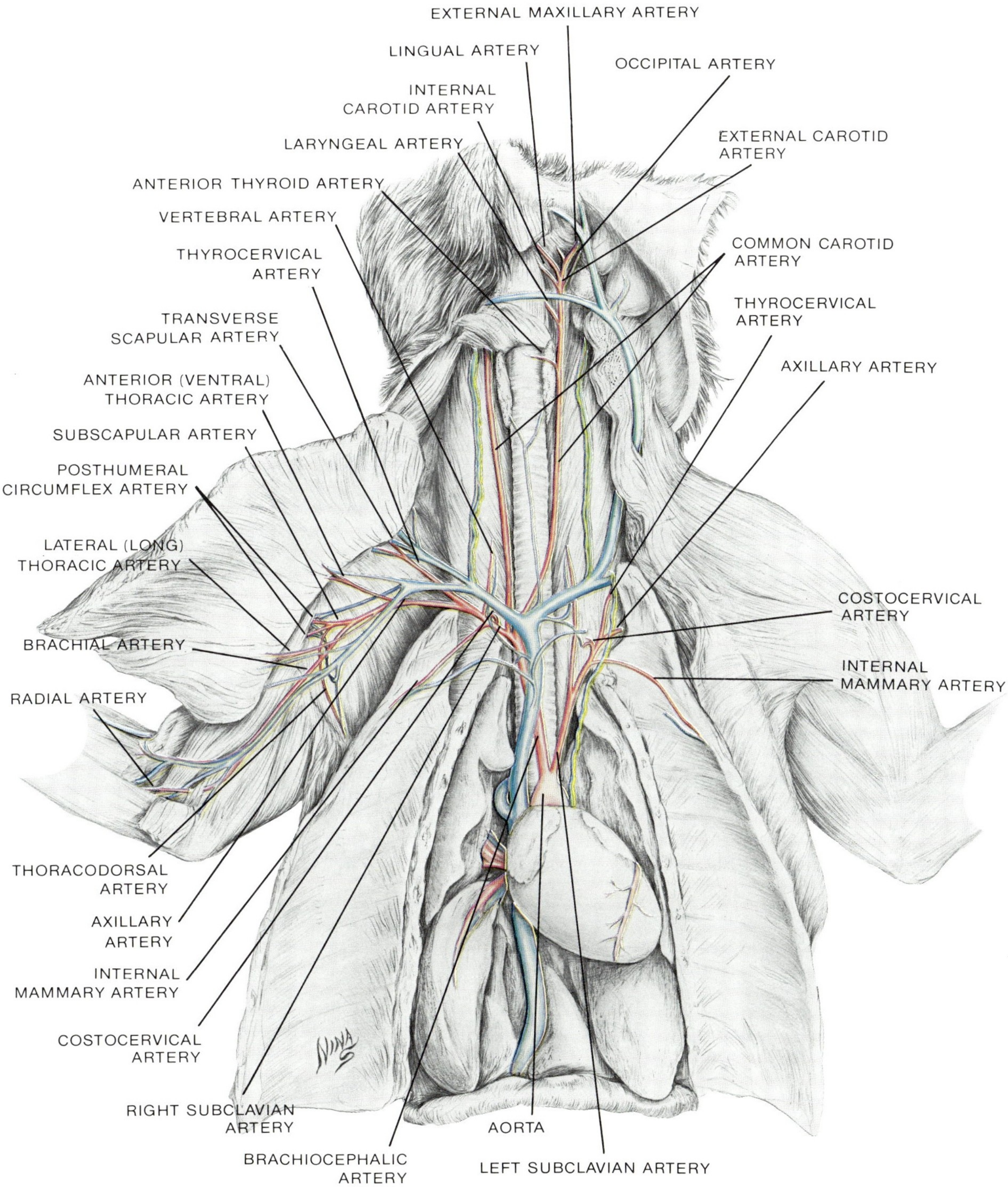
EXTERNAL MAXILLARY ARTERY
LINGUAL ARTERY
OCCIPITAL ARTERY
INTERNAL CAROTID ARTERY
LARYNGEAL ARTERY
EXTERNAL CAROTID ARTERY
ANTERIOR THYROID ARTERY
VERTEBRAL ARTERY
COMMON CAROTID ARTERY
THYROCERVICAL ARTERY
THYROCERVICAL ARTERY
TRANSVERSE SCAPULAR ARTERY
AXILLARY ARTERY
ANTERIOR (VENTRAL) THORACIC ARTERY
SUBSCAPULAR ARTERY
POSTHUMERAL CIRCUMFLEX ARTERY
LATERAL (LONG) THORACIC ARTERY
COSTOCERVICAL ARTERY
BRACHIAL ARTERY
INTERNAL MAMMARY ARTERY
RADIAL ARTERY
THORACODORSAL ARTERY
AXILLARY ARTERY
INTERNAL MAMMARY ARTERY
COSTOCERVICAL ARTERY
RIGHT SUBCLAVIAN ARTERY
AORTA
BRACHIOCEPHALIC ARTERY
LEFT SUBCLAVIAN ARTERY

Figure 27–7

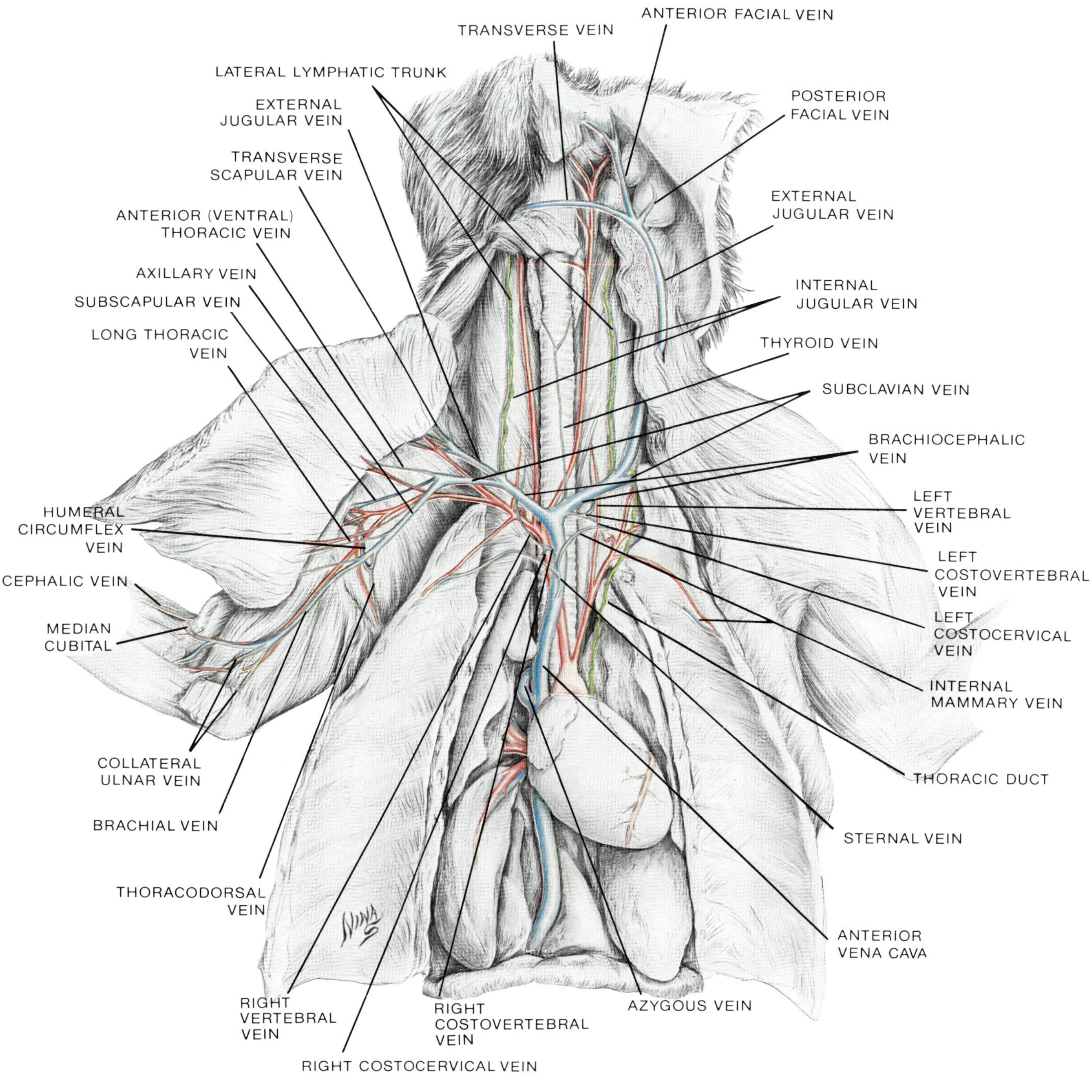
ANTERIOR FACIAL VEIN
TRANSVERSE VEIN
LATERAL LYMPHATIC TRUNK
EXTERNAL JUGULAR VEIN
TRANSVERSE SCAPULAR VEIN
ANTERIOR (VENTRAL) THORACIC VEIN
AXILLARY VEIN
SUBSCAPULAR VEIN
LONG THORACIC VEIN
HUMERAL CIRCUMFLEX VEIN
CEPHALIC VEIN
MEDIAN CUBITAL
COLLATERAL ULNAR VEIN
BRACHIAL VEIN
THORACODORSAL VEIN
RIGHT VERTEBRAL VEIN
RIGHT COSTOCERVICAL VEIN
RIGHT COSTOVERTEBRAL VEIN
AZYGOUS VEIN
ANTERIOR VENA CAVA
STERNAL VEIN
THORACIC DUCT
INTERNAL MAMMARY VEIN
LEFT COSTOCERVICAL VEIN
LEFT COSTOVERTEBRAL VEIN
LEFT VERTEBRAL VEIN
BRACHIOCEPHALIC VEIN
SUBCLAVIAN VEIN
THYROID VEIN
INTERNAL JUGULAR VEIN
EXTERNAL JUGULAR VEIN
POSTERIOR FACIAL VEIN
NINA S

Figure 27–9

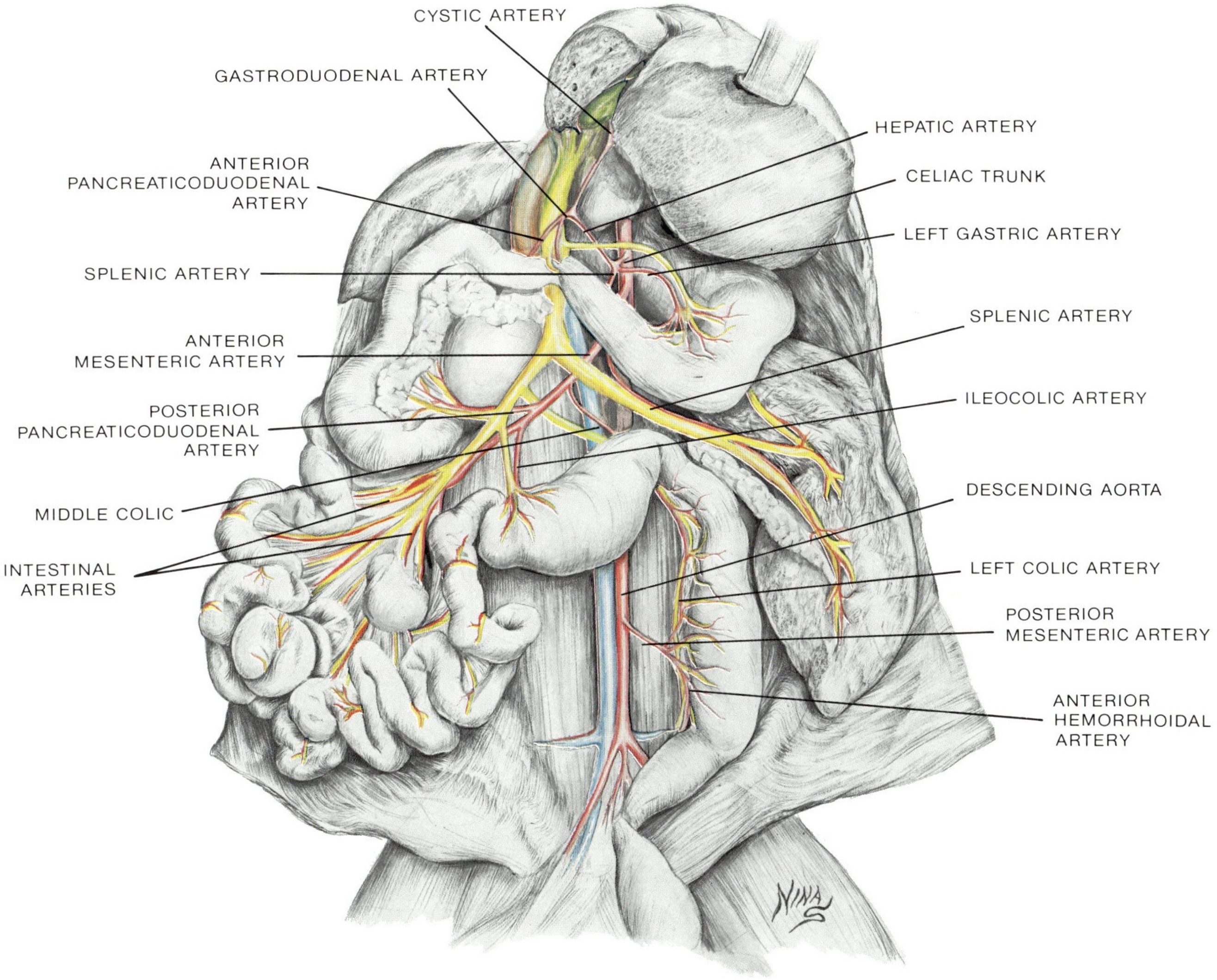
CYSTIC ARTERY
GASTRODUODENAL ARTERY
HEPATIC ARTERY
ANTERIOR PANCREATICODUODENAL ARTERY
CELIAC TRUNK
LEFT GASTRIC ARTERY
SPLENIC ARTERY
SPLENIC ARTERY
ANTERIOR MESENTERIC ARTERY
ILEOCOLIC ARTERY
POSTERIOR PANCREATICODUODENAL ARTERY
DESCENDING AORTA
MIDDLE COLIC
LEFT COLIC ARTERY
INTESTINAL ARTERIES
POSTERIOR MESENTERIC ARTERY
ANTERIOR HEMORRHOIDAL ARTERY

Figure 27–11

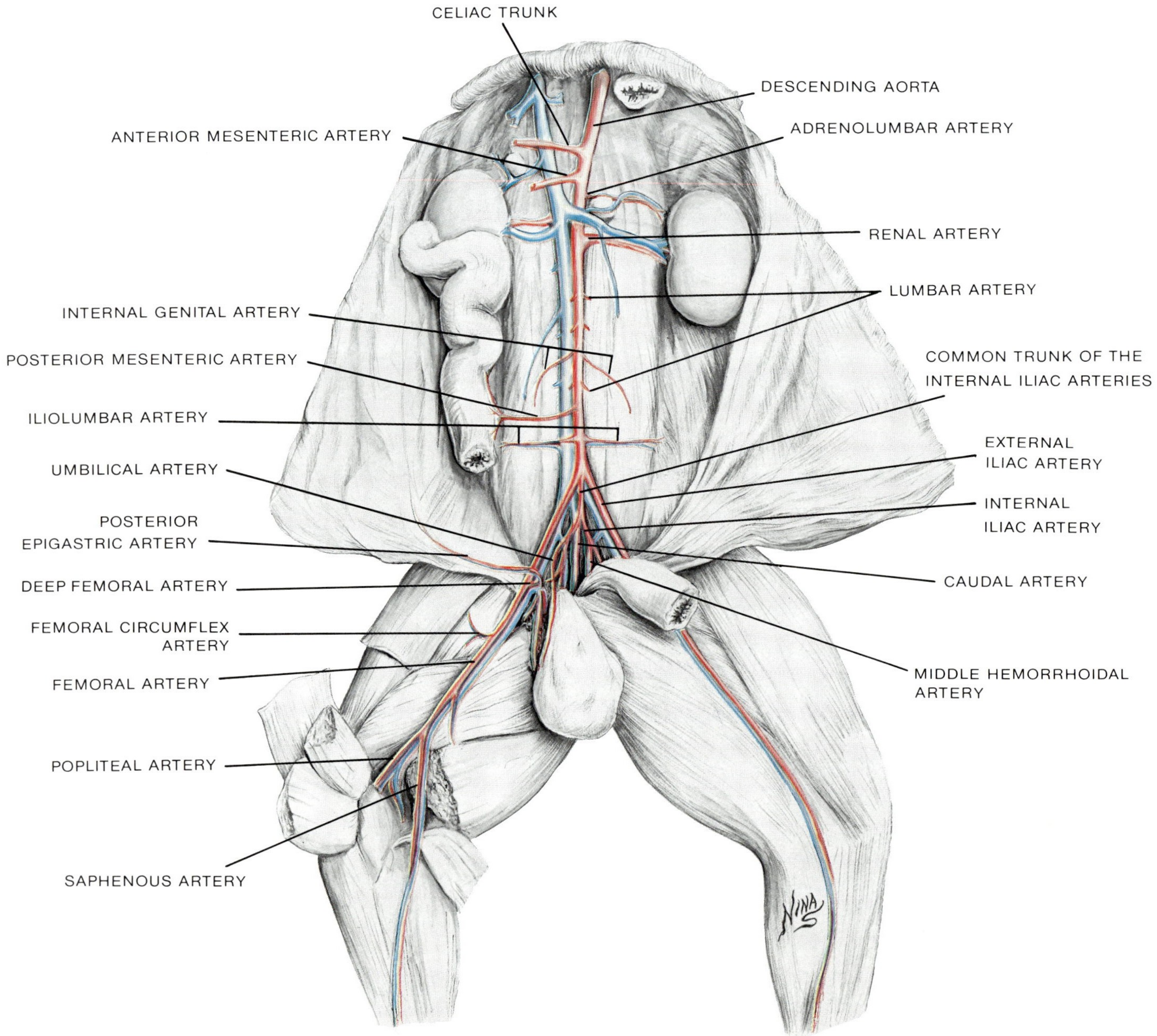
CELIAC TRUNK
DESCENDING AORTA
ANTERIOR MESENTERIC ARTERY
ADRENOLUMBAR ARTERY
RENAL ARTERY
LUMBAR ARTERY
INTERNAL GENITAL ARTERY
POSTERIOR MESENTERIC ARTERY
COMMON TRUNK OF THE INTERNAL ILIAC ARTERIES
ILIOLUMBAR ARTERY
EXTERNAL ILIAC ARTERY
UMBILICAL ARTERY
INTERNAL ILIAC ARTERY
POSTERIOR EPIGASTRIC ARTERY
CAUDAL ARTERY
DEEP FEMORAL ARTERY
FEMORAL CIRCUMFLEX ARTERY
MIDDLE HEMORRHOIDAL ARTERY
FEMORAL ARTERY
POPLITEAL ARTERY
SAPHENOUS ARTERY
NINA S

Figure 27–12

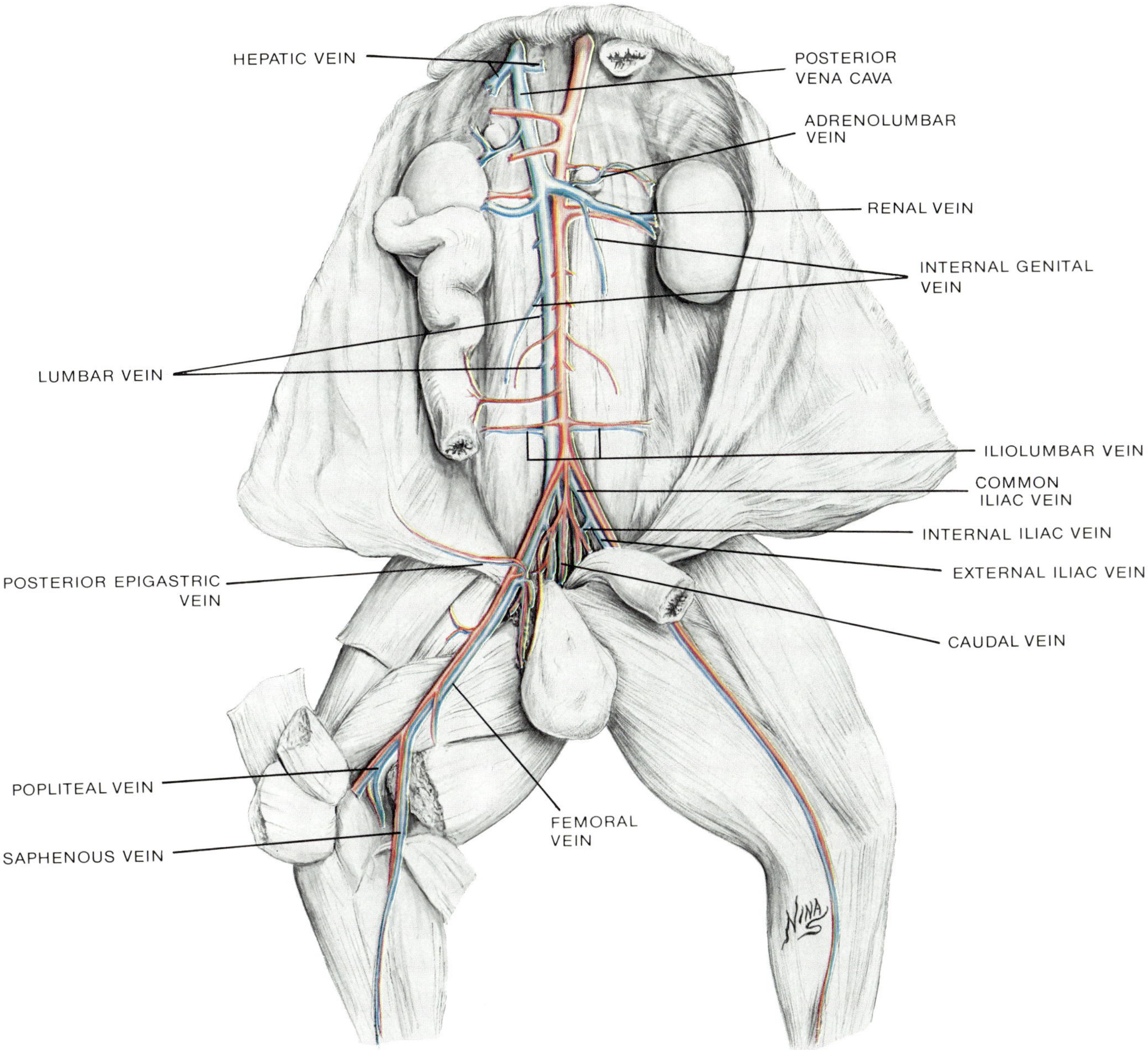
HEPATIC VEIN
POSTERIOR VENA CAVA
ADRENOLUMBAR VEIN
RENAL VEIN
INTERNAL GENITAL VEIN
LUMBAR VEIN
ILIOLUMBAR VEIN
COMMON ILIAC VEIN
INTERNAL ILIAC VEIN
EXTERNAL ILIAC VEIN
POSTERIOR EPIGASTRIC VEIN
CAUDAL VEIN
POPLITEAL VEIN
FEMORAL VEIN
SAPHENOUS VEIN
NINA S

Figure 27–13

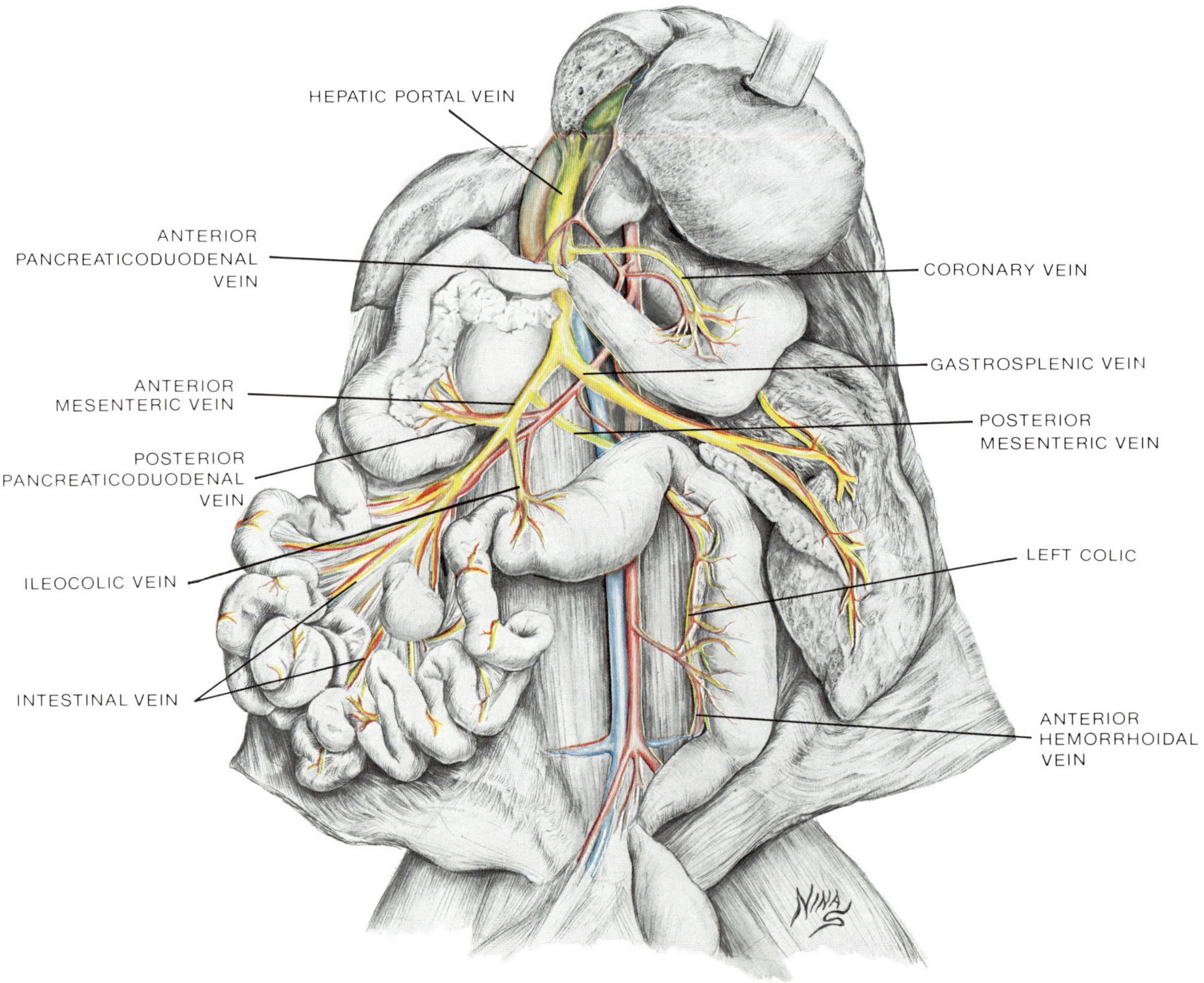
HEPATIC PORTAL VEIN
ANTERIOR
PANCREATICODUODENAL
VEIN
CORONARY VEIN
GASTROSPLENIC VEIN
ANTERIOR
MESENTERIC VEIN
POSTERIOR
MESENTERIC VEIN
POSTERIOR
PANCREATICODUODENAL
VEIN
LEFT COLIC
ILEOCOLIC VEIN
INTESTINAL VEIN
ANTERIOR
HEMORRHOIDAL
VEIN

Figure 27–14